AF581887

Fit-for-Purpose Land Administration-Providing Secure Land Rights at Scale. Volume 2: Country Implementation

Fit-for-Purpose Land Administration-Providing Secure Land Rights at Scale. Volume 2: Country Implementation

Editors

Stig Enemark
Robin McLaren
Christiaan Lemmen

MDPI • Basel • Beijing • Wuhan • Barcelona • Belgrade • Manchester • Tokyo • Cluj • Tianjin

Editors
Stig Enemark
Aalborg University
Denmark

Robin McLaren
University of Glasgow
United Kingdom

Christiaan Lemmen
University of Twente
The Netherlands

Editorial Office
MDPI
St. Alban-Anlage 66
4052 Basel, Switzerland

This is a reprint of articles from the Special Issue published online in the open access journal *Land* (ISSN 2073-445X) (available at: https://www.mdpi.com/journal/land/special_issues/FFPLA).

For citation purposes, cite each article independently as indicated on the article page online and as indicated below:

LastName, A.A.; LastName, B.B.; LastName, C.C. Article Title. *Journal Name* **Year**, *Volume Number*, Page Range.

Volume 2
ISBN 978-3-0365-2085-8 (Hbk)
ISBN 978-3-0365-2086-5 (PDF)

Volume 1-2
ISBN 978-3-0365-2087-2 (Hbk)
ISBN 978-3-0365-2088-9 (PDF)

Cover image courtesy of Stig Enemark

Contents

About the Editors

Stig Enemark is Professor Emeritus of Land Management at Aalborg University, Denmark, where he was Head of School of Surveying and Planning for 15 years. He holds an MSc in Surveying, Planning and Land Management from the Royal Agricultural Academy in Copenhagen and, before joining Aalborg University in 1982, he worked for ten years as a licensed surveyor in private practice. He is Honorary President of the International Federation of Surveyors, FIG, where he served as President 2007-2010. He is a well-known international expert in the areas of land governance, land administration and spatial planning, and related issues of education and capacity development. He has consulted widely in these areas especially in Eastern Europe, South East Asia and Sub-Sahara Africa. He has published a number of books and more than 400 scientific and professional papers.

Robin McLaren is an Honorary Fellow at the School of GeoSciences, University of Edinburgh where his research interests are focused on how crowdsourcing can be used to support land administration. He holds degrees from the University of New Brunswick and the University of Glasgow where he was awarded an honorary doctorate for contributions to geomatics and land administration. He is also the owner and director of Know Edge Ltd a UK based, independent management consulting company formed in 1986 specialising in the application of geospatial information and is a prominent consultant in land administration. He has been at the forefront of the GIS/LIS revolution and is recognised as an expert in Spatial Data Infrastructures and Land Policy. He works extensively with United Nations agencies, World Bank and EU on land policy/land administration / NSDI programmes in Africa, Far East, Asia and Europe.

Christiaan Lemmen is a full professor at ITC, the Faculty of Geo-information Science and Earth Observation, University of Twente, the Netherlands. He is a geodetic advisor with Kadaster International, the international branch of the Netherlands Cadastre, Land Registry and Mapping Agency. Christiaan is co-editor of the the Land Administration Domain Model (LADM), an international standard published as ISO–19152. He is a co-chair of the Land Administration Domain Working Group of the Open GeoSpatial Consortium. Christiaan is director of the OICRF, the Office International du Cadastre et du Regime Foncier (the international Office for Cadastre and Land Records), a documentation center for land administration, and a permanent institution of the International Federation of Surveyors (FIG). He holds a Ph.D. in Land Administration from Delft University of Technology, The Netherlands.

 land

MDPI

Editorial

Fit-for-Purpose Land Administration—Providing Secure Land Rights at Scale

Stig Enemark [1,*], Robin McLaren [2] and Christiaan Lemmen [3,4]

1 Department of Planning, Aalborg University, 9000 Aalborg, Denmark
2 Know Edge Ltd., Edinburgh EH14 1AY, UK; Robin.McLaren@KnowEdge.com
3 School for Land Administration Studies, ITC Faculty, University of Twente, 7500 AE Enschede, The Netherlands; chrit.lemmen@kadaster.nl
4 Kadaster International, Cadastre, Land Registry and Mapping Agency, 7311 KZ Apeldoorn, The Netherlands
* Correspondence: enemark@plan.aau.dk

This Special Issue provides an insight, collated from 26 articles, focusing on various aspects of the Fit-for-Purpose Land Administration (FFPLA) concept and its application. It presents some influential and innovative trends and recommendations for designing, implementing, maintaining and further developing FFP solutions for providing secure land rights at scale. The first group of 14 articles is published in Volume One and discusses various conceptual innovations related to spatial, legal and institutional aspects of FFPLA and its wider applications within land use management. The second group of 12 articles is published in Volume Two and focuses on case studies from various countries throughout the world, providing evidence and lessons learned from the FFPLA implementation process. However, in order to facilitate a more global understanding of the issues and their interrelationship, this editorial embraces both volumes. It should be noted, though, that the online version of this Special Issue presents the articles in a different order, namely in the chronology of their publication.

 check for updates

Citation: Enemark, S.; McLaren, R.; Lemmen, C. Fit-for-Purpose Land Administration—Providing Secure Land Rights at Scale. *Land* 2021, *10*, 972. https://doi.org/10.3390/land10090972

Received: 6 September 2021
Accepted: 9 September 2021
Published: 15 September 2021

Publisher's Note: MDPI stays neutral with regard to jurisdictional claims in published maps and institutional affiliations.

1. Evolution of the FFPLA Approach

The term "land administration" is rooted in cadastral and land registration systems originally developed for providing information about land value, land ownership and types of land use [1]. Historically, these systems were designed for slightly different purposes in various cultures, judicial systems and regions throughout the world. The key difference is whether the transaction alone is recorded (deeds systems) or the title itself is recorded and secured (title systems). The cultural and judicial aspects relate to whether a country is based on Roman law (deeds systems) or German or Anglo common law (title systems). This difference is also apparent in relation to the legacy of colonization.

A couple of decades ago, land administration emerged as a more generic term referred to as "the processes of determining, recording, and disseminating information about ownership, value, and use of land when implementing land management policies" [2,3]. This focus on information is still present, but, in recent years, the type and quality of information needed have changed and pushed the design of land administration systems (LASs) towards "an enabling infrastructure for implementing land policies and land management strategies in support of sustainable development" [4]. The operational components of this land governance infrastructure are the four functions of land tenure, land value, land use and land development. These four functions ensure the proper management of rights, restrictions, and responsibilities in relation to property, land use and natural resources, including the marine environment. However, the basis or the backbone of such solutions is the land tenure component related to the individual land parcel and establishing the relationship between people and land.

In most developed countries, security of tenure is taken for granted. Over centuries, these countries developed mature land institutions and laws that protect the people to land relationship and provide the services needed for supporting an efficient land market and effective land use management. However, an educated estimate indicates that for 70 percent of the world's population, this is not the case [5]. In most developing countries, people cannot register and safeguard their land rights, nor it is too costly. The majority of these people are the poor and the most vulnerable in society. Therefore, over recent years, LASs have developed to also capture and include more informal and social types of tenure. This is encouraged and supported through the development of concepts such as the continuum of land rights [6], the social tenure domain model [7], and aspects of responsible governance of tenure [8,9].

The key driver behind this evolution has been the overall global agenda focusing on poverty eradication, food security, gender equity, human rights, etc., as adopted by the Millennium Development Goals (MDGs) in 2000, and followed by the Sustainable Development Goals (SDGs) in 2015. This agenda has put a strong focus on security of land rights and provided targets and indicators for monitoring the progress of achieving the goals. Another key driver is technology development that has enabled easy access to new, innovative mapping and surveying techniques, such as satellite and drone imagery, mobile phones and handheld GPS, as well as techniques for the storage and management of huge datasets [10].

Over time, these evolutionary endeavors have been conceptualized in the FFPLA approach in 2014 designed to meet the challenges of providing secure land rights at scale [11,12]. This FFP approach indicates that is appropriate and of necessary standard for its main purpose . . . namely providing secure land rights at scale within a given jurisdiction. The concept, as illustrated in Figure 1, includes three interrelated frameworks that work together to deliver the FFP approach: the spatial, legal and institutional frameworks.

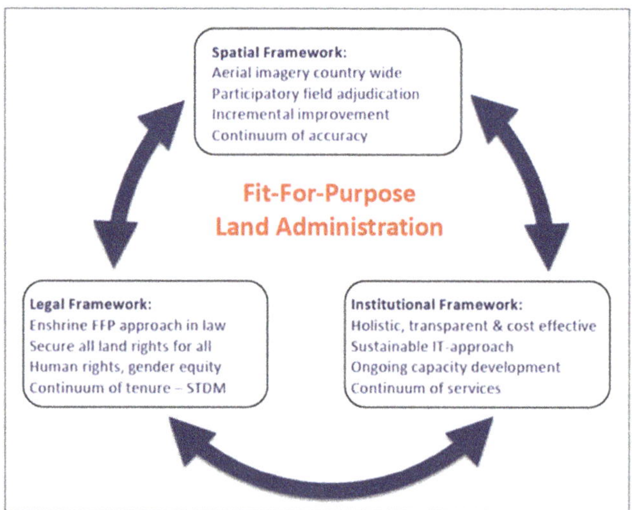

Figure 1. The FFPLA concept [12] (p. 17).

The spatial framework geospatially defines the way in which land is occupied and used. The scale and accuracy of this representation are not determined by rigid, high accuracy regulations, but instead by the users' requirements for effectively identifying the land parcels as a basis for securing the various kinds of legal and legitimate rights and tenure forms recognized through the legal framework. The institutional framework and partnerships are designed to manage these rights, the use of land and natural resources, and to deliver inclusive and accessible services. The approach is flexible, affordable, participatory, and the outcome is upgradeable over time [12].

2. The FFPLA Special Issue

In recent years, the FFPLA concept has been introduced in many countries throughout the world for providing secure land rights at scale, within a short timeframe and at affordable costs. Figure 2 shows the range of countries where FFPLA assessment and implementation are addressed in this Special Issue.

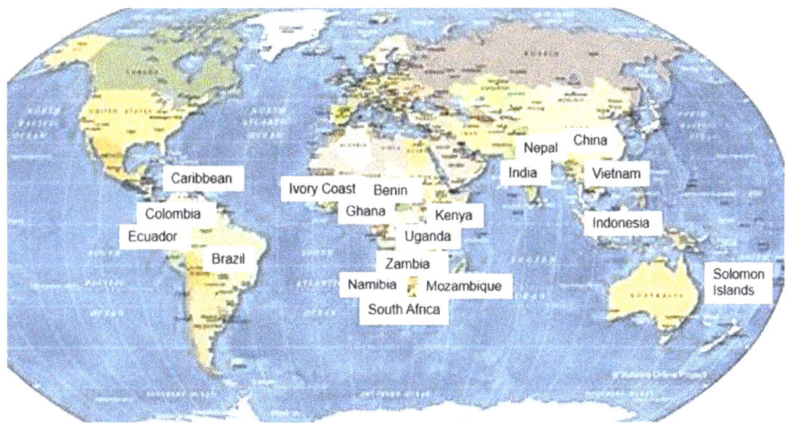

Figure 2. FFPLA Country assessments and implementations addressed in this Special Issue.

The first aim of this Special Issue is to present some recent innovations in the design and use of the FFPLA concept that are shaping new, more efficient approaches to FFPLA. This first group of 14 articles is published in Volume One. The second aim is to synthesize the experiences and lessons learned from country implementation in different cultures and jurisdictions throughout the world. This second group of 12 articles is published in Volume Two.

The conceptual innovations (Volume One) include issues such as:

(i) Assessing procedures of maintenance of conventional as well as unconventional systems;
(ii) Assessing adjudication and quality assurance for legal and geospatial data collected in the participatory processes of field work;
(iii) Applying innovative geospatial tools to FFPLA;
(iv) Using decentralization as a strategy for scaling FFPLA;
(v) Assessing the role of FFPLA for providing security of tenure in violent conflict settings;
(vi) Applying the FFP approach to wider land management functions and to urban resilience in times of climate change and the COVID-19 pandemic; and:
(vii) Exploring the role and opportunities of the private financial sector and public private partnerships within FFPLA.

These innovations are making the implementation of the FFPLA approach more efficient and widening the use in land management applications. Overall, these conceptual innovations are making the approach more attractive for countries to implement and allow the social and economic benefits to be realized more quickly for a sustainable future.

The experiences and lessons learned from country implementation (Volume Two) include cases such as:

(i) Assessing the development impacts of the processes used in China and Vietnam for providing secure land rights at scale;
(ii) Analyzing the strategy and implementation processes for applying an FFPLA approach in Indonesia, Nepal, Uganda and Mozambique;
(iii) Evaluating demonstrative cases of piloting a FFPLA approach and applying FFPLA tools for land recordation in Ghana, Kenya, Uganda, Zambia and Namibia;
(iv) Investigating the impact of applying the FFPLA approach to South Africa;

(v) Using a FFP approach for upscaling of land administration in Benin;
(vi) Applying the FFPLA approach in response to post disasters in the Caribbean; and:
(vii) Assessing FFPLA applications in Colombia and Ecuador.

This wide range of country cases clearly demonstrates that the FFPLA approach is applicable within different contexts by reflecting the specific cultural, legal and institutional settings. The pilot cases validate that the FFPLA methodology for recording land rights in the field is flexible and is working effectively.

3. Concluding Remarks

The main motivation for this special edition was to share experiences and research into the FFPLA approach to help accelerate its implementation at scale and quickly resolve the global insecurity of tenure crisis. The articles indeed illustrate the significant progress that has been achieved over the past decade. They provide some very encouraging lessons learned, as well as exciting, innovative technologies to inspire land professionals to achieve the challenging objectives of the SDGs.

These new highlighted opportunities for going to scale include a clearer understanding of how to decentralize roles and responsibilities and manage organizational change. It also includes a better comprehension of how to obtain political support, gain consensus and formulate national FFPLA strategies, and new insights into implementing robust and the sustainable maintenance of land rights. The articles provide examples of obtaining alternative sources of financing for FFPLA through new types of PPPs, and a pioneering use of private social enterprises for embracing FFP land financing to support the regularization and upgrading of informal settlements. A range of technical innovations is presented, including greater efficiencies derived from the use of machine learning to extract information from drone imagery. Finally, and very importantly, the articles provide a rich set of experiences from FFPLA national scale implementations, as well as pilot projects from developing countries in three continents.

The articles also indicate that the impact of the FFP approach is unfolding beyond its initial focus on security of tenure. UN agencies are widely adopting FFPLA as a tool to mitigate underlying land issues in violent conflict settings, and it is being embraced in wider, urban land management functions to support housing resilience, property valuations, and mitigate the impact of climate change and pandemics. The articles confirm that the FFPLA approach is growing in acceptance across the land professional community, is gaining considerable momentum and is a game changer in achieving key elements of the global agenda, the SDGs, and the globally accepted policies and guidelines around responsible governance of tenure. The FFPLA is already triggering a change in society towards greater social equity, leaving no one behind.

4. Overview of Contributions

This Special Issue includes 26 articles divided into two groups: conceptual innovations (Volume One) and country implementations (Volume Two). The full list of articles is presented in Tables 1 and 2, each of which is followed by a short synthesis of the individual articles.

Bennett et al. [13] use an impressive range of contemporary sources to review FFPLA approaches from the perspective of system maintenance. The "fit-for-purpose" era is producing a wide range of new social and technological innovations; however, large-scale and sustainable implementations still struggle with system maintenance. They present a consolidated model summarizing the story of the maintenance concept in land administration—in terms of key terminology, typologies, approaches, aspects and options. Then, they provide an overview of maintenance problems and related solutions. Finally, they identify that new solutions, as yet unpublished, and newly identified challenges, are emerging.

Table 1. Conceptual Innovations.

	Title	Country Focus	Application
Bennett et al.	Land Administration Maintenance: A review of the Persistent Problem and Emerging Fit-for-Purpose Solutions	Global	Methodologies of maintenance
Lengoiboni et al.	Initial Insights on Land Adjudication in a Fit-for-Purpose Land Administration	Global	Methodologies of adjudication
Augustinus and Tempra	Fit-for-Purpose Land Administration in Violent Conflict Settings	Sudan, Iraq, DRC, Honduras, Peru, Somalia,	Addressing land rights in conflict settings
Ho et al.	Decentralization as a Strategy to Scale Fit-for-Purpose Land Administration: An Indian Perspective on Institutional Challenges	India	Decentralization as a FFPLA tool
Mitchell et al.	The Benefits of Fit-for-Purpose Land Administration for Urban Community Resilience in a Time of Climate Change and COVID-19 Pandemic	Solomon Islands	FFPLA in support of Improving urban resilience
Kelm et al.	Applying the FFP Approach to Wider Land Management Functions	Global	The wider use of the FFPLA approach
Childress et al.	Fit-for-Purpose, Private-Sector Led Land Regularization and Financing of Informal Settlements in Brazil	Brazil	Applying a private sector led approach
Moran et al.	Exploring PPPs in Support of Fit-for-Purpose Land Administration: A Case Study from Côte d'Ivoire	Ivory Coast	Applying a PPP in support of FFPLA
Reydon et al.	The Amazon Forest Preservation by Clarifying Property Rights and Potential Conflicts: How Experiments Using Fit-for-Purpose Can Help	Brazil	Applying a FFP approach in support of forest preservation
Rocha et al.	Quality Assurance for Spatial Data Collected in Fit-for-Purpose Land Administration Approaches in Colombia	Colombia	Assessing the FFPLA data quality
Hall and Whittal	Do Design Science Research and Design Thinking Processes Improve the 'Fit' of the Fit-for-Purpose Approach to Securing Land Tenure for All in South Africa?	South Africa	Exploring the use of design science research and design thinking within FFPLA
Koeva et al.	Geospatial Tool and Geocloud Platform Innovations. A Fit-for-Purpose Land Administration Assessment	Rwanda, Kenya, Ethiopia, and Zanzibar	Assessing the use of geospatial tools in Africa
Chipofya et al.	SmartSkeMa: Scalable Documentation for Community and Customary Land Tenure	Global	Spatial documentation of community land tenure
Biraro et al.	Good Practices in Updating Land Information Systems that Used Unconventional Approaches in Systematic Land Registration	Global	Updating practices in unconventional land registration

Lengoiboni et al. [14] explore how primary ('ownership') and secondary, overlapping ('non-ownership') interests in land are being adjudicated and recorded in a FFPLA context. They prepared questionnaires and developed criteria that organizations invited to answer the questionnaires must meet. They define the components of land adjudication. Then, the processes used by the organizations to achieve these components are described and insights are gained on how land tenure and land rights are framed and how this influences the outcomes of what is recorded. Results show that the legal perspective of land rights intersects with the perspective of communities regarding legitimate rights.

Augustinus and Tempra [15] tackle the challenging subject of UN peace keeping in violent conflict settings and examine how FFPLA interventions have become an integral part of the dispute and conflict resolution. They discuss seven cases across multiple countries and conflict sites where UN-Habitat either supported, or was directly involved in, UN peace keeping. They identify that land governance is of importance because of the way in which land-related power dynamics play out across the conflict cycle; the UN uses its power and capacity to strengthen land governance. They recommend that the FFPLA approaches provide practical options to support peace-building operations by the UN and other stakeholders.

Ho et al. [16] analyze the socio-political and institutional consequences of using decentralization as a scaling strategy for implementing FFPLA through three case studies in India. They review how decentralization is coordinated and governed across multiple levels. Their cases demonstrate a reduced role for the state, and a need for increased collaboration across a diverse set of stakeholders, including a greater number of non-state actors at multiple levels. Although decentralization can work to effectively kickstart the implementation of FFPLA at scale, there is significant work required to ensure that implementation is "fit-for-people" to introduce a trustworthy system that redistributes power and distributes critical social justice.

Mitchell et al. [17] investigate the interlinkages between land tenure, climate vulnerability, and pandemics. Through research in Honiara, Solomon Islands, they consider how improving tenure security at scale through FFPLA can enhance climate resilience to mitigate vulnerability to both climate and pandemic impacts. They contend that this can be achieved at both the city and settlement levels by including tenure in vulnerability and risk assessments (VRA) and the development of resilience action plans. Their proposed FFPLA process, informed by participatory enumeration of the complexities of urban land tenure, can support scaling up efforts to improve tenure security, and deliver more effective and equitable climate resilience actions for vulnerable urban communities.

Kelm et al. [18] analyze how the FFP approach, which has predominantly been applied to the land tenure aspects, can be expanded into a wider set of land management functions. They test their hypothesis through three World Bank urban case studies focusing on land valuation, housing resilience and waste management. Machine learning techniques extract information from drone and street-level imagery to produce minimum viable product models. Their analysis has revealed that there is a common set of geospatial datasets that can be captured once and shared across many other land management functions in an urban environment. This will allow single land intervention projects to be holistically integrated into a wider program of land management functions.

Childress et al. [19] use the analysis of an innovative private social enterprise in Brazil, called Terra Nova, to demonstrate that the concept of FFP land regularization can be widened to include FFP land financing with relevance for wider efforts in informal settlement regularization and upgrading. Their analysis of parcel-level repayment and price data provides some evidence of the sustainability of the business model and the increase of property values of the regularized parcels (pre-COVID-19). Since 2001, Terra Nova has regularized over 20,000 parcels, primarily in São Paolo and Curitiba. They contend that the approach is widely replicable.

Morán et al. [20] introduce a new, innovative form of public private partnership (PPP) being piloted in the coffee growing areas of Côte d'Ivoire, that includes a partnership between the Government and a consortium of cocoa industry leaders and Meridia, the Dutch private land documentation firm. The private sector companies provide funding and service delivery, while the Government enables a political environment for interventions, provides in-kind contributions, collaborates in the execution of projects, and operates the land information system/registry to which the consortium's service connects. This PPP approach could potentially provide countries with alternatives to donor funding/loans.

Reydon et al. [21] blame deforestation in Brazil on the absence of cadastral mapping, land registration, and an effective regulation of property rights. This already involves some 200 million hectares, mostly on public land or undesignated land. This land is easy to grab, deforest and to be used for speculative purposes. The availability of well-defined land rights can reduce the process of deforestation. Participatory determination of land rights based on FFPLA methodologies promotes forest preservation. They hope that their methodology for determining the land rights of small landholders and of traditional population landholders will become mainstream. This will require some legal and institutional adjustments in order to improve the sustainability of the Amazon rainforest.

Rocha et al. [22] present a FFPLA quality assurance model for the evaluation of the quality of the geospatial data collected in the Municipality of Apartadó in Colombia. The FFPLA approach allowed the right holders to walk their parcel boundaries using a smartphone application connected to a GPS receiver to collect their boundary data points. The project evaluated how well the FFPLA dataset conformed to its product specification and was able to determine whether FFPLA data were of sufficient quality, specifically in the case of positional accuracy and logical consistency. The model supported the creation of a product quality life cycle and a quality model in the Colombian context.

Hull and Whittal [23] conduct a reflective retrospective of the processes of land administrative reform in South Africa to determine how land administration systems should be reshaped and new land tenure reforms to be developed. They adopt a thematic framework that innovatively combines FFPLA, design science research, and design thinking processes to help to unlock sensitive and empathetic innovations in land administration systems reform initiatives that will deliver restorative justice. These approaches should be embedded into FFPLA at the start. They admit that this is new and untested and they encourage case studies explicitly implementing these additional approaches in land administration reform.

Koeva et al. [24] assess a series of innovative tools recently developed within the framework of a European Commission Horizon 2020 project. The tools they review are designed to effectively apply FFPLA approaches and are based on requirements from FFPLA projects in Rwanda, Kenya, Ethiopia and Zanzibar. The study was conducted under the appealing title "its4land". The tools include software that implements the smart sketch mapping concept, a workflow for data acquisition based on unmanned aerial vehicles, and a boundary delineator tool based on semi-automatic feature extraction from UAV images. The 'Publish and Share' platform enables the integration of all the outputs of tool sharing and publishing of land information through geocloud web services.

Chipofya et al. [25], with a reference to the 'landmarkmap database', point out that fifty percent of habitable land on this planet is held by indigenous communities. There are no proper tools to document these rights quickly and effectively. Existing software and facilities for documentation of these rights still assume parcel-oriented thinking with statutory rights. The Smart Sketch Map (SmartSkeMa) allows people to document their land rights using concepts from their everyday experiences. SmartSkeMa supports both the legibility of customary land tenure to government authorities and the preservation of the customs within which the tenure relations operate.

Biraro et al. [26] study the maintenance of data in land information systems for which the data were obtained using unconventional approaches. The paper proves that there are good, recommendable practices in the selected countries, including infrastructure for updating; simplified systems; reasonable registration fees; decentralized services; accessible and secured digital databases; awareness raising about registration; availability of a legal framework; incentives to motivate people to report transactions on time; and trained staff and political support. The authors conclude that efforts are still needed to shorten updating procedures, introduce data-sharing platforms, ensure financial and technical sustainability, and reduce the number of involved institutions.

Table 2. Country Implementations.

	Title	Country Focus	Application
Byamugisha	Experiences and Development Impacts of Securing Land Rights at Scale in Developing Countries: Case Studies of China and Vietnam	China, Vietnam	Securing land rights at scale in China and Vietnam
Martono et al.	The Legal Element of Fixing the Boundary for Indonesian Complete Cadastre	Indonesia	Applying FFPLA in Indonesia
Panday et al.	Securing Land Rights for All through Fit-for-Purpose Land Administration Approach: The Case of Nepal	Nepal	Applying FFPLA in Nepal
Musinguzi et al.	Fit for Purpose Land Administration: Country Implementation Strategy for Addressing Uganda's Land Tenure Security Problems	Uganda	Applying FFPLA in Uganda
Chigbu et al.	Fit-for-Purpose Land Administration from Theory to Practice: Three Demonstrative Case Studies of Local Land Administration Initiatives in Africa	Ghana, Kenya, Namibia	Applying FFPLA approaches in Africa
Antonio et al.	Transforming Land Administration Practices through the Application of Fit-for-Purpose Technologies: Country Case Studies in Africa	Uganda, Kenya, Zambia	Applying the STDM in Africa
Mekking et al.	Fit-for-Purpose Upscaling Land Administration—A Case Study from Benin	Benin	Applying FFPLA in Benin
Balas et al.	The Fit for Purpose Land Administration Approach-Connecting People, Processes and Technology in Mozambique	Mozambique	Applying FFPLA in Mozambique
Williams-Wynn	Applying the Fit-for-Purpose Land Administration Concept to South Africa	South Africa	Assessment of applying FFPLA in South Africa
Griffith-Charles	Application of FFPLA to Achieve Economically Beneficial Outcomes Post Disaster in the Caribbean	Caribbean Islands	Applying FFPLA in the Caribbean
Becerra et al.	Fit-for-Purpose Applications in Colombia: Defining Land Boundary Conflicts between Indigenous Sikuani and Neighbouring Settler Farmers	Colombia	Applying FFPLA in Colombia
Todorovski et al.	Assessment of Land Administration in Ecuador Based on the Fit-for-Purpose approach	Ecuador	Assessment of applying FFPLA in Ecuador

Byamugisha [27] details, for the first time, the journeys that China and Vietnam embraced to register all land rights within their countries. This formidable task was triggered when the countries decollectivized agricultural production and allocated rural land to farming households in the 1980s and 1990s; about 1.5 billion rural arable land parcels in China and about 70 million in Vietnam. This was in addition to registering the urban land rights. In both countries, the registration of rural land was done in two rounds and the FFPLA approach was adopted. He distills an excellent set of lessons learned and challenges, and these should inform other countries embarking on similar security of tenure journeys to eradicate extreme poverty.

Martono et al. [28] aim to establish the distinction between physical and legal elements in determining cadastral boundaries in Indonesia. Interviews were conducted for this purpose, and six "cadastral elements" have been investigated and assessed: the parties that locate the boundary, the agreement between the adjoining landowners, the use of boundary markers, the role of the determination officer, the survey method, and the accuracy of the base map. Agreements could be obtained using aerial imagery instead of a field survey. Fixed boundary with exact coordinates based on prescribed survey methods, and with the accuracy of base map, as required by regulations, is not important to people in rural areas.

Panday et al. [29] analyze two pilot studies designed for testing the implementation of a FFPLA National Strategy in Nepal for providing security of tenure for 10 million land parcels currently outside the formal land registration system. They present the methodological workflow of this action-oriented research using one pilot study in an urban setting, including about 1500 spatial units of informal settlements, while the other is rural with 3400 arable land parcels. They explain how the results validate the FFPLA national implementation strategy approach designed for the specific and complex Nepal country context, and they argue that this methodology may be applicable for other low-income countries, where a large amount of the land is informally occupied.

Musinguzi et al. [30] explore, in great depth, the process of developing a national strategy for providing secure land rights at scale in Uganda, covering 23 million land parcels. They describe the current tenure types in Uganda and examine three representative pilot projects, in order to identify how lessons learned from these case studies informed a FFPLA implementation strategy in terms of building the spatial, legal and institutional frameworks. They highlight how pilot projects can provide opportunities for explaining benefits to obtain the necessary political, community and stakeholder support. In conclusion, they argue that a country implementation strategy, if developed as a result of a national dialogue and consensus among all stakeholders, is a promising way of advancing the FFPLA concept.

Chigbu et al. [31] provide evidence that the FFPLA approach represents an unprecedented opportunity to provide tenure security in Africa. They use three country case studies based on hands-on, local land administration projects to demonstrate how the features of the FFPLA guidelines were adopted. Support is provided for the understanding that high-precision measurements are not necessary for legal certainty. They conclude that local people, including youth and women, can be used for data collection and cadastral mapping purposes that are both inexpensive and can be used as necessary documentation for the promotion of tenure security.

Antonio et al. [32] investigate whether the STDM (social tenure domain model) tool facilitates the improvement of land tenure security. The STDM provides a flexible way to meet local needs in capturing human–country relationships. The tool is internationally recognized as practical, fast, and affordable. Their study shows that the STDM can be effectively applied to establish the spatial framework for land administration and to facilitate the implementation of land tenure security of poor communities. Focusing on one pillar of FFPLA can influence positive changes in other frameworks of the FFPLA concept through the use and application of technology, such as the STDM tool.

Mekking et al. [33] present a case study from Benin, with a focus on upscaling the FFPLA approach. At present, only 60,000 of the estimated 5 million plots are registered. For a parcel of 500 m^2, the cost of a title amounts to 540 USD, which is unaffordable for the vast majority of the population. The Benin Government wants legal security for all, and the FFPLA approach offers the opportunity to achieve this. The core of their approach is the introduction of a tenure system based on presumed ownership in parallel to the existing title system. Right holders then have the option to move from "presumed" ownership to state-guaranteed ownership.

Balas et al. [34] provide evidence from Mozambique that the FFPLA approach works. The former "Terra Segura" programme lasted four years and provided only 220,000 parcels and boundaries of 400 communities out of a set target of 5 million plots. The average costs were 50 USD per parcel and 10,000 USD per community. This is too expensive. A fundamentally different FFPLA approach was needed. The FFPLA-MOZ approach was developed and resulted in a better performance and a cost per parcel of 15 USD and 2000 USD for a community boundary definition. From the end of 2017 to March 2020, almost 1.4 million parcels and 826 community delimitations have been processed.

Williams-Wynn [35], the Surveyor General, investigates whether South Africa can adopt FFPLA to provide security of tenure for the five million land occupants that exist outside the formal land tenure system. As Surveyor General, he uses South Africa as a case study to demonstrate how adjustments to institutional, legal and spatial frameworks will develop a fully inclusive, sufficiently accurate land administration system that fits the purpose for which it is envisioned. He is optimistic that the adoption of the FFPLA approach with political support, trust built through community participation, and endorsement of the approach by land professionals will provide security of tenure that is beneficial to all.

Griffith-Charles [36] reviews the experiences with adjudication and titling being undertaken by countries of the Caribbean, with specific examples from Trinidad and Tobago, Barbados, and Jamaica, and others. Her assessment identifies that many countries had spent a lot of time, money, and effort, but were still without a predicted time of completion. The unhurried progress in some countries can be accelerated through the adoption of the FFPLA principles. She reasons that an essential aspect of achieving economically beneficial results is for a country to first identify and publicize a clear vision and objective of what is to be achieved, which requires land-related solutions to be efficient and inclusive.

Becerra et al. [37] introduce a FFPLA approach to support conflict resolution related to overlapping land claims. Indigenous people in Coumaribo, Colombia encounter land-related conflicts with newly arrived and established farmers. The methodology involves both parties independently surveying their land claims. This results in representations of the claims in georeferenced polygons, making any overlaps visible. In a public inspection, the results of the field measurements are displayed, with the presence of the cadastral authority. Discussing the results with all stakeholders helps to clarify the conflicts, to reduce the conflict to specific, relatively small, geographical areas, and to define concrete steps towards solutions.

Todorovski et al. [38] present an assessment of the existing land administration in Ecuador based on the spatial, legal and institutional frameworks—and related principles—of the FFPLA approach. This assessment is used to make recommendations for the improvement of the existing land administration to make the Government's plans for the implementation of a country-wide land administration system more feasible. They identify principles in a developed score table with a low and medium alignment that need to be addressed and adapted to a FFPLA approach; specifically with interventions in the current requirements for the precise measurement of fixed boundaries and a large number of text attributes collected in rural areas.

Author Contributions: S.E.: visualization, conceptualization and writing—original draft preparation. R.M.: conceptualization, writing—review and editing. C.L.: conceptualization, writing—review and editing. All authors have read and agreed to the published version of the manuscript.

Funding: The funding of the publication costs for this editorial, as well as the 26 articles included in this Special Issue, has kindly been provided by the School of Land Administration Studies, Faculty of ITC from the University of Twente, in combination with Kadaster International, The Netherlands.

Institutional Review Board Statement: Not applicable.

Informed Consent Statement: Not applicable.

Data Availability Statement: Not applicable.

Acknowledgments: The authors, as guest editors, would like to acknowledge the contributions of the authors of the individual articles included in this Special Issue, as well as all the external reviewers involved in assessing the scientific level of the manuscripts. We would also like to acknowledge the support of colleagues of academic and government originations that supported this work, in particular the School of Land Administration, Faculty of ITC from the University of Twente, in combination with Kadaster International, The Netherlands.

Conflicts of Interest: The authors declare no conflict of interest.

References

1. International Federation of Surveyors (FIG). *The FIG Statement on the Cadastre*; FIG Publication No. 11; FIG: Copenhagen, Denmark, 2015.
2. United Nations Economic Commission for Europe (UNECE). *Land Administration Guidelines: With Special Reference to Countries in Transition*; UNECE: Geneva, Switzerland; New York, NY, USA, 1996.
3. Dale, P.; McLaughlin, J. *Land Administration*; Oxford University Press: Oxford, UK, 1999.

4. Williamson, I.; Enemark, S.; Wallace, J.; Rajabifard, A. *Land Administration for Sustainable Development*; ESRI Academic Press: Redlands, CA, USA, 2010.
5. McLaren, R. *How Big Is Global Insecurity of Tenure?* Geomares Publishing: Lemmer, The Netherlands, 2015.
6. Global Land Tool Network (GLTN). *Secure Land Rights for All*; UN-Habitat: Nairobi, Kenya, 2008.
7. FIG/GLTN. *The Social Tenure Domain Model*; FIG Publication No. 52; FIG: Copenhagen, Denmark, 2010.
8. Food and Agriculture Organization of the United Nations (FAO). *Voluntary Guidelines on the Responsible Governance of Tenure of Land, Fisheries and Forests in the Context of Food Security*; FAO: Rome, Italy, 2012.
9. Zevenbergen, J.; De Vries, W.; Bennett, R. (Eds.) *Advances in Responsible Land Administration*; CRC Press: Boca Raton, FL, USA, 2016.
10. World Bank. *New Technology and Emerging Trends: The State of Play for Land Administration*; World Bank: Washington, DC, USA, 2017.
11. Enemark, S.; Bell, K.C.; Lemmen, C.; McLaren, R. *Fit-for-Purpose Land Administration*; FIG Publication No 60. Joint Publication; International Federation of Surveyors (FIG), World Bank: Copenhagen, Denmark, 2014.
12. Enemark, S.; McLaren, R.; Lemmen, C. *Fit-for-Purpose Land Administration—Guiding Principles for Country Implementation*; GLTN; UN-Habitat: Nairobi, Kenya, 2016.
13. Bennett, R.M.; Unger, E.-M.; Lemmen, C.; Dijkstra, P. Land Administration Maintenance: A Review of the Persistent Problem and Emerging Fit-for-Purpose Solutions. *Land* **2021**, *10*, 509. [CrossRef]
14. Lengoiboni, M.; Richter, C.; Aspersen, P.; Zevenbergen, J. Initial Insights on Land Adjudication in a Fit-for-Purpose Land Administration. *Land* **2021**, *10*, 414. [CrossRef]
15. Augustinus, C.; Tempra, O. Fit-for-Purpose Land Administration in Violent Conflict Settings. *Land* **2021**, *10*, 139. [CrossRef]
16. Ho, S.; Choudhury, P.R.; Haran, N.; Leshinsky, R. Decentralization as a Strategy to Scale Fit-for-Purpose Land Administration: An Indian Perspective on Institutional Challenges. *Land* **2021**, *10*, 199. [CrossRef]
17. Mitchell, D.; Barth, B.; Ho, S.; Sait, M.S.; McEvoy, D. The Benefits of Fit-for-Purpose Land Administration for Urban Community Resilience in a Time of Climate Change and COVID-19 Pandemic. *Land* **2021**, *10*, 563. [CrossRef]
18. Kelm, K.; Antos, S.; McLaren, R. Applying the FFP Approach to Wider Land Management Functions. *Land* **2021**, *10*, 723. [CrossRef]
19. Childress, M.; Carter, S.; Barki, E. Fit-for-Purpose, Private-Sector Led Land Regularization and Financing of Informal Settlements in Brazil. *Land* **2021**, *10*, 797. [CrossRef]
20. Morán, A.G.; Ulvund, S.; Unger, E.-M.; Bennett, R.M. Exploring PPPs in Support of Fit-for-Purpose Land Administration: A Case Study from Côte d'Ivoire. *Land* **2021**, *10*, 892. [CrossRef]
21. Reydon, B.; Molendijk, M.; Porras, N.; Siqueira, G. The Amazon Forest Preservation by Clarifying Property Rights and Potential Conflicts: How Experiments Using Fit-for-Purpose Can Help. *Land* **2021**, *10*, 225. [CrossRef]
22. Rocha, L.A.; Montoya, J.; Ortiz, A. Quality Assurance for Spatial Data Collected in Fit-for-Purpose Land Administration Approaches in Colombia. *Land* **2021**, *10*, 496. [CrossRef]
23. Hull, S.; Whittal, J. Do Design Science Research and Design Thinking Processes Improve the 'Fit' of the Fit-for-Purpose Approach to Securing Land Tenure for All in South Africa? *Land* **2021**, *10*, 484. [CrossRef]
24. Koeva, M.; Humayun, M.I.; Timm, C.; Stöcker, C.; Crommelinck, S.; Chipofya, M.; Bennett, R.; Zevenbergen, J. Geospatial Tool and Geocloud Platform Innovations: A Fit-for-Purpose Land Administration Assessment. *Land* **2021**, *10*, 557. [CrossRef]
25. Chipofya, M.; Jan, S.; Schwering, A. SmartSkeMa: Scalable Documentation for Community and Customary Land Tenure. *Land* **2021**, *10*, 662. [CrossRef]
26. Biraro, M.; Zevenbergen, J.; Alemie, B.K. Good Practices in Updating Land Information Systems that Used Unconventional Approaches in Systematic Land Registration. *Land* **2021**, *10*, 437. [CrossRef]
27. Byamugisha, F.F.K. Experiences and Development Impacts of Securing Land Rights at Scale in Developing Countries: Case Studies of China and Vietnam. *Land* **2021**, *10*, 176. [CrossRef]
28. Martono, B.; Aditya, T.; Subaryono and, S.; Nugroho, P. The Legal Element of Fixing the Boundary for Indonesian Complete Cadastre. *Land* **2021**, *10*, 49. [CrossRef]
29. Panday, U.S.; Chhatkuli, R.R.; Joshi, J.R.; Deuja, J.; Antonio, D.; Enemark, S. Securing Land Rights for All through Fit-for-Purpose Land Administration Approach: The Case of Nepal. *Land* **2021**, *10*, 744. [CrossRef]
30. Musinguzi, M.; Enemark, S.; Mwesigye, S.P. Fit for Purpose Land Administration: Country Implementation Strategy for Addressing Uganda's Land Tenure Security Problems. *Land* **2021**, *10*, 629. [CrossRef]
31. Chigbu, U.E.; Bendzko, T.; Mabakeng, M.R.; Kuusaana, E.D.; Tutu, D.O. Fit-for-Purpose Land Administration from Theory to Practice: Three Demonstrative Case Studies of Local Land Administration Initiatives in Africa. *Land* **2021**, *10*, 476. [CrossRef]
32. Antonio, D.; Njogu, S.; Nyamweru, H.; Gitau, J. Transforming Land Administration Practices through the Application of Fit-for-Purpose Technologies: Country Case Studies in Africa. *Land* **2021**, *10*, 538. [CrossRef]
33. Mekking, S.; Kougblenou, D.V.; Kossou, F.G. Fit-for-Purpose Upscaling Land Administration—A Case Study from Benin. *Land* **2021**, *10*, 440. [CrossRef]
34. Balas, M.; Carrilho, J.; Lemmen, C. The Fit for Purpose Land Administration Approach-Connecting People, Processes and Technology in Mozambique. *Land* **2021**, *10*, 818. [CrossRef]
35. Williams-Wynn, C. Applying the Fit-for-Purpose Land Administration Concept to South Africa. *Land* **2021**, *10*, 602. [CrossRef]

36. Griffith-Charles, C. Application of FFPLA to Achieve Economically Beneficial Outcomes Post Disaster in the Caribbean. *Land* **2021**, *10*, 475. [CrossRef]
37. Becerra, L.; Molendijk, M.; Porras, N.; Spijkers, P.; Reydon, B.; Morales, J. Fit-for-Purpose Applications in Colombia: Defining Land Boundary Conflicts between Indigenous Sikuani and Neighbouring Settler Farmers. *Land* **2021**, *10*, 382. [CrossRef]
38. Todorovski, D.; Salazar, R.; Jacome, G. Assessment of Land Administration in Ecuador Based on the Fit-for-Purpose Approach. *Land* **2021**, *10*, 862. [CrossRef]

 land

MDPI

Review

Experiences and Development Impacts of Securing Land Rights at Scale in Developing Countries: Case Studies of China and Vietnam

Frank F. K. Byamugisha

Independent Consultant, Washington, DC 20433, USA; fkbyamugisha@gmail.com

Abstract: This paper reviews experiences and development impacts of a selected number of developing countries in Asia and Africa that have used emerging land registration approaches to rapidly secure land rights at scale. Rapid and scalable registration is essential to eliminate a major backlog of the world's unregistered land, which stands at about 70 percent. The objective of the review, based on secondary data, is to draw lessons that can help accelerate land registration across many countries. While the focus is on China and Vietnam, the findings are buttressed by those from previous reviews in Ethiopia and Rwanda. The registration approaches used in these four countries were found to be cost-reducing, fast, inclusive and scalable enough to secure land rights for all within one generation. They also had significant positive impacts on land tenure security and investment. In addition, they indirectly along with other economic reforms contributed to rapid economic growth and a reduction in extreme poverty. The experience from these Asian and African countries offers important lessons including the need for strong political commitment and to develop flexible legal and spatial frameworks that fit the purpose of land registration, instead of the rigid technical standards set by land professionals.

Keywords: securing land rights; land registration; development impacts; fit-for-purpose land administration

Citation: Byamugisha, F.F.K. Experiences and Development Impacts of Securing Land Rights at Scale in Developing Countries: Case Studies of China and Vietnam. *Land* **2021**, *10*, 176. https://doi.org/10.3390/land10020176

Academic Editor: Christiaan Lemmen

Received: 31 December 2020
Accepted: 5 February 2021
Published: 9 February 2021

1. Introduction

The role of property rights in the economic development of the Western world has been well documented by economic historians and development economists including North and Thomas [1] and Rosenberg and Birdszell [2]. Well-defined and enforced property rights are key to economic development. Property rights in land are defined and enforced within formal governance structures called land administration systems. In Western European countries, virtually all their land, more than 95 percent, is registered in land administration systems, as reported by Schmid and Hertel [3]. However, the vast majority of the world's land, about 70 percent, remains unregistered and administered outside formal land administration systems [4]. To accelerate economic development and eradicate extreme poverty, especially in developing countries, it is important to register land quickly, at scale and in sustainable land administration systems.

In many developing countries, systems of land registration were initiated more than a century ago based on Western-style approaches that used rigid, high accuracy and skill-intensive standards of land surveying that are too costly to scale-up and maintain. Consequently, national registration coverage of land parcels (and owners) has been low, and limited primarily to urban areas and selected high-value rural areas. Moreover, subsequent transactions have not always been registered due to high registration charges relative to expected benefits, thus rendering the land registration records outdated. These conventional land registration approaches and their associated high costs have been at the root of the low levels of formal documentation and administration of land rights in many developing countries, including those in sub-Saharan Africa, where only about 10 percent of rural land is registered [5].

With new cost-reducing technologies emerging, an increasing number of countries has documented their land rights faster and at scale, including former Soviet Union countries after 1990 [6] and Thailand after 1984 with its 20-year land titling program [7]. A handful of countries have followed suit in Asia, Latin America and Africa, but they have struggled to document the more challenging lands in urban informal settlements, state lands, pastoral rangelands, woodlands and forestlands.

To register land at scale, including the challenging areas indicated above, new approaches have been developed including the fit-for-purpose approach to land administration. The approach has been embraced by key development players such as the International Federation of Surveyors (FIG) and World Bank [8] as well as the United Nations Food and Agriculture Organization (UN FAO)-led *Voluntary Guidelines on the Responsible Governance of Tenure of Land, Fisheries and Forests in the Context of National Food Security (VGGTs)*. Global Land Tool Network (GLTN)/United Nations Human Settlement Programme (UN HABITAT) supported the development of guidelines for its implementation at country level [9]. The emerging cost-reducing, affordable, fast and scalable approaches hold considerable promise to accelerate registration and improve land administration. However, a thorough evaluation of implementation experience is required to draw lessons that can inform adoption across many countries, building on some previous evaluations [5–8,10–12]. This article is intended to supplement these evaluations.

In addition to technical evaluations, socio-economic evaluations of land registration initiatives have also been done to assess their development impacts. A review of this literature suggests a positive impact of documentation of individual land rights on investment and productivity although gains have on average been more modest in Africa but stronger in Latin America and Asia [13,14]. The weaker gains in Africa have been attributed to the pre-existing context there, primarily the predominant customary land tenure having been relatively secure before formalization (hence weaker productivity gains from formalization); and the operating environment there (with less developed financial markets, and weak infrastructure and complementary investments) being inadequate to support a robust response of investment and productivity. Overall, these socio-economic evaluations suggest that documentation of land rights has significant impacts on investment and productivity, but context and complementary factors matter.

This paper reviews experiences of two Asian countries, China and Vietnam, in securing land rights at scale. The two populous nations have many rural arable land parcels, 1.5 billion in China and 70 million in Vietnam (see Table 1). They initiated country-wide documentation of land rights in the 1980s and 1990s using principles similar to those of fit-for-purpose approach to land administration long before it was articulated and formalized. Yet, there has not been a comprehensive and structured review of how the land registration was done and its results and development impacts. The objective of this paper is to review their land registration experiences, draw lessons and identify remaining challenges. The findings from the review are cross-checked against those from previous reviews of experience in two African countries, Ethiopia and Rwanda, to strengthen the conclusions and to broaden the global relevance of the lessons learnt.

Table 1. Population and rural land parcels of case study countries.

Country	Population Est. 2020 (Millions)	No. of Rural Arable Land Parcels	
		Year	Millions
China	1439	2008	1500
Vietnam	97	2002	70
Ethiopia	115	2019	50
Rwanda	13	2012	12
Total	1664		1632

Source: Estimates of population are based on data from the World Bank Database [15]. Rural land parcels data for China, Vietnam, Ethiopia and Rwanda were obtained from [12,16–18].

2. Approach and Methodology

This is a desk review that addresses the following research questions. How was land registration done in case study countries and what was the national coverage by land types and rights? How do the underlying principles to land registration relate to those in the fit-for-purpose land administration? What were the development impacts notably on investment, productivity, economic growth, poverty reduction and gender equity? To address these questions, the review covers four areas: (i) the legal framework to register land rights; (ii) the documentation, registration and certification of land rights; (iii) the development of unified and sustainable registration systems; and (iv) the development impacts of securing land rights at scale. The first two areas are important in assessing whether the legal framework was flexible enough and the registration approaches participatory, affordable, fast and scalable to secure land rights for all. The results would determine if the guiding principles of fit-for-purpose land administration are applicable retrospectively to the approaches that were used in those countries. The third helps in assessing whether the registration systems are unified or adequately coordinated to avoid high costs of establishment and maintenance, and to preserve quality, integrity and consistency of land registration data. The fourth area is important to justify investments in securing land rights.

Research for this paper was based on secondary data obtained through internet search and also from the records and databases of the World Bank where the author used to work and currently works as a consultant. Additional data was obtained from (a) government and donor-funded program and project documents including design, appraisal and evaluation documents, (b) publications by researchers and development practitioners, and (c) publications and databases of global development agencies and other actors. The method of research is qualitative based on case studies. The assessment of land registration and tenure security impact on economic growth is done by relating the former to the latter through intermediate outcomes such as investment and agricultural productivity since there is no direct relationship between land tenure security and economic growth. Past attempts to estimate their relationship have faced major problems of attribution [19]. Hence, we have assessed the impact of land registration on economic growth based on previous empirical research findings on registration impacts on investment and productivity. This has been supplemented by an assessment of economic growth (and poverty reduction) trends over the period of land registration interventions. Assessment of gender equity was based on reported land registration data disaggregated by gender.

3. Assessment and Results of Case Studies

3.1. China

Since the establishment of the People's Republic of China in 1949, the country relied on a centrally planned socialist economic system for its development until 1978 when it began a program of gradual but fundamental reform of the economic system toward a market-based mixed economy that continues to the present day. But even under the reform program to date, land has remained under state ownership in urban areas and under collective ownership in rural areas where farmland is contracted to households [20]. Hence, the registration of contractual rights to land plays an important role in confirming the land rights of users. With land tenure being central to the reform of the socialist economic system, there have been a strong political commitment to land tenure reform at all levels of government.

Land issues being local by nature, the decentralized system of government from central to provincial, municipality, prefecture-city and county down to commune level has enhanced participatory engagement especially in the registration and certification of land rights [21]. In addition, the securing of land rights has benefited from the fundamental reform of the socialist economic system in that the legal framework has had to be developed afresh and flexibly to cater for a multiplicity of reforms geared to achieving a market-based mixed economy. This provided opportunities to try out pragmatic approaches to land registration with principles similar to those of fit-for-purpose land administration [9].

3.1.1. The Legal Framework to Register Land Rights

From 1978, when it initiated major reforms to dismantle the collective system of production in favor of production by households through a Household Responsibility System (HRS), China's land tenure reform has followed a steady path to develop a comprehensive and flexible legal framework for land administration [20].

Rural Land Rights

Under the HRS and with rural land still collectively owned, rural households accessed land for farming through contract from local commune authorities, initially for five years, then extended in 1984 to 15 years, and to 30 years in 1993 when transfers of contracts for value among households within the collective were permitted but with prior consent from the collective. In addition, restrictions were imposed on periodic readjustment of contracted farmland which had been necessary to accommodate population growth within communes [20]. In the revisions to the Land Administration Law in 1998 [22], farming households were granted long-term and guaranteed land use rights which, in 2013, were confirmed to stay unchanged for a long time. The law also required farmers to receive a 30-year land use contract certificate [23]. In 2002, the land rights over farmland were strengthened by permitting households to transact them freely within their collective and also to sub-lease to households outside the collective. Contractual terms for grassland were set at 30–50 years, and for forestland at 30–70 years while they were undefined for residential rural land but with buildings on it owned in perpetuity. However, residential land could not be mortgaged or transferred to urban residents, to households outside the collective or used for non-agricultural development [24].

In 2007, the land use rights were further strengthened by clarifying that they were no longer contractual rights but property rights whose 30-year term were extendable upon expiration; it also laid out and strengthened arrangements for property registration [25]. The strengthened provisions for registration of farmland were reiterated and expanded in 2010 to also cover forestland and residential land, and to establish a rural land registration system. In 2013, the central authorities set a target of completing registration of rural land rights within five years. In the Decision of the Communist Party in November 2013 and in the newly-revised Rural Land Contract Law in 2019, the mortgaging and guaranteeing of contracted farmland as collateral were confirmed [24].

Rural land, collectively owned, is contracted out for farming (farmland) while grassland and forestland are contracted out for other rural development enterprises. The remaining land is construction land which is under two categories; land on which residential houses are built for the households (residential or homestead land) and also used to meet demand for village towns and enterprises; and for urban expansion. Construction land for residential houses, village towns and enterprises is allocated by collective authorities while land for urban expansion is expropriated by local governments with payment for compensation, and added to the stock of urban construction land for urban expansion or contracted out for private sector development [21].

Urban Land Rights and Buildings

Urban land is state-owned and managed by local governments. Land for construction is used by local governments for urban development, including infrastructure and other public services, while the remaining land is contracted to institutions and individuals for residential, industrial and commercial development. The contracting of land, through land market auctions, is for 70 years for residential purposes; 50 years for industrial development projects; and 40 years for commercial, tourism, or recreational purposes. The contractual terms are renewable, transferable and mortgageable and, for individuals, inheritable [26].

Prior to 1988, there were no individual rights to urban land in China. The state had a monopoly on housing, and distributed houses to households through their work units and enabled them to obtain accommodation at low rents. Since there was only public landownership at that time, there was little need to introduce a full-fledged land registration system. In 1988, China introduced a policy of privatizing its housing and of creating a housing market. It changed its constitution to allow individual urban land use rights and building ownership, and their transferability. While rules, laws and guidelines for registering urban private housing were initiated as early as 1982, they were only gradually developed and consolidated in the Real Estate Administration Law of 1994 which premised legal ownership of private housing on registration and award of a certificate for building ownership and another award of a land use rights certificate for land on which the property is built [27]. Subsequent laws and regulations were promulgated to fine tune the details and processes of registration of real estate transactions including mortgaging, starting with the Property Law of 2007. The latter, in particular, strengthened property rights by tackling deficiencies in the immovable property registration system by providing a national vision and framework to guide local registration offices [26].

3.1.2. Documenting, Registering and Certifying Land Rights

Rural Land Rights

There are two sets of documentation required to register and certify rural land rights. The first is textual or alphanumeric data that records rights in land. The second is a spatial framework in the form of cadastral or map data that show boundaries and extent of land over which these rights apply. Until the adoption of the 2007 Property Law which required and provided for the registration and certification of property rights, the registration of rural land rights (contractual farmland, grassland and forest land; residential or homestead land; and rural construction land) was based mostly on textual data, without a requirement for a spatial framework. Registration was required by the 1998 Land Administration Law to confirm the 30-year contractual term with certificates, referred to as standardized and notarized contracts [23]. The applications for certification of contractual rights for various kinds of rural land were submitted to the responsible agencies for processing and registration at county level as required by their statutes: to Ministry of Agriculture for registration of farmland and grassland; and to the State Bureau of Forestry for registering forestland. The Ministry of Land and Resources (MLR) was responsible to register other rural collective lands and to oversee the registration of other lands [28]. The registration and certification was based mostly on textual ownership data, without using a precise spatial framework [21,29].

After adopting the 2007 Property Law and based on its provision for national standards for registration, MLR in 2007 issued detailed measures to provide not only textual data on contractual rights but also spatial data that indicate the geo-position, boundaries and size of land covered by the rights as well as the methods of recording and indexing, and the powers of review by the registrar [21].

In 2009, the central government initiated measures to undertake pilot projects in the country, building on a land registration experiment in Anhui province started in 2005 [23]. Each province was ordered by the central government to continue promoting these land registration projects. The provincial governments were given the responsibility to organize and lead, the city-level governments to organize and coordinate, and the county and township governments to implement these projects ensuring that they are completed in every village in their jurisdictions [21].

The projects piloted an area by area systematic and participatory approach using commune leaders and members (with little training) to demarcate, adjudicate and register land use rights. They used textual data of contracted rural land rights and spatial data mostly from high-resolution satellite imagery (0.5 m resolution), supplemented by more precise ground survey data in case of high value land, to document and register contractual land use rights. They also introduced digital land records [30,31]. The piloting continued through 2012 and, in 2013, it was scaled up to cover the whole country. The procedures used in the scale-up included: publicity and mobilization campaigns, and short-training of local non-professionals in land demarcation and adjudication by surveyors and other land professionals; survey of households and confirmation of their membership in and contractual rights obtained from the collective; ortho-rectified aerial imagery (0.4 or 0.5 m resolution) supplemented where necessary with detailed survey of boundaries using a variety of survey instruments; public display and verification of field survey and adjudication results; registration of land rights; and issuing of certificates of land use rights [21,31]. While the piloting and scaling up of registration was the responsibility of county and township governments for all villages in their jurisdictions, it's the land professionals in MLR in the counties and townships that were responsible for training and supervision of non-professionals (mainly commune leaders and members) to do the demarcation and adjudication, while also undertaking quality control [21,28].

The scaling up of registration in 2013 was done rapidly while also expanding the capacity and national coverage of local Land Administration Offices under MLR (restructured and renamed Ministry of Natural Resources in 2018) to support first time registration and to also handle subsequent land transactions in a computerized environment. By the end of 2018, China had virtually completed first-time registration and certification of rural contracted land, which was 1.48 billion mu (one mu is equivalent to 0.165 acres) or about 98.87 million hectares, accounting for 89 percent of the measured area of contracted land [21].

From 2015, the scaling up of land registration was done hand-in-hand with the implementation of a nationwide, unified real property registration system which was completed in mid-2018. It included organizational consolidation of registration responsibilities under MLR and the development of a national digital information platform to share land information and enhance collaboration across all stakeholders [32].

Urban Land Rights and Buildings

While the legal framework to support registration of urban land rights was initiated in 1982, actual registration commenced in 1988 when an economic policy of privatizing housing and a housing market were introduced, and constitutional changes made to allow transferability of privatized land use rights. The main requirement for registration was textual documentation of land use rights. An additional requirement for a spatial framework was introduced later with the adoption of the 2007 Property Law which also provided for national standards to register urban land rights (urban construction land; and contractual land use rights and the associated developments). The Ministry of Housing and Urban-Rural Development (formed in 2008; replacing Ministry of Construction) followed up in 2008 with the issuing of detailed measures of registration including requirements for textual data on land use rights and spatial data. The required spatial data were geo-position, boundaries and area of the immovable property captured in updated cadastral maps based on orthophotos (0.2 to 0.4 m resolution), high resolution satellite imagery (0.5 m resolution), drones and detailed ground surveys [26].

Unlike the documentation of rural land rights which followed an area by area systematic registration approach, the registration of urban land rights was done sporadically based on applications from rights holders. It should be noted that the registration of building ownership was done separately from land use rights, with building ownership registered at the local Authority for Housing while land use rights and related transactions such as mortgages were registered at the local Land Administration Authority, resulting in the issuing of separate building ownership and land use rights certificates, respectively. With commencement of registration of contractual urban land rights in 1988, pressure was put on both the local Authority for Housing and the local Land Administration Offices to expand their capacity and urban coverage to handle the registration demand which increased with the boom in real estate that ensued in the 1990s and 2000s following housing privatization and development of the housing market [33]. The registration of urban land and properties was based on cadastral maps with scales of 1:500 and 1:1000 in the larger cities and 1:2000 in the smaller cities and other developed areas [26,34]. They also initiated the digitization of land records and to operate in a computerized work environment [35]. The registration of land and buildings, combined with the incorporation of peripheral and urban villages (informal settlements) in urban master plans [26], reduced the percentage of the urban population living in slums from about 44 in 1990 to about 25 in 2018 [36].

3.1.3. Developing a Unified and Sustainable Registration System

The Property Law of 2007 provided a vision of a unified land and property registration system to overcome fragmentation in registration of property rights which was leading to inefficiencies and increased incidences of error. Fragmentation had three sources: property registration being done at three separate levels of government; building ownership being registered separately from land use rights; and different rural land types (farmland, grassland and forestland) being registered in separate systems and agencies. But the law lacked regulations and implementation measures to achieve the vision, and hence was not implemented until 2015 when they were put in place [33]. But some jurisdictions could not wait that long and so some municipalities, provinces and prefecture-cities came up with their own local regulations either modifying or supplementing the national regulations while others made much more profound change to the national framework [33]. For example, Shanghai Municipality in 2009 overhauled its Regulations on Registration of Real Estate to implement a unified registration of land and buildings thus issuing unified certificates of right to land and buildings [35]. A few other municipalities followed suit including Chongqing and Tianjin, in 2004 and 2006, respectively. On the other hand, Beijing municipality, decided to continue with registration of building ownership while discouraging registration of land use rights except for an exclusive list of residential developments. Some prefecture cities also did likewise including Dalian of Liaoning Province, Qingdao of Shandong Province, Xiamen of Fujian Province and Kaifeng of Henan Province [33].

In November 2013, MLR was assigned the responsibility to guide and supervise the implementation of the unified property registration system throughout China. Accordingly, MLR established in August 2014 a new department, the Bureau of Real Estate Registration and put it in charge of implementing the unified property registration system [37]. In November 2014, it issued Interim Regulations on Real Property Registration and commenced its implementation in March 2015 with a plan to establish a nationwide, unified real property registration system in three years [33]. Through 2015 and 2016, Real Estate Registration Bureaus were established in various provinces to implement the unified registration system [38]. Based on the experience gained by some municipalities and prefecture-cities that had implemented unified registration systems on their own, a national digital land information platform was developed to support the integration of real property registration data of buildings, urban land, farmland, woodland, grassland and forestland of varying quality, under one system [35]. According to the Ministry of Natural Resources, the unified real property registration system came into effect on 18 June 2018. It started connecting "3001 property registration stations in 335 cities and 2853 counties serving more than 300,000 enterprises and individuals on average each day, according to latest statistics" [32]. With the national unified real property registration system in place and operational, the quality and efficiency of registration services are expected to improve throughout the country as they did in Shanghai and other jurisdictions which implemented unified property registration initiatives on their own before the introduction of the national program in 2015.

3.1.4. Development Impacts

Implementation of the HRS had a significant positive impact on investment and agricultural growth as reported by Lin [39]. Also, according to impact studies, the accompanying first round certification of contractual rights to farmland had a significant positive impact on investment and productivity for the farmers whose land was not subjected to frequent reallocation by commune authorities; the impact was achieved through increased investment incentives, renting-out land and migrating out of farming to more rewarding economic activities [23,40,41]. The second-round certification, both piloting and scaling up, also significantly promoted investment incentives among farmers who had not had land reallocation experience but negatively affected those who had experienced big reallocations [42].

Notwithstanding the dampening effects of land reallocations on investment response to certification of contractual land rights, the legal clarification and registration of land rights together with other economic reforms and public investments introduced after 1978 had a profound impact on China's economy as reported by some researchers [43]. Over the 40 years of economic reforms (1980–2019), China's economy achieved an average annual real growth rate of 9.4 percent measured in GDP constant prices, based on calculations using data from the IMF World Economic Outlook Database [44]. Land registration and tenure security contributed to the growth indirectly mainly by stimulating private investment while the greater contribution to growth came from the other major economic reforms and public investments [43]. The rapid and sustained economic growth contributed to reducing the proportion of people living below the poverty line (US$1.25 per day) from over 85 percent in early 1980s to about 0.5 percent (using the poverty line of US$1.90 per day) in 2016, according to data from the World Bank Poverty and Equity Data Portal [45].

3.2. *Vietnam*

After its 1954 independence from the French, which left it divided into two parts (North and South), Vietnam emerged from the "Vietnam war" in 1975 with a reunified country. In 1986, it introduced sweeping economic reforms (the "Doi Moi" policy) and a move from a socialist economic system towards a market-oriented economy, including the dismantling of collective production and the allocation of land use rights to households [46]. Like in China, where land tenure was central to the fundamental reform agenda, there was a strong political commitment to land tenure reform at all levels of government in Vietnam. And reform implementation, including land registration, was participatory, aided by "peoples committees" which operate at every decentralized level of government from provincial to district and down to the commune level [46]. In addition, the legal framework was developed afresh and flexibly to cater for a multiplicity of sweeping economic reforms. This created opportunities to try out pragmatic approaches to land registration that are underpinned by principles similar to those of the recently formalized fit-for-purpose land administration [9].

3.2.1. The Legal Framework to Register Land Rights

Since 1986 and the "Doi Moi" policy, Vietnam progressively moved towards a market economy, with two major changes in rural land rights. In 1988, the collective system of production was dismantled in favor of production by households. In the agricultural sector, Resolution 10 of the 1988 Land Law granted land use rights (for 10 to 15 years) to individual households, with land allocation done by commune authorities. Land use rights could not be traded or exchanged until adoption of the 1993 Land Law which assigned five rights to land users-to transfer, exchange, lease, inherit and mortgage land use rights.

Ownership was vested in the entire people, according to the 1992 Constitution, with the state managing the land on their behalf [47].

The duration of rural land use rights was increased to 20 years for annual crops (increased further to 30 years in a 1998 revision) and to 50 years for perennial crops and forestland. The law also enabled authorities to allocate land for urban residential use on a stable long-term basis while land used for income production could be allotted or leased for short periods based on business production plans. It also provided for registration of land use rights.

These land rights were extended and clarified by the 1998 amendment to the Land Law to allow sub-leasing and Vietnamese entrepreneurs to contribute land rights to joint ventures with foreign companies. Further amendments in 2001 simplified procedures in urban areas to allow foreign investors to lease land for renewable periods of 50 to 70 years [47].

The 2003 Land Law provided a legal boost to the emergence of a market for land use rights. It also provided further equality on land rights between domestic and foreign investors, including Vietnamese permanent residents abroad, to buy property associated with their land use rights, and between husband and wife in certification of land use rights [48].

The 2013 Land Law extended lease terms for all agricultural land to 50 years for both annual and perennial crops, and broadened the scope and duration of land rights by allowing landholders to transfer land as gifts to others or as shares in joint ventures [49].

The land rights for rural residential and urban land were also reformed. For residential land (for houses) in rural and urban areas, the allocation to households and individuals is for "long term use", basically indefinite duration. The 1992 constitution and the Civil Code give citizens the right of housing ownership in both rural and urban areas [50]. The Housing Law and Real Estate Business Law of 2014 extended "land-use rights" to foreign investors, allowing title holders to conduct property transactions, including mortgages [51].

3.2.2. Documenting, Registering and Certifying Land Rights

Following introduction of the "Doi Moi" policy, Vietnam started registering land rights and issuing land use rights certificates (LURCs) to households in 1994 on the basis of the 1993 Land Law [46]. This was done based on textual data on land rights, mainly the land allocation approval package of documents including the application for LURCs by individual households, evidence of support by neighbors and approvals by the commune and district peoples committees and by the district land allocation and registration committees. Cadastral maps of varying quality (some with inadequate scale; others outdated) [52] indicating location, boundaries and size of land, were also used in some cases [47,53].

The registration process of land rights involves all levels of government, from central down to commune level, and non-professionals (with short training) especially members of commune and district peoples committees. At central government level, the General Department of Land Administration (GDLA), established in 1994, is responsible for registration and issuing LURCs for rural land. After the Ministry of Natural Resources and Environment (MONRE) was established in 2002, its departments absorbed the functions of GDLA. GDLA was re-established in 2009 and took over daily oversight of land administration under MONRE's guidance [46,54]. The Ministry of Construction had responsibility to issue building ownership certificates in urban areas. At regional and local level, the peoples committees are responsible for implementing land administration according to the laws, working with the Department of Natural Resources and Environment (DONRE) including land administration offices [46,54].

In 1994, the GDLA arranged with Ministry of Construction to sponsor the joint issuing of land use rights and building ownership certificates for urban land. This was formalized through amendments to the Land Law in 1998 to develop a unified system to register land use rights and ownership of the attached properties; the amendments laid out the associated processes and procedures for registration. At provincial and city levels, the respective housing departments supervised by GDLA and Ministry of Construction were merged under the supervision of GDLA thereby enabling GDLA to take responsibility for issuing not only LURCs but also the Building Ownership and Land Use Certificates (BOLUCs) [53].

The registration of rural land was very rapid but slow for urban land. By the end of 2000 (after only six years of implementation), 90 percent of rural land users had been issued with LURCs while in urban areas, only 16 percent of land users had been issued LURCs [53]. Rural land registration was compulsory and undertaken systematically, area by area, while registration in urban areas was voluntary and done on request. Low registration in urban areas was attributed also to high land allotment fees (up to 20 per cent of the land values) accentuated by complex procedures of registration and incomplete cadastral mapping [50]. The issuing of LURCs in rural and urban areas was based largely on textual data, with hardly any precise cadastral data. The requirements were consistent with the urgency and national scale of the task to register an estimated 70 million land parcels in rural areas [17].

Following the adoption of the 2003 Land Law, Vietnam embarked on registering land rights using integrated textual and spatial data based mostly on orthophotos (0.4 m resolution) and high resolution satellite imagery (0.5 m resolution). In addition, an initiative was made to start issuing LURCs on a parcel (instead of household) basis to overcome challenges of registering subsequent transactions [55].

The cumulative total number of LURCs and BOLUCs issued after the adoption of the 1993 Land Law was about 30 million in 2006, according to reports by GDLA, indicating an annual average of 2.5 million LURCs/BOLUCs. As of December 2007, about 82 percent of agricultural land area, 62 percent of urban residential land area, 76 percent of rural residential land area and 62 percent of forest land area had been covered with LURCs [52]. The high volume of registration notwithstanding, the quality of registration documents was inadequate, with cadastral maps mostly inaccurate and outdated [52]. It was estimated by the World Bank that, to complete the nationwide registration including updating cadastral maps (at scales of 1:500 to 1:2000 in urban areas; and 1:2500 to 1:10,000 in rural areas), about 20 million of LURCs/BOLUCs would need to be issued or re-issued [52,56]. Much of this work was supported by a World Bank project, from 2008 to 2015, including the registration and issuing of 3 million LURCs, out of which 80 percent were certificates reissued with an upgraded spatial framework based mostly on orthophotos (0.4 m resolution) and high resolution satellite imagery (0.5 m resolution), supplemented by ground surveying [57]. Field adjudication and demarcation work was done mostly by local people, mainly members of communes and peoples committees, trained and supervised by qualified surveyors from GDLA supported by contract staff, while the back-office document processing was done entirely by GDLA land professionals supported by contract staff [57]. GDLA had about 12,000 land professionals (called cadastral staff) supporting the land registration program in the entire country as of 2015 [54]. Approximately 62 percent of the LURCs issued under the project were registered in the names of women or joint spouses [57].

Using support from development partners led by the World Bank, Vietnam was able to ramp up the quality and quantity of registration of land use rights from 2009 to 2015. As of January 2015, Vietnam had virtually completed its nationwide program of good quality land registration, with the following outputs and coverage of first time land registrations [54]:

- 20.2 million LURCs for agricultural land, representing 90% total area;
- 2.0 million LURCs of forestland, representing 98% total area;
- 13.0 million LURCs of rural residential land, representing 94% total area;
- 5.3 million LURCs of urban residential land, representing 97% total area; and
- 0.3 million LURCs for special purpose land, representing 85% total area.

According to MONRE, as of September 2018, the government had issued LURCs for 96.9 percent of land in Vietnam [51]. The registration of urban land and property, combined with the formalization of urban informal settlements, reduced the percentage of the urban population living in slums from about 61 in 1990 to about 14 in 2018 [36].

3.2.3. Developing a Unified and Sustainable Registration System

Following the 2003 Land Law that required a unified registration system in terms of organizational arrangements and operations, Vietnam established unified land registration offices at the provincial level in all its 58 provinces (and 5 municipalities) and in one third of their office branches in the country's 650 districts by mid-2007 [52]. While the organizational unification went well and rapidly, the unification of operations lagged much behind as both spatial and textual data at provincial, district and commune levels remained outdated and thus inconsistent at different levels, and also digitally unlinked [52]. To accelerate the unification of land registration while also developing and modernizing land administration, MONRE outlined a 15-year comprehensive program which included a "Strategy for Information Technology Application and Development for Natural Resources and Environment to Year 2015 with a vision to 2020" approved by the Prime Minister. The Strategy included modernization of land administration by the year 2020 with emphasis on several areas including modernizing the system to collect and update land information and establishing a nationwide unified land database [52]. To implement the program, the government mobilized internal and external financing exceeding US$100 million, including a US$70 million World Bank loan to finance completion of first time land registration, upgrading land rights documentation, completing the unified land registration system and modernizing land registration and administration infrastructure.

As of 2019, Vietnam had developed a unified, comprehensive and decentralized land registration system covering all types of land in the country. At the central level, MONRE has developed and is overseeing reliable procedures and standards for land registration that are being implemented at provincial, district and commune levels. Cadastral data, both textual and spatial, has been updated and is mostly digitally linked with a Land Information System (LIS) software in land registration offices at provincial and district levels that are well-equipped with modern IT equipment. At commune-level, access points with internet connectivity have been established in many of the commune offices.

While a unified land registration system is in place, there are still gaps [58]. It is not comprehensive enough and does not include key areas of land administration such as land valuation and land use planning. Nor does it have links with other relevant computerized information systems within and outside government. To meet these needs, the government is developing a uniform and comprehensive system, the National Land Information System (NLIS), which will include a national land database, a national land information portal and a unified electronic land registration system within a unified framework. It will also include interoperable and standardized data modelling for land data exchange as well as necessary ICT infrastructure to support links with other relevant information systems within and outside government [58].

3.2.4. Development Impacts

The implementation of the 1993 Land Law and the issuing of LURCs led to significant increases in agricultural investment [47] and allowed households to pursue non-farming activities [55]. It also increased agricultural productivity and efficiency in the overall economy by promoting the leasing of land to enable its full use as more rural households moved out of farming to more rewarding activities in the evolving non-agricultural sectors of the economy [59].

Over the 34 years of implementation of securing land rights at scale and undertaking other major economic reforms (Moi Doi) and public investments (1986–2019), Vietnam's economy grew at an average annual real growth rate of 6.6 percent as measured by GDP in constant prices, according to calculations based on data from the IMF World Economic Outlook Database [44]. Land registration and tenure security contributed to the growth indirectly mainly by increasing private investment and agricultural productivity while the greater contribution to growth came from the other major economic reforms and public investments [60]. The rapid and sustained economic growth led to reducing the proportion of the population living below the poverty line (US$1.90 per day) from 51.9 percent in 1992 to 1.9 percent in 2018, according to data from the World Bank Poverty and Equity Data Portal [45].

4. Discussion

This evaluation of land registration experiences of China and Vietnam has produced notable results and lessons that could help other developing countries to register land at scale. To strengthen the lessons and conclusions from the review while also geographically broadening their global relevance, the findings from the review are supplemented by those from previous evaluations in Africa, notably for Ethiopia and Rwanda, which have had considerable success in registering their land at scale. The discussion below focuses on selected findings to draw lessons especially from the emerging land registration approaches that are cost-reducing, affordable, fast and scalable.

4.1. Flexible Legal Frameworks

Flexible and responsive legal frameworks enabled the adoption of successful approaches to secure land rights at scale. While initial legislations in both China and Vietnam focused on decollectivizing production, the legal frameworks were flexible enough to allow the adoption of spatial and institutional frameworks as well as new technologies that enabled the registration and certification of land use rights at scale and the efficient operation of land registration systems. In Africa, where Ethiopia and Rwanda have played a lead role in securing land rights at scale, the two countries have also had flexible legal frameworks to accommodate successful registration approaches and efficient operation of land registration systems [8,11,12,61–66].

4.2. Land Registration Without a Precise Spatial Framework

Registration at scale without a precise spatial framework was attained, upgradeable and had positive impacts on investment and productivity. When they decollectivized their farming and allocated land use rights to households, both China and Vietnam documented land use rights and issued lease certificates to their farmers based mostly on textual data, without a precise spatial framework, in their so-called first round of registration. The certification was done very rapidly to assure farmers of their newly granted land use rights, achieving virtually full national coverage (a large share of 1.5 billion rural arable land parcels in China and about 70 million in Vietnam) within 5–7 years, at a low cost (less than US$1.5 per parcel in the case of Vietnam). Despite lack of a precise spatial framework and the low cost of registration, the certification provided the required tenure security for 20 years (1993–2012) in the case of China, and 15 years (1994–2008) in the case of Vietnam.

Impact evaluation studies found that land certification led to significant increases in tenure security, agricultural investment and productivity in both countries. However, in China, the investment response was undermined by frequent administrative reallocation of land in case of farmers that experienced it. Ethiopia also certified rural land use rights (and most urban land outside informal settlements) without a precise spatial framework over 30 million rural land parcels (during 1998–2009), at a cost of US$1 per parcel [61,65]. Like in China and Vietnam, the certification of rural land in Ethiopia had a significant positive impact on investment and productivity, according to impact studies [62,63]. While documentation of land rights without a spatial framework was a good decision given the urgency to confirm the land rights of millions of households and the difficulties of accessing technology at the time, it was an interim measure which was later followed by the addition of a spatial framework when access to aerial imagery, especially from high resolution satellites and drones, became easier and cheaper. Rwanda took advantage of the accessibility and reduced cost of aerial imagery when it implemented its land registration program as indicated in the next section.

4.3. Land Registration with a Precise Spatial Framework

Registration at scale using mostly orthophotos and high resolution satellite imagery was done cost-effectively. In their second-round registration, both China and Vietnam upgraded their approach to registration of rural land use rights by adding a spatial framework based mostly on orthophotos (0.4 m resolution) or high resolution satellite imageries (0.5 m resolution), supplemented by ground surveys where physical land boundaries were not visible or land values were high enough to justify the cost. China piloted the second round registration in 2009–2012, scaled it up from 2013 and used it to register 89 percent of all the contracted rural lands by the end of 2018, within five years of scaling up. Vietnam, on the other hand, started implementing second round registration of rural land around 2009, and had, within 10 years, issued land use certificates for around 97 percent of its land as of September 2018.

While information on the average cost of registration per parcel is not available for China and Vietnam, in the case of Ethiopia which has been implementing second level certification of rural land since 2014 and has demarcated 20 million land parcels in 6 years using a similar spatial framework, the average cost per parcel has been around US$8.5 [66]. Rwanda, which registered the entire country's rural and urban land within five years (2009–2013) using only orthophotos (0.4 m resolution) and high-resolution satellite imageries (0.5 m resolution), did so at an average cost of US$8 per parcel [12]. It must be noted that, to register land at scale, rapidly and at a reduced cost, all the four countries maximized the use of orthophotos and high resolution satellite imageries (0.4 or 0.5 m resolution) as well as non-professionals, trained for a short time. While doing so, they minimized the use of high accuracy ground survey instruments, and used land professionals sparingly mainly to provide training and supervision of non-professionals and the management of processes [12,21,31,46,57,66].

Registration of urban land, in both China and Vietnam, was sporadic, and not systematic. It was initially done using low quality cadastral maps, either of inadequate scale, incomplete or outdated. The cadastral mapping was later improved with updated maps of larger scales of 1:500 to 1:1000 in the larger cities and 1:2500 in the smaller cities [34,56]. In Africa, Ethiopia has followed a similar trend while Rwanda registered all its rural and urban land systematically using cadastral maps generated from aerial imageries (0.4 or 0.5 m resolution) at an average cost of US$8 per parcel [12,66].

4.4. Unified Land Registration Systems

Unified registration systems which are efficient, transparent and protective of data quality have been mostly developed. Fragmented registration systems had not only been costly to build and maintain, they had also been cumbersome and time-consuming. For example, both China and Vietnam had used an approach in which land was registered in one registration system while the attached buildings were registered in a separate system. In addition, China had used separate registration systems for farmland (together with grassland), woodland, forestland and urban land. The development of unified registration systems, involving organizational, operational and digitization components, took about 15 years from vision to full implementation in both Vietnam and China. Much of the implementation was completed by end of 2018 in the case of China while in Vietnam it was still on-going as of November 2020. The design and implementation involved flexibility in legislating, organizing, staffing and in designing land information systems. Implementation required three things: consolidating the responsibilities of real property registration under one organization; integrating or linking real property data bases under one information platform for easy sharing within and outside government; and improving business processes and services. The most challenging activities were the development of national digital land information platforms and systems to support the integration or linking and sharing of real property registration data that, in the case of China, brought together the different registration data for buildings, urban land, farmland, grassland, woodland and forestland of varying quality. But due to experiences gained from earlier implementation initiatives, development and implementation of a national unified real property registration system integrating or linking all real property registration data was done within the planned time frame (end of 2018) at least for China. Vietnam on the other hand completed implementation of the unified registration system but is still developing a broader land administration information system to accommodate all core land administration processes as well as links with other information systems within and outside government.

As for the two African countries, Rwanda possesses a well-developed digital land administration information system with links to other information systems within and outside government. On the other hand, Ethiopia has developed one for rural land but has ways to go in developing another for urban land and links between them, or integrating the two to form a unified national land administration information system [12,64].

4.5. Development Contributions of Secured Land Rights

Reforms and investments to secure land rights at scale contributed to high economic growth and a rapid reduction of poverty. In all the case study countries, previous evaluation studies confirmed positive impacts of the land tenure reforms (legal clarification and certification of rights) on land tenure security, investment and productivity except for Rwanda where the productivity response time was too short to allow quantifiable results [12]. But context matters. The land tenure reforms in all the case study countries were part of broader economic reforms to transform from central planned to market-based systems (in China, Vietnam and Ethiopia) or, in the case of Rwanda, to resettle, rebuild and recover from civil war, displacement and genocide [12]. The countries grew their economies strongly over the periods of land tenure reform, with China, Vietnam, Ethiopia and Rwanda recording average annual real economic growth rates of 9.4%, 6.6%, 8.2% and 7.7%, respectively, calculated using data from the IMF World Economic Outlook Database [44]. As earlier noted, while land tenure reforms contributed indirectly to the growth particularly through enhanced investment incentives and productivity, a much greater contribution came from complementary economic reforms and public investments as documented for China by Garnaut, Song and Fang [43], for Vietnam by Le [60], for Ethiopia by World Bank [67] and for Rwanda by Crisafulli and Redmond [68].

The combined impact of land tenure and other major economic reforms and public investments on poverty reduction was equally impressive, with China and Vietnam mostly eradicating extreme poverty while Ethiopia and Rwanda reduced considerably the proportion of people living below the poverty line (US$1.90 per day) from 72.3 percent in 1995 to 32.6 percent in 2016 and from 69 percent in 2005 to 56.5 percent in 2016 respectively, according to data from the World Bank Poverty and Equity Database [45]. Land rights for women were enhanced at least in the cases of Ethiopia, Rwanda and Vietnam. Ninety (90) percent of land rights certificates were issued to women in sole or joint ownership in Ethiopia [66], 68 percent in Rwanda [12] and 62 percent in Vietnam [57].

4.6. Lessons Learnt and Challenges

There are important lessons to learn from these Asian and African land registration experiences. At least five lessons stand out from the land registration experience of China and Vietnam, reinforced by the more recent experience of Ethiopia and Rwanda, to secure land rights at scale.

1. **Strong political commitment**. Political commitment is crucial to securing land rights at scale. As land registration was part of fundamental economic reforms to move from a socialist planned economic system to a market-based economic system in China, Vietnam and Ethiopia and part of comprehensive reforms to recover from civil war and genocide in Rwanda, there was strong political commitment to land registration from the highest to the lowest level of government. The political commitment carried the day.

2. **Flexible legal framework**. Flexibility in legal frameworks facilitates the adoption of some emerging registration approaches that are fast, affordable and scalable to register land rights for all.
3. **Registration based on imprecise spatial framework**. Registering land rights mostly on the basis of textual data, without a precise spatial framework, and with limited use of land professionals is a low-cost, affordable, scalable and upgradable approach best suited for first round registration especially for low value rural land as was demonstrated by China, Vietnam and Ethiopia in the 1990s and 2000s. But with rapidly declining costs and easier access to aerial imagery, the option of adding a spatial framework (based on aerial imagery) to textual data is increasingly becoming realistic.
4. **Registration using mostly aerial imagery for a spatial framework**. Adding a spatial framework based mostly on orthophotos and high resolution satellite imagery as an upgrade from first round registration has proven to be a scalable cost-effective land registration approach as it minimizes the use of expensive ground survey equipment while maximizing the use of non-professionals (with short training) in place of surveyors and land lawyers to demarcate and adjudicate land. In fact, Rwanda covered the registration of all its rural and urban lands using only aerial imagery (orthophotos of 0.4 m resolution and satellite imagery of 0.5 m resolution) while China, Vietnam and Ethiopia supplemented aerial imagery with ground surveying only in areas where either physical boundaries were not visible or land values were high with a potential to cause contestation of land boundaries.
5. **Developing a unified digital registration system**. Developing a unified digital registration system is feasible and improves efficiency, transparency, protection of data quality and integrity, and facilitation of information sharing. Rwanda has done it and it has been beneficial. China, Vietnam and, to a less extent, Ethiopia (for rural land) have also done and it has started to pay off.

There are also at least three challenges.

1. **Keeping land registration data updated.** In Vietnam, Ethiopia and Rwanda, registration of subsequent transactions after the first round of certification was difficult because registration forms used in the latter were in a format that could not allow the recording of subsequent transactions. In the second round of land certification (the tail end of first registration in the case of Rwanda), the forms were redesigned and digitized, and systems were developed to enable registration of subsequent transactions. Another challenge related to the maintenance of land registration, at least in the case of Rwanda where the issue was closely monitored, is that registration of subsequent land transactions has remained relatively low for rural land mainly because registration charges have been high relative to the value of land. The government of Rwanda has been considering options to address it including reducing registration charges for rural land while increasing those for urban land to effect cross-subsidization since urban land is of higher value. In China and Vietnam, there are no reported issues of registration of subsequent transactions presumably because land values in both rural and urban areas are high enough to cover land registration charges. As for Ethiopia whose land registration coverage includes a lot of rural low-value land, the registration of sub-sequent land transactions has not been an issue so far presumably because many of the rural land offices have not started charging registration fees given that the new national rural land administration information system is still being installed and many land offices have not been covered yet [64].
2. **Addressing land tenure insecurity and informality in urban slums**. While registration of urban land in the four case study countries has gone well, notwithstanding the poor quality of cadastral mapping in the case of Ethiopia [65], the formalization of informal settlements and registration of land rights in urban slums are still inadequate despite great improvements made by Vietnam and China. For example, the percentage of the urban population living in slums declined from about 44 in 1990

to about 25 in 2018 in China and from 61 in 1990 to 14 in 2018 in Vietnam compared to the decline from 96 in 1990 to 64 in 2018 in Ethiopia and from 96 in 1990 to 42 in 2018 in Rwanda [36]. The prevalence of slums especially in the African countries of Ethiopia and Rwanda remains a serious development challenge.

3. **Registering customary land rights**. While China and Rwanda have virtually no customary or communal land tenure systems left, Vietnam and Ethiopia still have them but the latter has successfully piloted approaches which are being scaled up to register land rights especially for pastoralist groups [69]. Vietnam, on the other hand has registered most of the customary lands which are found in the mountainous and forest areas occupied by indigenous ethnic communities which account for about 13 percent of Vietnam's total population [70]. Vietnam's customary land tenure systems are recognized under state laws and, before the lands are registered and land use rights certificates (LURCs) issued, the land-owning communities are organized into legal entities on the basis of Bylaws [70]. The main challenge customary land tenure systems face in Vietnam, like in many other countries especially in Africa, is how to organize the land-owning groups into legal entities and to strengthen their capacity to manage their land, forestry, pastoral and other natural resources [5].

5. Conclusions

When China and Vietnam decollectivized agricultural production and allocated rural land to farming households in the 1980s and 1990s, they were faced with a formidable task of registering land rights covering about 1.5 billion rural arable land parcels in China and about 70 million in Vietnam. This was in addition to the need to registering urban land rights. In both countries, the registration of rural land was done in two rounds. The first round was done rapidly and much of it within 7–8 years after land allocation, covering about 90 percent of the allocated land parcels in both countries. While the registration was done at county level in China and at district level in Vietnam, the field work of adjudication and demarcation was done mainly by commune authorities (and peoples committees in Vietnam) after short training done by land professionals from land administration departments, who also managed the registration processes. The processes were participatory involving not only the holders of land use rights but also the neighbors. They were also cost-effective, costing in the case of Vietnam, about US$1.5 per parcel.

The second round in China started with piloting from 2009 to 2012, and scaling up from 2013 to 2018 while in Vietnam, implementation of the second round started in 2009 and was completed around 2018 as well. The second round was done in both countries mainly to add or upgrade cadastral maps and to digitize and develop unified land registration systems. The cadastral maps were based mostly on orthophotos and high resolution satellite imageries. While the field work of adjudication and demarcation was done by local non-professionals mainly commune leaders and peoples committees supported by land professionals (called cadastral officers in Vietnam) from land agencies, the documentation and digitization was done only by government staff from land agencies supplemented by contracted staff, in the case of Vietnam. The work in Vietnam involved 12,000 cadastral staff from government and many more in the case of China. When the second round of registration ended in 2018, China had registered 89 percent of the contracted rural lands while Vietnam had issued land use rights certificates covering about 97 percent of its rural land. Grassland, woodland, forestland, collectively owned land and state lands were all registered in both countries. In addition, urban land was registered over time and completed by 2018 in both countries. The registration of urban land was based on applications from holders of land use rights (sporadic method), and the cadastral maps used were updated over time.

The land registration processes in both China and Vietnam were underpinned by flexible legal and spatial frameworks that accommodated the technologies and standards used during the two rounds of registration. They were also participatory involving the holders of land rights, their neighbors and local leaders. In addition, they were inclusive, enabling the registration of rights of all landholders. The registration processes were also upgradable and scalable, and used skilled land professionals and costly survey equipment sparingly while maximizing the use of non-professionals and aerial imageries. In short, the registration approaches used were flexible, participatory, cost-reducing, affordable, fast, inclusive and scalable. These are the same kind of principles that underpin the fit-for-purpose land administration approach, notwithstanding the fact that they were alluded to in the VGGTs in 2012 and used long before the approach was formalized in the guiding principles for country implementation published in 2016 [9]. Rwanda, which completed its land registration program recently, and Ethiopia, whose land registration program is ongoing, have applied similar principles with success, as indicated in the last chapter. In all the four countries, there was strong political commitment from government from central level down to local level, and land registration was part of comprehensive economic reform programs. The findings from this review of land registration experience in China and Vietnam, buttressed by those from previous reviews for Ethiopia and Rwanda, suggest that developing countries can secure their land rights at scale within a generation if they adopt similar registration approaches. They also suggest that the fit-for-purpose land administration guiding principles for country implementation indicated above [9] would be useful for developing countries considering to engage in providing secure land rights at scale.

It should be noted that in China, Vietnam and the two African countries of Ethiopia and Rwanda, previous evaluation studies confirmed positive impacts of the land registration programs on investment and productivity except for Rwanda where the productivity response time was too short to allow quantifiable results. While the relationship between land tenure security and economic growth is indirect and its measurement has been confounded by attribution problems [19], the positive contributions of land tenure to private investment and agricultural productivity suggest that the land registration programs together with the other major economic reforms and public investment in the four countries contributed considerably to the economic development of these countries. During the periods when land registration programs were implemented, the economies of China, Vietnam, Ethiopia and Rwanda recorded average annual real economic growth rates of 9.4%, 6.6%, 8.2% and 7.7%, respectively. The impact on poverty reduction was equally impressive, with China and Vietnam mostly eradicating extreme poverty while Ethiopia and Rwanda reduced considerably the proportion of people living below the poverty line (US$1.90 per day) from 72.3 percent in 1995 to 32.6 percent in 2016 in the case of Ethiopia, and from 69 percent in 2005 to 56.5 percent in 2016 in the case of Rwanda. In addition, land rights for women were enhanced at least in the cases of Ethiopia, Rwanda and Vietnam. Ninety (90) percent of land rights certificates were issued to women in sole or joint ownership in Ethiopia, 68 percent in Rwanda and 62 percent in Vietnam.

The experience of these Asian and African countries in securing their land rights at scale offers lessons that other developing countries can apply to secure land rights for all in one generation. The most two crucial lessons are: political commitment to securing land rights at scale; and using flexible legal and spatial frameworks that fit the purpose of land registration, instead of the rigid technical standards set by land professionals. However, there are also three challenges to note from the experience of these four countries. The first challenge is keeping land registration records updated. This is an important issue in Rwanda where registration of subsequent land transactions has remained relatively low for rural land mainly because registration charges have been high relative to the value of land. The government of Rwanda has been considering options to address it including reducing registration charges for rural land while increasing those for urban land to effect cross-subsidization since urban land is of higher value. Keeping land records updated is critical to the sustainability and integrity of land registration systems.

The second challenge is addressing urban slums, notwithstanding the success achieved in registering urban land. The percentage of the urban population living in slums in 2018 was 25 for China, 14 for Vietnam, 64 for Ethiopia and 42 for Rwanda. The prevalence of slums especially in the African countries of Ethiopia and Rwanda remains a serious development challenge. The third challenge is registering customary and community land rights. Among the four countries covered by this review, China and Rwanda have virtually no customary land tenure systems remaining while Vietnam and Ethiopia still have them, and they are recognized under state laws. In Vietnam, customary communities mostly in the mountain areas have been formalized and issued land use rights certificates covering their agricultural and forest lands. The challenge they face, like many other customary tenure communities especially in Africa, is how to organize themselves into legal entities and to strengthen their capacity in managing their land, pastoral, forestry and other natural resources.

The persistence of the three challenges highlighted above, namely, keeping land registration data updated, addressing land tenure insecurity and informality in urban slums and registering customary land rights, suggests a need for intensifying evaluation of implementation experiences in these areas to learn lessons that can help in addressing these challenges more effectively.

Funding: This work received no external funding.

Institutional Review Board Statement: Not applicable.

Informed Consent Statement: Not applicable.

Conflicts of Interest: The author declares no conflict of interest.

References

1. North, D.; Thomas, R. *The Rise of the Western World*; Cambridge University Press: Cambridge, UK, 1973; ISBN 9780511819438.
2. Rosenberg, N.; Birdszell, L.E. *How the West. Grew Rich: The Economic Transformation of the Industrial World*; Tauris, I.B., Ed.; Basic Books Co., Ltd.: London, UK, 1986; ISBN 1-85043-016-0.
3. Schmid, H.C.U.; Hertel, C. *Real Property Law and Procedure in the European Union: General Report, Final Version*; European University Institute Florence/European Private Law Forum Deutsches Notarinstitut Wurzburg: Wurzburg, Germany, 2005; Available online: https://www.eui.eu/Documents/DepartmentsCentres/Law/ResearchTeaching/ResearchThemes/EuropeanPrivateLaw/RealPropertyProject/GeneralReport.pdf (accessed on 4 November 2020).
4. Williamson, I.; Enemark, S.; Wallace, J.; Rajabffard, A. *Land Administration for Sustainable Development*; ESRI Press Academic: Redlands, CA, USA, 2010; ISBN 978-1-58948-041-4.
5. Byamugisha, F.F.K. *Securing Africa's Land for Shared Prosperity: A program to Scale up Reforms and Investments*; Africa Development Forum Series; World Bank: Washington, DC, USA, 2013; ISBN 978-0-8213-9810-4.
6. Törhönen, M.-P. *Keys to Successful Land Administration: Lessons Learned in 20 Years of ECA Land Projects*; World Bank Group: Washington, DC, USA, 2012; Available online: https://fig.net/resources/proceedings/fig_proceedings/fig2017/papers/p04g/P04G_torhonen_8760.pdf (accessed on 30 October 2020).
7. Burns, A. *Thailand's 20 Year Program, To Title Rural Land*; Background Paper Prepared for the World Development Report, No. 31363; World Bank: Washington, DC, USA, 2005; Available online: http://documents1.worldbank.org/curated/en/223631468777611336/pdf/313630TH0land111WDR20050bkgd0paper1.pdf (accessed on 30 October 2020).
8. Enemark, S.; Bell, K.C.; Lemmen, C.; McLaren, R. *Fit.-For.-Purpose Land Administration*; Joint FIG/World Bank Publication, FIG Publications No 60; FIG Office: Copenhagen, Denmark, 2014; Available online: https://www.fig.net/resources/publications/figpub/pub60/Figpub60.pdf (accessed on 20 October 2020).
9. Enemark, S.; McLaren, R.; Lemmen, C. *Fit.-For.-Purpose Land Administration: Guiding Principles for Country Implementation*; GLTN/UN Habitat: Nairobi, Kenya, 2016.
10. Navajas, H.; Byamugisha, F. *Global Land Tool Network (GLTN), Phase 2: End of Phase Evaluation*; Evaluation Report 2/2018; GLTN/UN Habitat: Nairobi, Kenya, 2018.
11. English, C.; Locke, A.; Quan, J.; Feyertag, J. *Securing Land Rights at Scale: Lessons and Guiding Principles from DFID Land Tenure Regularization and Land Sector Support. Programmes. LEGEND*; Technical Report; DFID: London, UK, 2019. [CrossRef]
12. Ngoga, T.H. *Rwanda's Land Tenure Reform: Non-Existent to Best Practice*; CABI: Wallingford, UK, 2018.
13. Deininger, K.; Feder, G. Land registration, governance and development: Evidence and implications for policy. *World Bank Res. Obs.* **2009**, *24*, 233–266. [CrossRef]
14. Lawry, S.; Samii, C.; Hall, R.; Leopold, A.; Hornby, D.; Mtero, F. The impact of land property rights interventions on investment and agricultural productivity in developing countries: A systematic review. *J. Dev. Effect.* **2017**, *9*, 61–81. [CrossRef]
15. World Bank Population Database. Available online: https://data.worldbank.org/indicator/SP.POP.TOTL (accessed on 25 January 2021).

16. Rabley, P.; Yuen, E. In China, GIS-Based Land Registry Aims to Protect Farming Rights and Enhance Food Security. ESRI, ArcNews, Spring Issue. 2009. Available online: https://www.esri.com/news/arcnews/spring09articles/in-china.html (accessed on 25 January 2021).
17. Marsh, S.P.; MacAulay, T.G. Farm size and land use changes in Viet Nam following land reforms. In Proceedings of the 47th Conference of the Australian Agricultural and Resource Economics Society, Fremantle, WA, Australia, 12–14 February 2003; Available online: https://ideas.repec.org/p/ags/aare03/57919.html (accessed on 25 January 2021).
18. Ministry of Foreign Affairs of Finland. REILA Deliverable: Systematic Land Registration. Available online: https://www.niras.com/media/10911267/reila2_landregistration_compressed.pdf (accessed on 22 October 2020).
19. Roth, M.; McCarthy, N. *Land Tenure, Property Rights, and Economic Growth in Rural Areas*; USAID Issue Brief: Washington, DC, USA, 2013; Available online: https://www.land-links.org/wp-content/uploads/2016/09/USAID_Land_Tenure_Economic_Growth_Issue_Brief-061214-1.pdf (accessed on 28 January 2021).
20. Liu, S. The structure of and changes to China's land system. In *China's 40 Years of Reforms and Investments: 1978–2018*; Garnaut, R., Song, L., Fang, C., Eds.; ANU Press, Australian National University: Canberra, Australia, 2018; Chapter 22; ISBN 9781760462246.
21. Zhou, Y.; Li, X.; Liu, Y. Rural land system reforms in China: History, issues, measures and prospects. *Land Use Policy* **2019**, *91*, 104330. [CrossRef]
22. People's Republic of China. Land Administration Law. 1998. Available online: https://www.cecc.gov/resources/legal-provisions/land-administration-law-of-the-peoples-republic-of-china-0 (accessed on 1 February 2021).
23. Deininger, K.; Jin, S. The impact of property rights on households' investment, risk coping, and policy preferences: Evidence from China. *Econ. Dev. Cult. Chang.* **2003**, *51*, 851–882. [CrossRef]
24. Li, L.; Tan, R.; Wu, C. Reconstruction of China's farmland rights system based on the 'trifurcation of land rights' reform. *Land* **2020**, *9*, 51. [CrossRef]
25. National People's Congress. Property Law of the People's Republic of China. 2007. Available online: http://www.china.org.cn/china/LegislationsForm2001-2010/2011-02/11/content_21897791.htm (accessed on 8 February 2021).
26. World Bank; Development Research Center of the State Council, the People's Republic of China. *Urban. China: Toward Efficient, Inclusive, and Sustainable Urbanization*; World Bank: Washington, DC, USA, 2014.
27. People's Republic of China. Land Administration Law on the Urban. Real Estate. 1994. Available online: http://www.asianlii.org/cn/legis/cen/laws/aoturel381/ (accessed on 1 February 2021).
28. Ministry of Land and Resources, PRC. The Main Functions of the Ministry of Land and Resources of the Peoples' Republic of China, Undated. Available online: http://cfguide.cn/govshow/Ministry-of-Land-and-Resources_16.htm (accessed on 1 February 2021).
29. Ho, P. Who owns China's land? Policies, property rights and deliberate institutional ambiguity. *China Q.* **2001**, *166*, 387–414. [CrossRef]
30. Barlow, C.H. Chinese rural land registration and certification program: A case study in modeling costs and benefits for a systematic registration campaign. In Proceedings of the World Bank Annual Conference on Land and Poverty, Washington, DC, USA, 18–20 April 2011.
31. Ye, J.; Feng, L.; Jiang, Y.; Lang, Y.; Roy, P. 2016 China rural land use rights survey and research—17 provinces survey results and policy recommendations. *Manag. World* **2018**, *34*, 98–108.
32. Ministry of Natural Resources. The State Council of the Republic of China. 2018. Available online: http://english.www.gov.cn/stat.e_council/ministries/2018/06/18/content_281476190239424.htm (accessed on 14 October 2020).
33. Xu, L. The new real property registration structure in China: Progress with unanswered questions. *Glob. J. Comp. Law* **2016**, *5*, 91–117. [CrossRef]
34. State Bureau of Survey and Mapping, PRC. Development of surveying and mapping in China during 1997–2000. Country Report Submitted by State Bureau of Surveying and Mapping, The People's Republic of China. In Proceedings of the Fifteenth United Nations Regional Cartographic Conference for Asia and the Pacific, Kuala Lumpur, Malaysia, 11–14 April 2000; Available online: https://unstats.un.org/unsd/geoinfo/RCC/docs/rccap15/E_CONF_92_L.2.pdf (accessed on 1 February 2021).
35. Yang, G.; Yang, Z.; Han, Z. Research on basic database for real property unified registration. *MATEC Web Conf.* **2018**, *176*, 03003. [CrossRef]
36. World Bank/UN Habitat Urban Population in Slums Database. Available online: https://data.worldbank.org/indicator/EN.POP.SLUM.UR.ZS?locations=1W (accessed on 26 January 2021).
37. China to Launch Unified Property Registration. *China Daily*, 21 August 2014. Available online: http://www.chinadaily.com.cn/china/2014-8/21/content_18458759.htm (accessed on 14 October 2020).
38. Zhang, N.; Wang, C. China's New Real Estate Registration System Launched. *Mondaq*, 9 February 2016. Available online: https://www.mondaq.com/china/real-estate/464906/china39s-new-real-estate-registration-system-launched (accessed on 14 October 2020).
39. Lin, J.Y. Rural reforms and agricultural growth in China. *Am. Econ. Rev.* **1992**, *82*, 34–51.
40. Ye, J.; Jiang, Y.; Prostman, R. 2005 survey of rural land use rights in China 17 findings and policy recommendations from the provinces. *Manag. World* **2006**, *7*, 77–84.
41. Deininger, K.; Jin, S. Land rental markets in the process of rural structural transformation: Productivity and equity impacts from China. *J. Comp. Econ.* **2009**, *37*, 629–646.

42. Hong, W.; Luo, B.; Hu, X. Land titling, land reallocation experience, and investment incentives: Evidence from rural China. *Land Use Policy* **2020**, *90*, 104271. Available online: https://www.sciencedirect.com/science/article/abs/pii/S0264837718314030 (accessed on 20 October 2020). [CrossRef]
43. Garnaut, R.; Song, L.; Fang, C. (Eds.) *China's 40 Years of Reforms and Investments: 1978–2018*; ANU Press, Australian National University: Canberra, Australia, 2018; ISBN 9781760462246. Available online: https://b-ok.global/book/5580749/762aa9 (accessed on 28 January 2021).
44. IMF World Economic Outlook Database. Available online: https://www.imf.org/en/Publications/WEO/weo-database/2020/October (accessed on 25 January 2021).
45. World Bank Poverty and Equity Database. Available online: https://data.worldbank.org/topic/poverty (accessed on 25 January 2021).
46. Nang, L.V. Land Administration and Land Registration and the Relationship among Government Bodies in Viet Nam. In Proceedings of Twentieth United Nations Regional Cartographic Conference for Asia and the Pacific, Jeju, Korea, 6–9 October 2015; Available online: https://unstats.un.org/unsd/geoinfo/RCC/docs/rccap20/27_paper_LuuVanNang.pdf (accessed on 20 October 2020).
47. Do, Q.T.; Iyer, L. Land Rights and Economic Development: Evidence from Vietnam. May 2003. Available online: https://ssrn.com/abstract=445220 (accessed on 9 October 2020).
48. USAID. USAID Country Profile: Property Rights and Resource Governance. Viet Nam. 2013. Available online: https://www.land-links.org/wp-content/uploads/2016/09/USAID_Land_Tenure_Vietnam_Profile.pdf (accessed on 2 November 2020).
49. Socialist Republic of Viet Nam. Vietnam Land Law 2013—Law No. 45/2013/Qh13. Available online: https://vietanlaw.com/vietnam-land-law-2013/ (accessed on 2 November 2020).
50. Japan Bank for International Cooperation. Urban Development and Housing Sector in Viet Nam, 1999. JBIC Research Paper No. 3. December 1999. Available online: https://www.jica.go.jp/jica-ri/IFIC_and_JBICI-Studies/jica-ri/publication/archives/jbic/report/paper/pdf/rp03_e.pdf (accessed on 19 October 2020).
51. US Department of State. 2019 Investment Climate Statements: Vietnam. Available online: https://www.state.gov/reports/2019-investment-climate-statements/vietnam/ (accessed on 2 November 2020).
52. World Bank. *Viet Nam Land Administration Project, Project Appraisal Document*; Report No. 39867-VN; World Bank: Washington, DC, USA, 2008.
53. Quan, D.A. *Two Land Registration Systems. The Land Law of Viet Nam and of Sweden*; Lund University: Lund, Sweden, 2011; Available online: http://lup.lub.lu.se/search/ws/files/5712795/4024268.pdf (accessed on 20 October 2020).
54. Nang, L.V.; Viet Nam Land Administration and Land Registration. Viet Nam General Department of Land Administration. October 2015. Available online: https://unstats.un.org/unsd/geoinfo/RCC/docs/rccap20/27_Presentation_Viet%20Nam_Luu%20Van%20Nang.pdf (accessed on 20 October 2020).
55. Do, Q.T.; Iyer, L. Land titling and rural transition in Vietnam. *Econ. Dev. Cult. Change* **2008**, *56*, 531–579. [CrossRef]
56. Vo, D.H.; Thuc, L.Q. Development of surveying and mapping technology in Vietnam. In Proceedings of the New Technology for a New Century, International Conference, FIG Working Week, Seoul, Korea, 6–11 May 2001; Available online: https://www.fig.net/resources/proceedings/fig_proceedings/korea/abstracts/pdf/session3/dang-le-abs.pdf (accessed on 26 January 2021).
57. World Bank. *Viet Nam Land Administration Project. Implementation Completion and Results Report*; Report No. ICR3590; World Bank: Washington, DC, USA, 2015.
58. Gil, S. An overview analysis in Asia with some current LIS initiatives. In Proceedings of the Presentation Made at the International Conference on Global Land Information System, Entebbe, Uganda, 20–21 February 2020.
59. Deininger, K.; Jin, S. Land sales and rental markets in transition: Evidence from rural Vietnam. *Oxf. Bull. Econ. Stat.* **2008**, *70*, 67–101.
60. Le, H. Economic reforms, external liberalization and macroeconomic performance in Viet Nam. *Int. J. Finance Econ.* **2019**, *176*. Available online: http://www.internationalresearchjournaloffinanceandeconomics.com/ISSUES/IRJFE_176_08.pdf (accessed on 28 January 2021).
61. Gebremeskel, T. FIG-WB Forum on Land Administration and Reform in Sub-Saharan Africa: Ethiopia. In Proceedings of the FIG Working Week, Abuja, Nigeria, 6–10 May 2013.
62. Deininger, K.; Ali, D.A.; Alemu, T. Impacts of land certification on tenure security, investment and land market participation: Evidence from Ethiopia. *Land. Econ.* **2011**, *87*, 312–334. [CrossRef]
63. Melesse, M.B.; Bulte, E. Does land registration and certification boost farm productivity? Evidence from Ethiopia. *J. Agric. Econ.* **2015**, *46*, 757–768. Available online: https://www.onlinelibrary.wiley.com/doi/abs/10.1111/agec.12191 (accessed on 3 November 2020). [CrossRef]
64. Zein, T.; Gebremeskel, T.; Tenno, A.T.; Reda, Y.; Abera, T. The innovative national rural land administration information system of Ethiopia. In Proceedings of the 2019 World Bank Conference on Land and Poverty, Washington, DC, USA, 25–29 March 2019; The World Bank: Washington, DC, USA, 2019.
65. Land Equity. *A Review of Urban. Legal Cadastre of Government of Ethiopia: Issues and Policy Recommendations Report*; Report No.138777; World Bank Group: Washington, DC, USA, 2016.
66. LIFT. Land Tenure and Certification. Available online: https://liftethiopia.com/programme-themes/land-tenure-and-certification/ (accessed on 24 December 2020).

67. World Bank. *Ethiopia's Great Run: The Growth Acceleration and How to Pace It*; World Bank Group: Washington, DC, USA, 2015; Available online: https://openknowledge.worldbank.org/handle/10986/23333 (accessed on 28 January 2021).
68. Crisafulli, P.; Redmond, R. *Rwanda, Inc.: How a Devastated Nation Became an Economic Model, for the Developing World*; Palgrave Macmillan: New York, NY, USA, 2014; ISBN 978-0-230-34022-0.
69. Bekure, S.; Mulatu, A.; Dagnew, A.; Negassa, D.; Haddis, Z.; Gebremeskel, T. Formalizing pastoral land rights in Ethiopia: A breakthrough in Oromia National Regional State. In Proceedings of the 2018 World Bank Conference on Land and Poverty, Washington, DC, USA, 19–23 March 2018; World Bank: Washington, DC, USA, 2018.
70. FAO; MRLG. Challenges and Opportunities of Recognizing and Protecting Customary Tenure Systems in Viet Nam. 2019. Available online: http://www.fao.org/3/CA1037EN/ca1037en.pdf (accessed on 30 January 2021).

 land

Article

The Legal Element of Fixing the Boundary for Indonesian Complete Cadastre

Dwi Budi Martono *, Trias Aditya, Subaryono Subaryono and Prijono Nugroho

Doctoral Study Program in Geomatics Engineering, Department of Geodetic Engineering, Faculty of Engineering, Universitas Gadjah Mada, Jalan Grafika No. 2, Yogyakarta 55284, Indonesia; triasaditya@ugm.ac.id (T.A.); ssubar@ugm.ac.id (S.S.); prinug@ugm.ac.id (P.N.)

* Correspondence: dwi.budi.martono@mail.ugm.ac.id; Tel.: +62-811342769

Abstract: In 2017, the Indonesian government implemented the systematic land registration (PTSL) process, projected to be finished by 2025. However, this process faces some challenges in the spatial and legal data collection process, resulting in the Indonesian cadastral system still being incomplete. For instance, during the three years of its implementation, out of about 135 million parcels, only 49.5% have been registered. Therefore, the level of completeness needs to be improved. This research aims to assess the compliance of the fixed boundary process' legal elements, such as the parties that locate the boundary, agreement between the adjoining landowners, and boundary markers. This is a piece of qualitative research in which the data were obtained through interviews from questionnaire surveys to land administration policymakers. Subsequently, the research carried out regulation assessments to develop a country-context cadastre typology of the current cadastral mapping activities. Data were obtained from the results of the PTSL campaign in the Madiun regency. The result showed that the high percentage (i.e., 96.61%) of legal elements regarding the boundary agreement in a rural area could be used as a potential enabler towards achieving completion of the Indonesian cadastre.

Keywords: complete cadastre; legal element; fixing boundary; eligible landowner; agreement; boundary marker

Citation: Martono, D.B.; Aditya, T.; Subaryonoand, S.; Nugroho, P. The Legal Element of Fixing the Boundary for Indonesian Complete Cadastre. *Land* **2021**, *10*, 49. https://doi.org/10.3390/land10010049

Received: 17 November 2020
Accepted: 1 January 2021
Published: 7 January 2021

1. Introduction

The United Nations Economic Commission for Europe (UNECE) is one of the five regional commissions under the United Nations Economic and Social Council's jurisdiction to promote economic cooperation and integration. In 1996, this organization introduced land administration terminologies [1]. Furthermore, in the 2000s, several studies relating to the land registration domain reported that the complete cadastral system is an engine that supports its administration, which concisely encompasses land tenure, value, use, and development functions [2–4]. However, since the Dutch colonialization, Indonesia's cadastral system has remained incomplete, although approximately 49.5% of land parcels have been registered [5]. This incompleteness creates a significant barrier to the construction of modern land administration [1].

A complete cadastral domain is not something new, as reported in the 2014 International Federation of Surveyors (FIG) statement on the Cadastre [6] and also in the Bathurst Declaration on Land Administration for Sustainable Development [7]. In the Indonesian context, a complete cadastre is described as an essential component of land reform [8]. The Bogor Declaration (1996) stated that it is considered an underlying infrastructure that supports land reform administration. The high percentage of unregistered lands deters the integration of this system. It was suggested that the Indonesian cadastre needs to include all types of land, namely forests, urban and rural areas. The cadastral framework (which is a map) is a fundamental layer within the National Spatial Data Infrastructure (NSDI) [9].

The government has initiated a process to accelerate land registration through complete systematic land registration (*Pendaftaran Tanah Sistematis Lengkap* or PTSL). It is national strategic program to register all land parcels in Indonesia starting from 2017 and projected to be finished by 2025. PTSL aims to increase tenure security for all land and to improve economic access. The PTSL standard is based on the "continuum of land rights" philosophy [10]. Its technical book shows that the cadastral mapping of a village includes undisputed land parcels ready for certification (K1). Disputed and undisputed land parcels can be identified using both spatial and legal status of ownership. This includes the disputed land parcels (K2), undisputed land parcels that are not ready for certification (K3), and existing certified land parcels (K4). Land certificate is evidence that the parcel is registered. K4 is a registered land that is inappropriately mapped or lacks spatial information [11] so that its spatial accuracy needs to be improved.

However, the results from PTLS were unable to attain the expected completeness (spatial and legal data). Spatial data are a description of the location, boundaries, and area of the land parcel, while legal information is the legal status, landholders, including the rights of other parties and its encumbrances [12].

The studies relating to cadastre, land registration, and land management to attain sustainable development contributes to the theoretical maturity of the land administration [3,13–15]. The "Land Administration Domain Model (LADM)" [10,16–18], in accordance with the Social Tenure Domain Model (STDM) as its sub-version [19–22], is a prominent example. LADM deals with standardization and facilitates the efficient set-up of land administrations to integrate various spatial data. Subsequently, the "Fit-for-Purpose Land Administration (FFP LA)" [13,14] is also popular. FFP LA focuses on cost-effective data collected quickly, which meet society's needs and are incrementally improved [13,14,23]. Globally, it is believed that the critical bottleneck for the provision of a complete cadastre that serves as infrastructure for an effective land administration system lies in the process of collecting spatial data [24–27]. In addition, land boundaries are a crucial component, and from a legal-societal perspective, it is defined as where a person's interests of the land end and that of the next begins. It is often manifested spatially by visible artifacts such as hedges, stone walls, ditches, or land-use changes [28].

Recent publications and studies reported that the use of a visible boundary concept is adequate and sufficient for incorporating the remaining unregistered parcels, particularly in rural, semi-urban, and forest areas [13,14,29–32]. In this regard, visible marks are evident, and fixed boundaries are used where relevant or for specific purposes, and they are paid for by the landowners or stakeholders. There are several ways of documenting boundaries, from a verbal description of words, a "metes and bounds" approach, depiction on an index map, capturing corners in coordinates, etc. [15]. Boundary determination is a critical aspect of the construction of a land ownership record, which maintains its integrity, reliability, and accuracy. However, Arruñada (2018) supports the voluntary type, rather than the mandatory demarcation [33].

The research aims to identify the distinction between physical (not the spatial) and legal elements in a fixed boundary, as reported by Arruñada (2018). The study describes some of the characteristics of the Indonesian cadastral system and the reason it is still incomplete. A cadastre typology and methodology were proposed to assess the compliance of boundary surveys, in addition, it also contributes to an improved national cadastral system. This research is divided into five sections starting with materials and methods followed by results obtained from the investigation, the assessment outputs, discussion, conclusions, and recommendation.

2. Materials and Methods

This research adopted a case study methodology [34,35] as well as qualitative and quantitative methods, as shown in Figure 1. The subsequent section describes the approaches implemented as follows, (A) an interview was carried out to identify the cadastral system in Indonesia, (B) analysis of the literature and regulations that describe the six cadastral elements in the land registration regime, (C) assessment of the cadastral measurements of PTSL in urban areas using a multivariate clustering tool which resulted in cadastral typologies, and (D) to analyze the issue of incompleteness in this system.

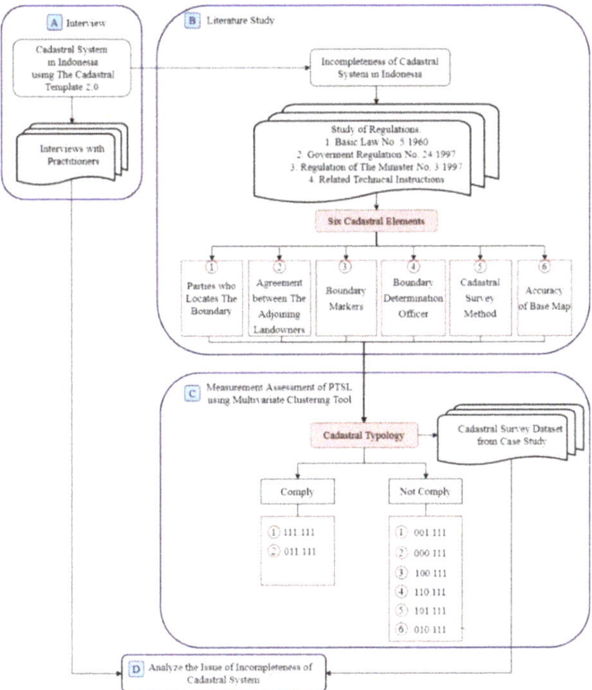

Figure 1. The workflow of the assessment of the legal elements in fixing the boundary in Indonesia.

2.1. Materials

The cadastral template 2.0 was adopted to identify this system [36–39]. A description of Indonesian cadastre, based on The Cadastral Template 2.0 is already available from http://cadastraltemplate.org/indonesia.php. It was completed in 2003. The recording at FIG is not updated. The questionnaire—shown in the blue box (A) in Figure 1—was used to interview five senior executives of the Ministry of Agrarian Affairs and Spatial Planning/National Land Agency (ATR/BPN).

The second material (B) is centered on several regulations in ATR/BPN, including the Basic Agrarian Law No. 5 of 1960 (hereafter UUPA) [40], Indonesian Government Regulation No. 24 of 1997 on Land Registration (hereafter PP No. 24/1997) [12], Regulation of Minister for Agrarian Affairs/Head of National Land Agency Regulation No. 3 of 1997, the Implementation of Government Regulation No. 24 of 1997 (hereafter PMNA No. 3/1997) [41], and its technical instructions. Finally, the third material includes cadastral survey datasets (C).

2.2. Methods

This research adopted a qualitative method by interviewing the respondents. This was used to identify the cadastral system according to the decision-makers' perspective in ATR/BPN. The interviews are used as a basis for providing a description of the current administrative structure, rules, and procedures. The literature review was carried out to determine the current and best practice for the fixed boundary demarcation. A multivariate clustering tool was used to group the parcels in ArcGIS Pro into cadastral typology, generated using government regulations, PMNA, and technical guidelines.

The fixed boundary is mandatory in the Indonesian cadastre. However, strict technical requirements are challenging, thereby leading to land registration's slow progress [13,42]. Articles 14 to 22 of PP No. 24/1997 stated six elements related to boundary demarcation, although there are 2 aspects, namely, spatial (some parties identify land parcels, measure and draw their boundaries) and legal (seeking the consent of neighbors) [33].

The six cadastral elements were used to assess the level of compliance during the survey of the PTSL project in 6 villages located at Madiun District, East Java Province. The proportions of cadastral measurements that do not comply with the regulation were determined in this study. A list of cadastral elements and scores are shown in Table 1. A score of assessment using binomial number 1 for "comply" and 0 for "not comply" was used to evaluate the compliance [43,44].

Table 1. Assessment of the Level of Compliance Score of Cadastral Survey at *Pendaftaran Tanah Sistematis Lengkap* (PTSL) Project (Adopted from PP No. 24/1997 and PMNA No. 3/1997) [12,41].

No.	Cadastre Elements	Score
1	Parties that locate the boundary	
	1.1. Landowner or proxy	1
	1.2. Occupant or possessing party	1
	1.3. Local officer *	1
	1.4. Interested or concerned parties	0
	1.5. No one	0
2	Agreement between the adjoining landowners	
	2.1. Yes	1
	2.2. No	0
3	Boundary markers	
	3.1. Concrete fence or wall	1
	3.2. Dike of paddy field, river, ditch, or road	1
	3.3. Monuments from wood or bamboo	1
	3.4. Plant fence or bamboo	0
	3.5. No boundary	0
4	Determination officer	
	4.1. Government Surveyor	1
	4.2. Licensed Surveyor	1
	4.3. Others	0
5	Survey method	
	5.1 Terrestrial	1
	5.2 Photogrammetric	1
	5.3 Others	1
6	Accuracy of the Base Map (Planimetric Accuration $\geq$ 0.3 mm x Scale)	
	6.1. $\leq$30 cm in the urban area	1
	6.2. $\leq$75 cm in the rural area	1
	6.3. $\leq$3 m in the plantation area	1
	6.4. Others	0
	6.5. No Base Map	0

* Note: Local officer represents the village officer who has supplementary function and authority to locate the parcel boundaries in their village in the systematic land registration.

This study assessed the cadastral measurement requirement for boundary determination, both legally and spatially, and 7552 of plots were evaluated using the six elements [33]. This research only focuses on three legal elements (parties that locate the boundary, agreement between the adjoining landowners, and boundary markers) in a rural area. Eight cadastral typologies, obtained from six villages were used as a case study, as shown in Table 2. The intention was to examine the possibility of implementing the FFP approach in rural locations where 78.9% of the population resides [45]. The implementation of its approach will address the challenge regarding the requirement for fixing the boundaries, as stipulated in the current regulation.

Table 2. Cadastral Survey Typology.

No.	Cadastral Typology	Remark
1	111.111	comply
2	011.111	comply
3	001.111	not comply
4	000.111	not comply
5	100.111	not comply
6	110.111	not comply
7	101.111	not comply
8	010.111	not comply

The PTSL surveyed parcel in the case study areas is distinguished with cadastral typology. It is classified as "comply" if the elements met the regulation, as shown in Table 1. It was discovered that eligible landowners' participation during the first land registration was an issue [11]. The eligible landowners are the parties who occupy and have control of the land legally or legitimately that is not contested by others. Legal owners lead to legal demarcation and they are part of the legal demarcation process. Their absence leads to legal demarcation, such as the party that locates the boundary, agreement between the adjoining landowners and the boundary markers. Zevenbergen [15], reported that the absence of right-holders in the case study areas is due to various reasons, including those domiciled in other regions, those that are unwilling for fear of rising taxes, and difficulties in proving ownership.

As shown in Table 1, the first digit refers to the parties that locate the boundary, while the subsequent one is the second element, and the fourth ended in the sixth element. Furthermore, there are three passive values, from the fourth to the sixth, with a score of 1. The three first digits are legal, and the three-second numbers are spatial demarcations, respectively. The absence of right holders or their proxies and the activeness of local officers [11] results in only the values of the first, second, and third elements to becoming active. However, when the cadastral element is unfulfilled, it tends to alter the scores. This research focuses on the legal element of demarcation, which refers to the social aspect of fixing the boundary. The social element leads to physical boundaries that may or may not represent the legal boundary location [46]. Spatial demarcation, which consists of the following elements, cadastre of determination officer, survey method, and the base map's accuracy, were not examined in this study. It was assumed that the spatial demarcation of PTSL complies with regulation as provided by the PTSL contract.

2.3. Assessment of Compliance

The proportion of compliance with the required cadastral elements was carried out using samples obtained from the case study area. A compliance analysis between the cadastral measurements of PTSL and regulation was carried out by calculating the proportion of those complied by the total sample. The outputs of this research were used to develop a strategy to improve the overall cadastral system's effectiveness, which serves as a basic layer for the land administration system. The proportion of surveyed parcels that complies with the cadastral elements required in Government Regulation No. 24/1997 is calculated using the percentage formula, which is stated as follows:

$$\%\, Comply = \frac{\Sigma\, Comply}{Total\ sample} \times 100\%$$

3. Results

3.1. Cadastral System in Indonesia

Indonesia is an archipelago country with a total landmass of 190 million ha. Conversely, 66.3% of the area is forest administered by the Ministry of Environment and Forestry (KLHK), and the remaining (33.7%) is managed by the Ministry of Agrarian Affairs and

Spatial Planning/National Land Agency (ATR/BPN). There are approximately 126 million parcels identified in 2016 and approximately 3.5 million new ones [47] appear yearly.

Historically, the cadastral system aimed to provide legal certainty of land rights. Subsequently, since 1813 (during the colonial era) to the beginning of the republican period, the Land Registry Services were part of the judicial or legal arrangement in the Ministry of Justice. The Cadastral Offices provided services, whereas the Ministry of Finance was in charge of land taxes [5]. In 1988, the National Land Agency (BPN) was established to integrate land registration and cadastre services. It was reported that for more than seventeen decades, the cadastral system served the judiciary rather than the administrative sector [13,14]. In 2014 the Ministry of Agrarian Affairs and Spatial Planning (ATR) was established to administer the functions of land use, previously supervised by the Ministry of Public Works, conversely, The Ministry of ATR acts as the head of BPN.

The Indonesian cadastral system provides spatial and legal data. The land office organizes the public registers (*Daftar Umum*), which consist of cadastral maps (*Peta Pendaftaran*), land registers (*Daftar Tanah*), surveying letters (*Surat Ukur*), land books (*Buku Tanah*), and name registers (*Daftar Nama*). The documents which serve as the basis for registration (*Warkah*) are given an identification number called DI (*Daftar Isian*) 208 and kept as an integral part of the *Daftar Umum*. The cadastral layer has not been integrated with other spatial data such as e-Government, community empowerment, and activities to attain sustainable development [36,48,49]. The cadastral system in Indonesia is reported in Appendix A.

Its main problems are incompleteness and the quality of the cadaster. These lead to the loss of public confidence in the land tenure systems, operational inefficiencies within land administration agencies, including costs and delays in land development processes [50]. Land computerization was implemented to digitalize land services. The challenges are as follows, firstly, some of the cadastral data have not been entered into the system (the *Komputerisasi Kegiatan Pertanahan* or KKP). Secondly, there is a validity issue regarding data quality. Thirdly, there is an analog document (papers-based documents). Fourthly, there are two group entry-levels of spatial data, un-plotted registered parcels (referred to as KW 4, 5, 6) and those which have been plotted (called KW 1, 2, 3). The classification of cadastral content entry levels into the KKP system is shown in Table A3, Appendix B. The quality of land parcels related to KW 1, 2, 3 differs in their accuracy and uncertainty level both relatively. In a position that needs to be improved, in addition, digital services are also still challenging.

The target and realization of PTSL from 2017 to 2019 are shown in Table A4 Appendix B. It implies that the higher realization versus the target does not guarantee the completeness of cadastre data. According to Enemark (2018), this target is only achieved by using an FFP approach. It was reported that FFP solutions tend to modify the current approach in Indonesia, which includes systematic registration with aerial mapping and participatory land adjudication, visual boundaries and areas calculated on the map, integrated land management based on a one map policy (OMP), and the use of locally trained land officers acting as trusted intermediaries [14]. The introduction of FFP land administration in Indonesia is supported by the President, who issued a land policy to boost tenure security and economic development [47]. However, its legal and institutional reform is still challenging.

3.2. The Fixed Boundaries

The term "boundary" refers to either the spatial objects that mark a parcel's limits or an imaginary line that divides two adjoining estates. It is also described as a surface marking or as defining the division between two legal interests, it also implies "the dividing line between contiguous parcels" [3]. Land boundaries are defined by laws and regulations, with several variations across countries and even states or provinces within a country [3].

This research distinguishes between spatial and legal demarcation [33]. According to the regulation, these characteristics are elaborated based on the six cadastral elements in the fixed boundary procedure as shown in the blue box (A) in Figure 1. The fixed boundaries are not an aspect of the technical exercise for surveying and mapping, rather it is legal. These activities, which include the party that locates the boundary, agreement between the adjoining landowners, boundary markers, boundary determination officer, cadastral survey method, and the accuracy of the base map, are shown in Figure 2. The differences between spatial and legal elements are distinguished by red and blue frames.

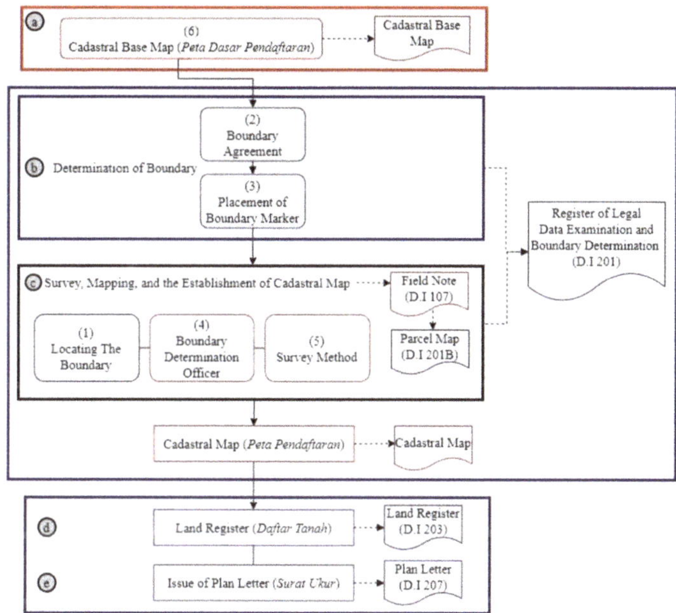

Figure 2. The workflow of spatial and legal elements in fixing the boundary. The spatial elements are marked by the red frames (4, 5, 6), and legal elements (1, 2, 3) by the blue frames.

The determination of boundaries is recorded in the register of legal data examination and boundary determination (*"Risalah Penelitian Data Yuridis dan Penetapan Batas* or *Daftar Isian 201"*). PMNA No. 3/1997 article 140 stipulates that the registers of spatial data are initiated with one, legal by two, and administration by three. Article 58 states that after the determination of the boundary and installation of its markers, the exercise of survey completion and mapping of the land parcels needs to be carried out, however, this separation is apparent.

3.3. Cadastral Typologies

This study provides a cadastre typology to assess the compliance of the boundary survey with that of cadastral regulation in Indonesia, which contributes to an improved understanding of the complexity, thereby forming the basis for adopting the best practices and a tool to improve national cadastral systems [37]. Furthermore, the continuum of accuracy is based on the cadastral typologies in terms of improving data quality.

This research only analyzes three legal elements (parties that locates the boundary, agreement between the adjoining landowners, and boundary markers). The study resulted eight cadastral typologies as follows:

1. Parties that locate the boundary (1), agreement between the adjoining landowners (1), boundary markers (1).
2. Parties who locate the boundary (0), agreement between the adjoining landowners (1), boundary markers (1).

3. Parties who locate the boundary (0), agreement between the adjoining landowners (0), boundary markers (1).
4. Parties who locate the boundary (0), agreement between the adjoining landowners (0), boundary markers (0).
5. Parties who locate the boundary (1), agreement between the adjoining landowners (0), boundary markers (0).
6. Parties who locate the boundary (1), agreement between the adjoining landowners (1), boundary markers (0).
7. Parties who locate the boundary (1), agreement between the adjoining landowners (0), boundary markers (1).
8. Parties who locate the boundary (0), agreement between the adjoining landowners (1), boundary markers (0).

The three-second numbers are a spatial demarcation (determination officer, survey method, and accuracy of base map) and were not investigated in the paper.

3.4. *Assessment of Compliance*

This research could be used to potentially facilitate the completion of the Indonesian cadastre, particularly in rural areas. In addition, it could be used to develop a strategy to improve the overall cadastral system's effectiveness as a basic layer of the land administration system.

3.4.1. The Participation of Eligible Landowners

The participation of eligible landowners during parcel registration is minimal, despite the compulsory legal requirement to register their lands in a systematic registration such as PTSL. Subsequently, the parties that locate the land boundary (the detailed number attached in Appendix C) are shown in Figure 3. Only 36.06% of the landowners or their proxy located the boundary of 7522 parcels. The majority of them, approximately 59.68%, believe in the use of local officers to locate their boundaries, while 0.62% are occupants. Since the systematic registration is compulsory, the local officer and the occupant tend to identify the boundary by regulation. The interested parties, e.g., the potential buyers, are ineligible to locate the boundary. This occurs when a local officer does not accompany the surveyor, and this was realized for approximately 2.17% of the data. Conversely, when the land has no occupant, it is difficult to locate the boundary, and this was realized for 1.47% of the data.

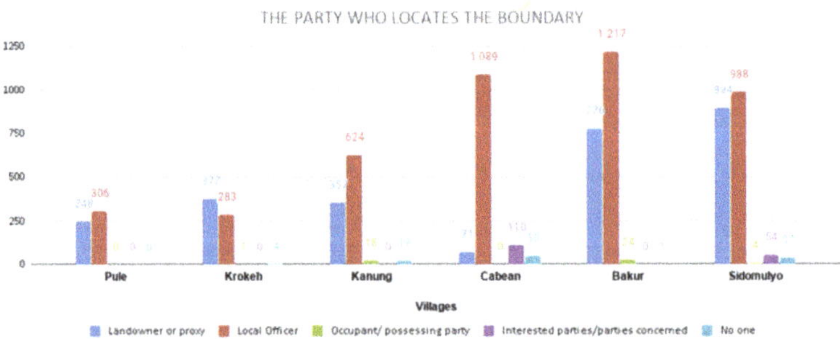

Figure 3. The party that locates the boundary.

The findings from PTSL in the case study area show that the minimum number of eligible landowners that were actively able to locate their boundaries disclosed essential factors that are crucial in completing the cadastral system in the future, and this is reported as follows

- Landholders feel safe, they do not directly locate the boundaries of their parcels. They trust both the local and land officers and based on the interview that was carried out, and it shows that fixed boundary concerning the exact coordinate is unimportant to them.

- The compulsory obligations stipulated in the regulations were not perceived as such by the right-holders. On the contrary, the obligatory enforcement was also not carried out, and the right holders who did not certify their land even at the systematic registration were not sanctioned.
- The parties must be eligible to locate the boundary and need to carry out such an act before ascertaining the truth through the procedure of determining the rights of the landowner. This logical sequence is important for the purposes of the land information system, which needs to be completed before accepting the general boundary. However, this assumes the boundary determination is based on agreements that are not between the actual owners, and rather, are between people that occupy or use the land, therefore making the legal basis of the agreement invalid.

The low level of participation of right-holders can be addressed through identification and delineation of boundaries using satellite/aerial imagery, as suggested in the FFP approach.

3.4.2. The Agreement between the Adjoining Landowners

Article 19 to 23 of PMNA No. 3/1997 regulates the agreement between the adjoining landowners before the survey and mapping activities. In addition, the majority of the respondents (96.61%) gave consent, as shown in Figure 4. The high number of agreements contradicts the low level of right-holders' participation. Most of them agree with adjacent right holders. Therefore:

- The boundary agreement does not significantly impede land registration.
- Only a few of them (3.39%) stated that they have not obtained boundary consensus. Based on the interview with landowners and local officer, they reported that boundary disputes are relatively rare. Most respondents stated that it is difficult to obtain approval from landowners that do not reside on the site.
- A total of 496 (6.57%) parcels signed agreements without proper boundary markers or even lacked boundaries.

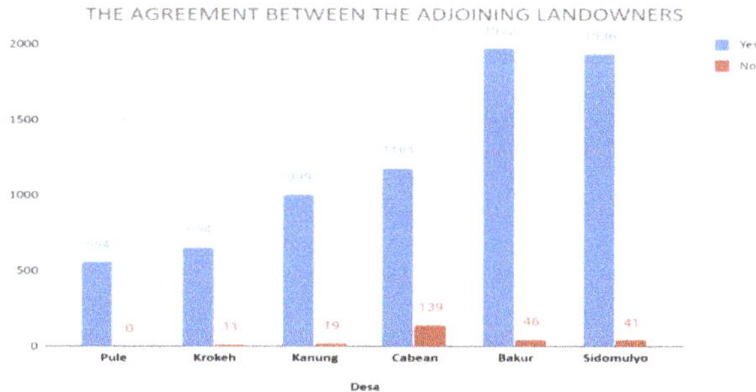

Figure 4. The agreement between the adjoining landowners.

Article 26 of PMNA No. 3/1997 regulates that the boundary of land parcels can be identified by aerial imagery. However, a field survey is mandatory, but this is too costly, time consuming, and capacity demanding.

Arruñada (2018) reported that irrespective of the method and precision used to define a land boundary, the neighbors need to be involved in this type of agreement.

3.4.3. The Boundary Markers

Based on PMNA 3/97, the boundary markers need to be installed by the right-holder, after obtaining an agreement from the actual landowner of the adjacent parcel. The various types of boundary markers are regulated in article 22. The majority of the parcels (92.47%) meet these requirements, as shown in Figure 5. Subsequently, only a few numbers of parcels do not comply (4.14% use plant fences or bamboo, and 3.39% have no borders at all). The majority of the landowners (65.81%) install permanent boundary markers (fixed boundary). It is surprising that the other nonpermanent boundary markers (34.19%) are plant fences or bamboo, the dike of a paddy field, rivers, ditches or roads, a monument made from wood or bamboo (which was unexpected). The landholders stated that fixed boundaries (permanent boundary markers) are considered inessential.

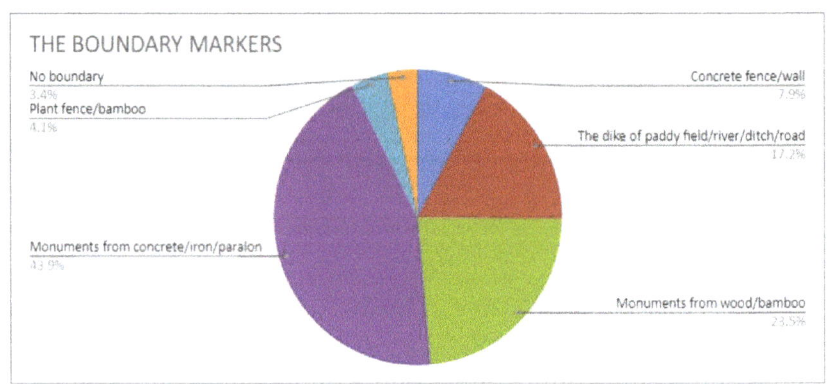

Figure 5. The boundary markers.

Based on the interview with the landowners and the local officer regarding the boundary agreement, it was discovered that they are satisfied. It was also reported that even though the boundary mark is "not fixed", they were not worried and assumed the marks got lost, damaged, or moved since the agreement with the adjoining landowners had been obtained.

The significant number of boundary monuments that did not comply (34.19%) with the fixed boundary regulation confirmed that the boundary monument is too costly for rural citizens. As suggested by the FFP approach, the regulation should accept general boundaries instead of fixed boundaries.

Appendix C provides information on the structure of the land parcels of the villages. It appears that large parts of a village district is made up of narrow, elongated land parcels, while other sections are subdivided by more square land parcels. The eastern part of Kanung Village consists of family houses. The cadastral elements addressed by the study will be improved if supplemented to include different boundary types, for example: the village boundary, road boundary, boundaries of sections of built-up land parcels, boundaries of sections of strip-shaped land parcels, boundaries of individual built-up land parcels, and boundaries of individual strip-shaped land parcels. This would require landholders of built-up land parcels to be much more able and willing to engage in boundary determination. However, this main scope of this research is the legal element of fixing the boundary that is regulated in the PP No. 24/1997 and PMNA No. 3/1997.

3.4.4. The Compliance of Cadastre Elements

As stated in the methods section, this research adopted three cadastral elements, namely, the party that locates the boundary, its markers, and the agreement between the adjoining landowners. In addition, the compliance of cadastre elements is shown in Table 3. The percentage formula described in Section 2.3 was used to calculate the proportion of surveyed parcels that comply with the cadastral elements, as stated in the Government Regulation No. 24/1997.

Table 3. The compliance of all data.

Village	The Party Who Locates the Boundary					The Green-ment between the Adjoining Landowners		The Boundary Markers						Remark		Total
	Landowner or Proxy	Local Officer	Occupant/ Possessing Part	Intetested Parties/ Parties Concered	No One	Yes	No	Concrete Fence/Wall	The Diky of Paddy/River/ Ditch/Road	Mounnments from Concrete/ Iron/Paralon	Mounnments from Wood/Bamboo	Plant Fence/ Bamboo	No Boundary	Comply	Not Comply	
Kanung	357	624	18	0	19	999	19	117	182	607	93	3	16	996	22	1018
Krokeh	377	283	1	0	4	654	11	83	28	324	191	39	0	615	50	665
Pule	248	306	0	0	0	554	0	32	10	291	182	39	0	515	39	554
Sidomulyi	894	988	4	54	37	1936	41	0	241	1422	173	5	136	1797	180	1977
Cabean	71	1089	0	110	50	1181	139	228	253	610	11	127	91	1021	299	1320
Bakur	776	1217	24	0	1	1972	46	135	586	1121	63	100	13	1859	159	2018
Total	2723	4507	47	164	111	7296	256	595	1300	4375	713	313	256	6803	749	7552

It was discovered that the majority of the parcels (90.08%) comply with the requirements. Further analysis from PTSL suggested a review of the field survey requirements to measure the boundaries that did not comply with the regulation. The proportion of non-compliance tends to be higher when the spatial elements in the boundary determination, e.g., determination officer, survey method, and base map accuracy, are also included. The distribution of land parcels that do not comply with the cadastral requirements and its cadastral typology is shown in Appendix C.

4. Discussion

The first land registration (PTSL) aims to register all land parcels in Indonesia by 2025. However, the requirement for fixing the boundaries, as stipulated in the regulation, is still a challenge. This research disclosed that the participation of eligible landowners in the first land registration and local officers' activeness during boundary determination are important factors of cadastral elements. This is consistent with the studies carried out by Aditya (2020) [11] and Zevenbergen (2002) [15].

Lack of compliance with the cadastral element causes the land office to lack confidence in implementing the PTSL completely and systematically as regulated. This is due to the fact that they are afraid of the consequences involved in issuing a parcel map—*Peta Bidang Tanah* (PBT)—that has "not comply" with legal requirements provided in the regulation. This skepticism is quite reasonable, the PBT with no certificate issued yet (K3) needs to be prepared for certification in the years ahead (K1). The dilemma was discussed as follows, supposing the PBT is used to issue a certificate, it implies some elements have not been fulfilled and probably deemed defective. Conversely, such deficiencies simply mean no eligible boundary locater, agreement, or markers. However, assuming it is refinanced to issue a new PBT, they tend to be afraid of double budgeting, and that published in PTSL becomes useless. The regulation in fixing the boundary consists of distinguishing between spatial and legal elements, as reported by Arruñada (2008) [33]. Legal elements have negative legal implications, assuming they are not fulfilled. After seventeen decades of the cadastral system in Indonesia, improvements are still needed, e.g., the judicial instead of administrative sector [36]. The interview and spatial analysis of PTSL show that fixing boundary determination slows down the progress of PTSL, even though the survey is carried out sporadically [5].

This research identifies that boundary determination using a fixed boundary is mandatory in the Indonesian cadastre [12,41]. Figure 2 shows that the six cadastral elements consist of the legal element (blue frame) and spatial element (red frame). The fixing boundaries are not merely related to technical surveying and mapping practices; rather there are closely related to the legal element. The legal aspect may be created by an agreement, followed later by the survey [46]. The six case studies in rural areas show that an agreement is reached in 96.61% of cases, and only 36.06% of the landowners' directly located boundaries are legally sufficient. Furthermore, the interview shows that the agreements could be obtained using aerial imagery instead of a field survey during the publication of the register of legal data examination and boundary determination. These data show that the fixed boundary concerning the exact coordinate, including survey method and accuracy of base map as required by regulation, is not important to people in the rural area.

Moreover, the provisions on the boundary determination when registering land is mandatory, as stated in PMNA 3/97. The aim is to develop a complete land information system, and ATR/BPN has initiated discussions using general boundaries for its land registration. Parts of the land information system may include general boundaries, and it could be argued that the Land Information System (LIS) could not be completed without them. The interviews with senior land officers revealed that the boundary determination between the non-landowners (e.g., people that occupy or use the land) is not valid. Therefore, land information is inaccurate.

The acceptance of the general boundary was started in England. Earlier attempts to introduce fixed boundaries in 1862 failed, although it was only after introducing the principle of general boundaries and avoiding expensive surveys (1875), adding selective compulsory registration (1897), simplifying the substantive land law (1925), and making conversion mandatory on transfers in designated areas that the system swung into action [15].

The purpose of this research is centered on the use of a visible boundary concept, particularly in rural and semi-urban areas. The results of the research are in line with other scientific papers dealing with methods for collecting cadastral parcel data based on visible boundaries in African countries with the purpose of increasing the speed of mapping and reducing the costs of land rights registration. Fixed boundaries are then used where relevant or necessary for any specific purposes or when required and paid for by the landowner or stakeholders [20,21,33,39–41].

5. Conclusions

Boundary demarcation consists of a six element cadastre in the legal and spatial elements recorded in the legal activity registers. The research in the case study area found a high percentage of the second element, an agreement between the adjoining landowners. This social aspect of fixing the boundary can be used as a potential method enabling the completion of the Indonesian cadastre by enshrining a legal and spatial framework of the FFP approach in Indonesian regulation.

The boundary determination using a fixed boundary is mandatory in the Indonesian cadastre. However, the research shows that the fixed boundary concerning the exact coordinate, including survey method and accuracy of the base map required by regulation, is not considered important to people in rural areas.

6. Recommendation

The ministry regulation should adopt aerial imagery in its regulation to build a complete spatial cadastre and also require government regulation to use the complete spatial cadastre as a basic layer in land administration system by integrating land tenure, value, use, and development functions.

Author Contributions: Conceptualization, D.B.M. and T.A.; methodology, D.B.M.; software, D.B.M.; validation, D.B.M., T.A.; formal analysis, D.B.M.; investigation, D.B.M.; resources, D.B.M.; data curation, D.B.M.; writing—original draft preparation, D.B.M.; writing—review and editing, D.B.M.; visualization, D.B.M.; supervision, T.A., S.S., and P.N. All authors have read and agreed to the published version of the manuscript.

Funding: The funding has kindly been provided from the School of Land Administration Studies, University of Twente, in combination with Kadaster International, The Netherlands.

Acknowledgments: The authors are grateful to the high officials of the Ministry of Agrarian and Spatial Planning/National Land Agency, Himawan Arief Sugoto, Secretary-General, R.M. Adi Darmawan, Director-General of Agrarian Infrastructure, Suyus Windayana, Director-General of Agrarian Legal Affairs, Virgo Eresta Jaya, Head of Information and Data Center, Gabriel Triwibawa, Head of Planning and Cooperation Bureau as well as Rangga Alfiandri Hasim for collecting the reference data during the research. The authors are also grateful to M. Sigit Widodo and Mulyadi for providing language help, as well as Stig Enemark for providing adequate assistance. Authors would also like to thank three anonymous reviewers for very helpful comments that have greatly improved the manuscript.

Conflicts of Interest: The authors declare no conflict of interest.

Appendix A

Table A1. List of Materials in Country Context Section (Adopted from the "Cadastral Template 2.0") [38].

Country Context	Statement
Purpose of Cadastral System	The cadastral system's primary purpose is to support the registration of land by providing legal security rights and facilitating the transfer of rights in the land tenure function.
Types of Cadastral System	Different institutions administer other cadastral systems to register forest area, fiscal cadastre, more discrete rights and responsibilities, such as mining rights and concessions, to carry out land activities. These cadastral systems generally use different systems for base data.
Cadastral Concept	The principal cadastral unit is a surveyed and boundary marked parcel, with spatial data, such as location, boundaries, and an area. The fixed boundary is mandatory. *Nomor Identifikasi Bidang* (NIB) is a parcel consisting of 13 digits, the first eight digits are the code of the province, district, sub-district, and *kelurahan*/village where it is located, and the last five digits are the parcel number. The administration area code is determined through the ministerial decree. The title rights unit is issued in accordance with the parcels defined in the survey system. Three distinct categories of land rights were registered, namely (a) the three real property rights specified in the *Undang-Undang Pokok Agraria* (UUPA), i.e., ownership right (*hak milik*), the right to build (*hak guna bangunan*), and the right of exploitation (*hak guna usaha*), (b) the pre-1960 land rights on former "western" and "Indonesian" land that has been "converted" to one of the three UUPA rights mentioned above, and (c) three additional types of land rights introduced after enactment of the UUPA, i.e., the right of use on state land (*hak pakai*), the "development right" (*hak pengelolaan*), and the strata title (*hak milik atas satuan rumah susun*). Properties differ from parcels and titles. The term "property" is used for administration and taxation purposes.
Content of Cadastral System	The principal cadastral components are spatial and legal data. To present these data, the land office organizes the public registers (*daftar umum*) consisting of registration maps (*peta pendaftaran*), land registers (*daftar tanah*), surveying letters (*surat ukur*), land books (*buku tanah*), and name lists (*daftar nama*). *Warkah* is defined as the documents used as evidentiary tools, and the basis for registration with identification kept as an integral part of the *daftar umum*. The *daftar umum*, in particular, *buku tanah* is a comprehensive and functional electronic database for checking all encumbrances, caveats, charges, or privileges affecting a registered property's encumbrances. Land offices maintain a hardcopy of *peta pendaftaran*, some of which are already in fully digital and stored at the Komputerisasi Kegiatan Pertanahan (KKP) system. *Peta pendaftaran* roles as a geographic information system (a fully digital geographic representation of the land plot)—an electronic database for recording boundaries, checking plans, and providing cadastral information on land ownership and maps, which are kept in a single database. Both the *buku tanah* and the *peta pendaftaran* use the same NIB as an identifier number to describe the same parcel. Information is accessed by mentioning the NIB, or land rights number (*nomor hak*), or based on the landmark on the *peta pendaftaran*, or the name of the right-holder, and not through the address.
Cadastral Map	Cadastral map in the form of *peta pendaftaran* illustrates the parcels of land for the purposes of recording/bookkeeping. It contains information on the shape, boundary, location, and NIB, as well as the existence of buildings when necessary. The parcel is the smallest unit surveyed and registered in the *peta pendaftaran*. The fixed boundary is mandatory. It is accomplished by boundary determination procedures stipulated in the regulation, which means that there needs to be an eligible landowner to locate the boundary markers, agreements between the adjoining landowners, determination officer, cadastral survey methods, and accuracy of the base map. The activity of fixing boundaries is not part of the technical exercise of surveying and mapping, rather it is legal. *Peta pendaftaran* is based on the cadastral base map—*peta dasar pendaftaran*—which is a map that contains technical base points and geographical elements, such as rivers, roads, buildings, and spatial boundaries of parcels. The *peta dasar pendaftaran* has a scale of 1: 1,000 or higher for urban areas, 1: 2,500 or higher for agricultural areas, and 1: 10,000 or smaller for large plantation areas. Planimetric accuracy is determined to be higher or equal to 0.3 mm on the map scale. The national coordinate system uses the Transverse Mercator projection with a zone width of 3 ° (three degrees) called TM-3. ° The central meridian zone TM-3 ° is located 1.5 ° (one point five degrees) to the east and west of the central meridian of the concerned UTM zone. The scale factor magnitude in the central meridian (k) used is 0.9999. Pseudo zero points used are east (x) = 200,000 m, and north (y) = 1,500,000 m. Earth's mathematical model as a reference field is a spheroid in the WGS-1984 datum with parameters a = 6,378,137 m and f = 1 / 29,825,722,357.
Example of a Cadastral Map	*Peta pendaftaran* can be seen at https://bhumi.atrbpn.go.id/ *Peta pendaftaran* comprises an incomplete thematic layer, such as spatial planning and overlaid land value zone map, that cannot be seamlessly interoperable. The role of *peta pendaftaran* is still limited to land tenure function services. The integration of the Land Information System at the national, provincial, or district/city level, and its role for other thematic mappings, are still discouraging.
Role of Cadastral Layer in SDI	The cadastral layer has not been integrated with other spatial data such as e-Government, community empowerment, and activities to attained sustainable development.
Cadastral Issues	The main problems currently addressed by the cadaster are the issues of incompleteness and the quality of content.

Table A1. *Cont.*

Country Context	Statement
Current Initiatives	Since 2017, the government has accelerated land registration through systematic land registration (PTSL), which was launched as mandated by the President through President Instruction No. 2/2018 and mandated to be completed by 2025. All land parcels are expected to be measured and mapped. It is categorized as existing certified land parcels (K4), undisputed land parcels not ready for certification (K3), disputed land parcels (K2), and undisputed land parcels ready for certification (K1). K4 means that land offices need to take action to improve the quality of land records. This is because previously, the land titles were either not mapped correctly or with no spatial information (known as floating titles). The computerization has been implemented in stages, both in the form of scanned documents and fully digital. However, several challenges still need to be resolved in the digital transformation of cadastral content. Firstly, not all cadastral content data have been completely entered into the KKP system, or divided into the six classifications of each entry-level. Secondly, the data quality is not entirely "valid" in accordance with the contents recorded in the analog document (papers). It is associated with various factors, such as incompleteness or analog data errors, data entry errors, double numbering, etc. The suitability status of electronic data with analog data for every parcel needs to be validated. Thirdly, there are two groups of entry-level spatial data, namely, the un-plotted registered parcels (called KW 4, 5, 6) and plotted (called KW 1, 2, 3). The quality of land parcels of KW 1, 2, 3 differs in their accuracy/uncertainty level, both relative and positional, which needs to be improved. https://www.atrbpn.go.id/

Table A2. List of Features Covered by Cadastral Country Principles and Statistics (Adopted from the "Cadastral Template 2.0") [38].

Cadastral Principles and Statistics	Categories
Type of registration system	Title registration.
Legal requirement for registration of land ownership	Compulsory and optional.
Approach for the establishment of cadastral records	Both systematic and sporadic.

Appendix B

Table A3. Classification of Cadastral Content Entry Levels into the KKP System.

Entry Levels (KW)	Parcel Plotted on *Peta Pendaftaran*	*Surat Ukur* (Spatial)	*Surat Ukur* (Spatial)	*Buku Tanah*
1	√	√	√	√
2	√	×	√	√
3	√	×	×	√
4	×	√	√	√
5	×	×	√	√
6	×	×	×	√

√ (Entered); × (not entered yet).

Table A4. Target and Realization of PTSL in the Year 2017–2019.

Year	Target	Realization (K1 + K2 + K3 + K4)
2017	5,000,000	5,069,513
2018	7,000,000	8,854,797
2019	9,000,000	8,963,415
Total	**21,000,000**	**22,887,725**

Appendix C

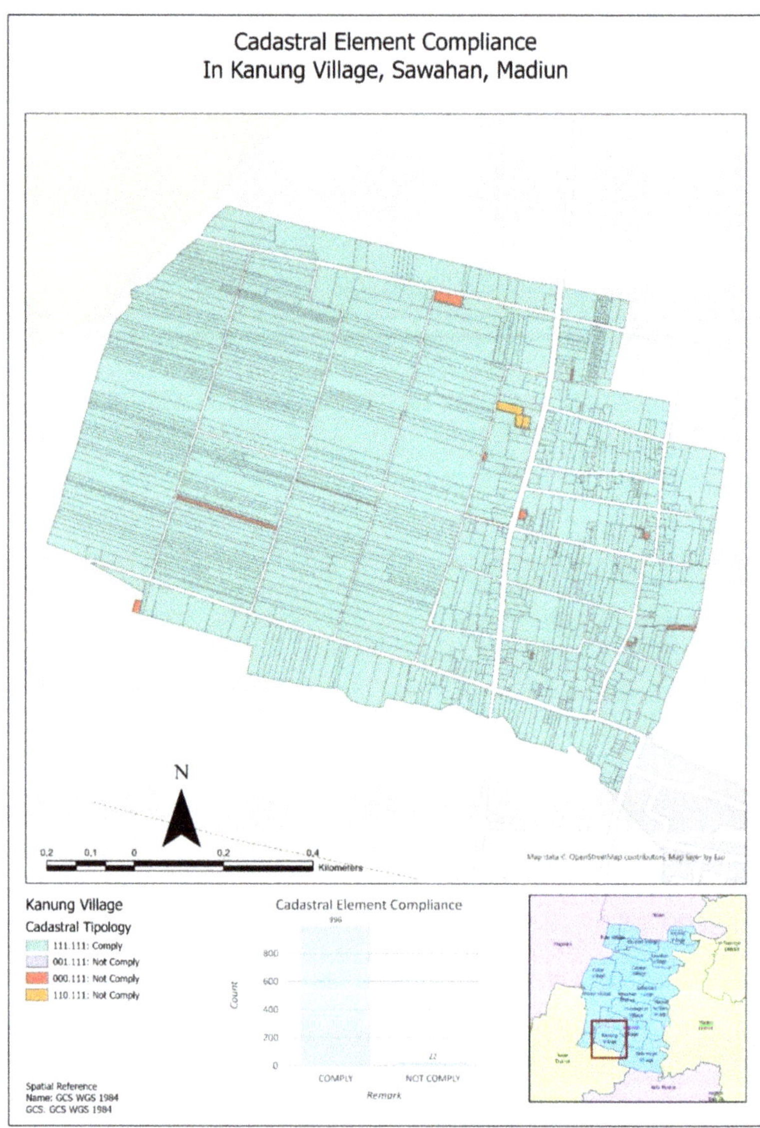

Figure A1. The distribution of non-compliance parcels and its cadastral typology in Kanung.

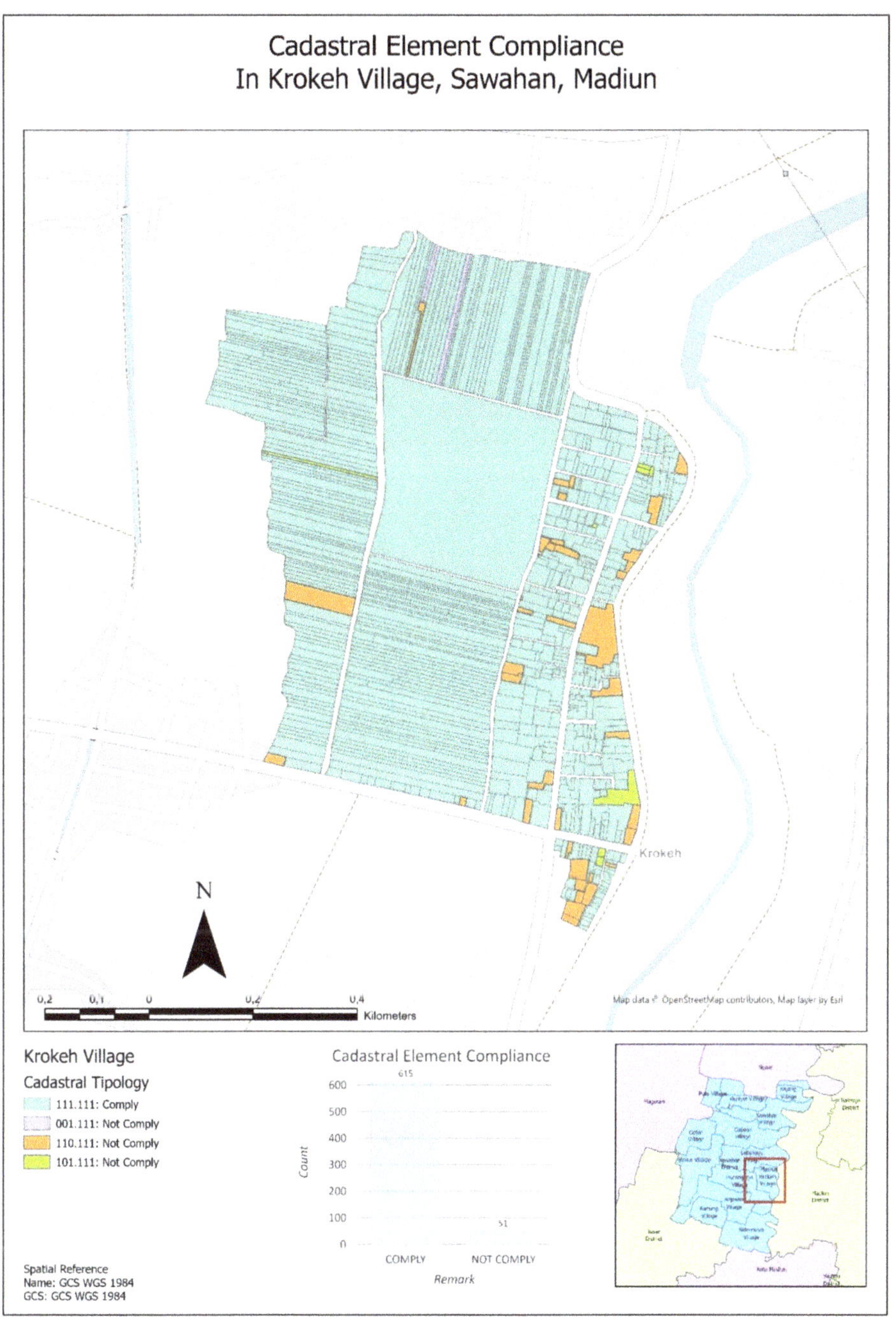

Figure A2. The distribution of non-compliance parcels and its cadastral typology in Krokeh.

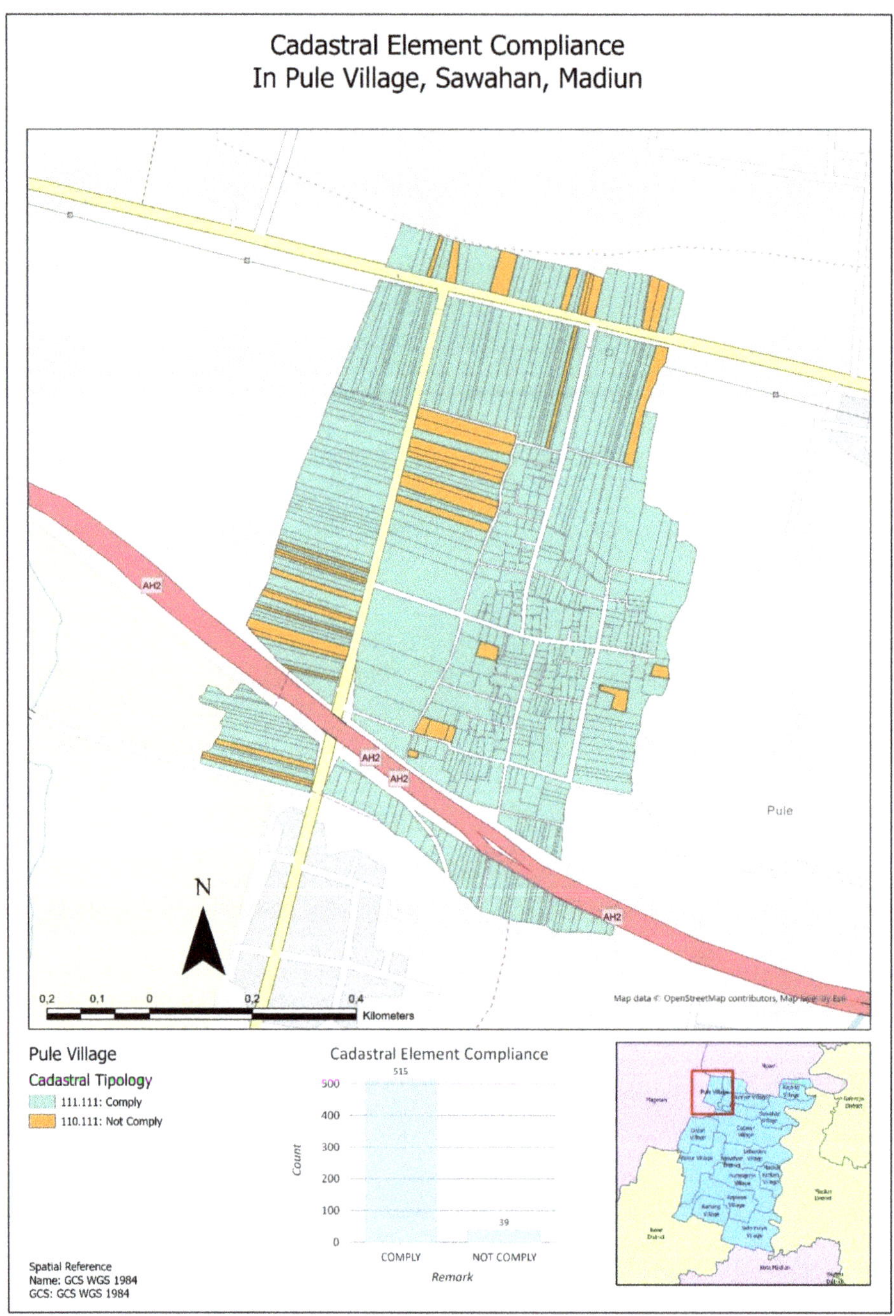

Figure A3. The distribution of non-compliance parcels and its cadastral typology in Pule.

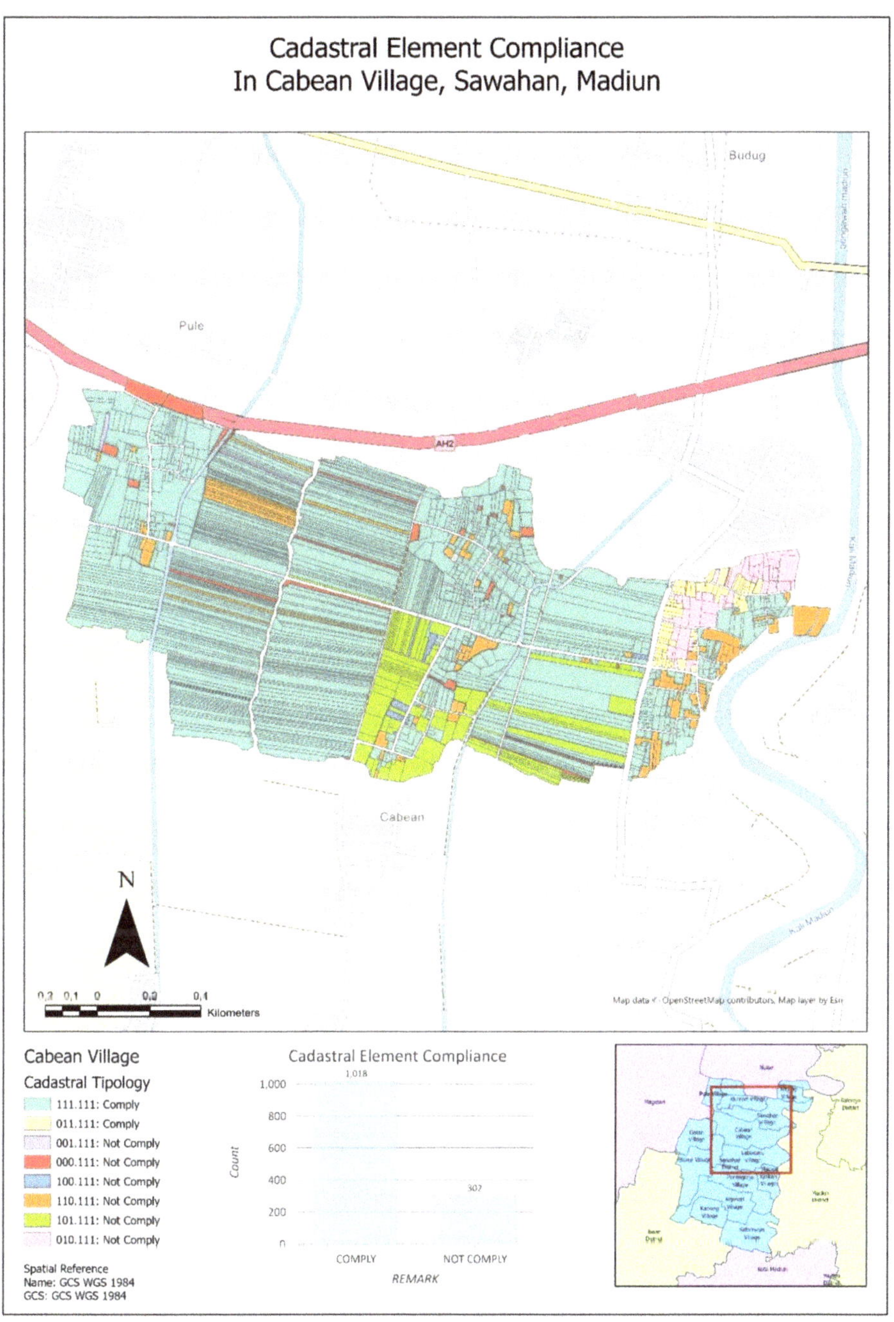

Figure A4. The distribution of non-compliance parcels and its cadastral typology in Cabean.

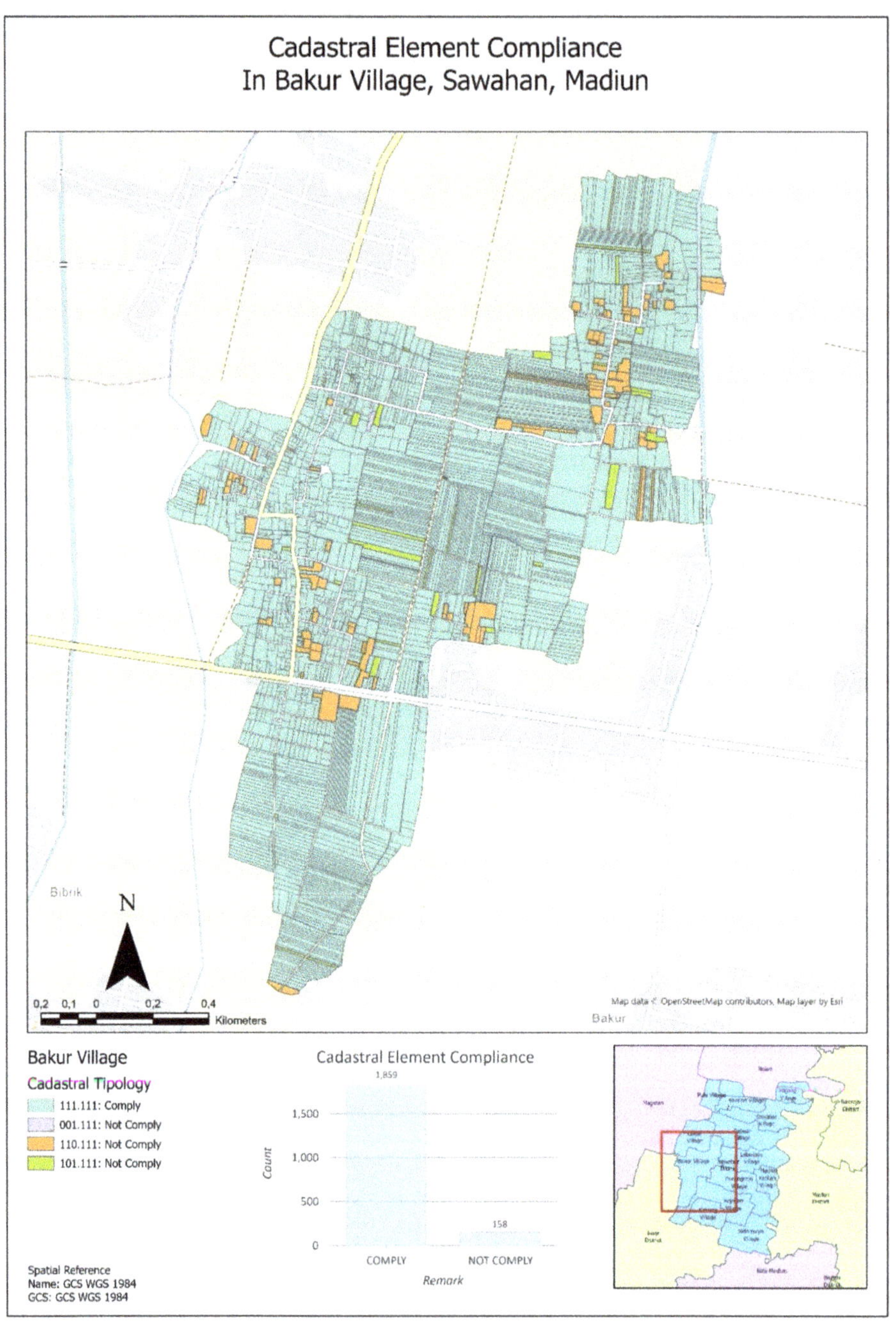

Figure A5. The distribution of non-compliance parcels and its cadastral typology in Bakur.

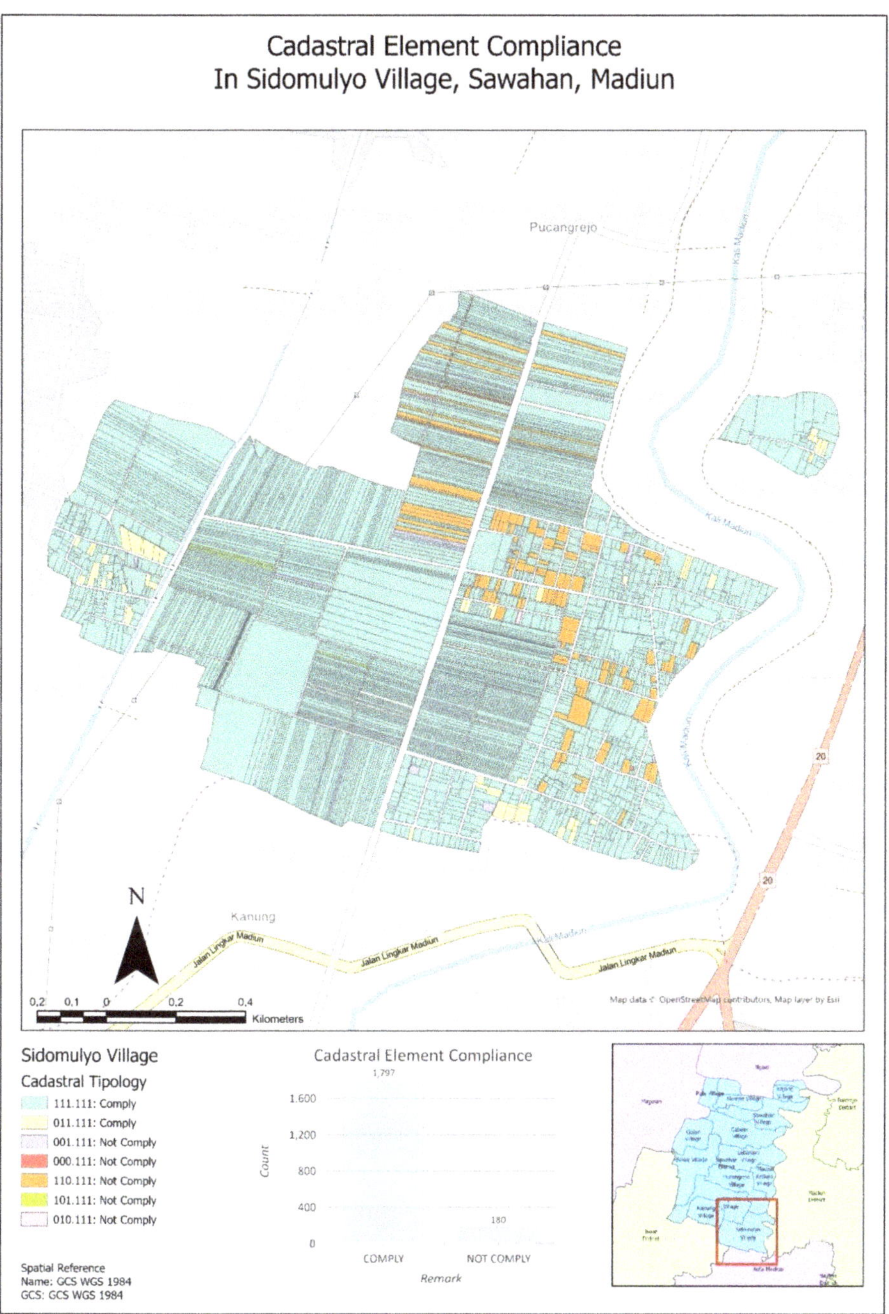

Figure A6. The distribution of non-compliance parcels and its cadastral typology in Sidomulyo.

References

1. Unece. *Land Administration Guidelines—With Special Reference to Countries in Transition*; Unece: New York, NY, USA; Geneva, Switzerland, 1996.
2. Enemark, S. From Cadastre to Land Governance. In Proceedings of the Annual World Bank Conference on Land Policy and Administration, Washington, DC, USA, 26–27 April 2010.
3. Williamson, I.; Enemark, S.; Wallace, J.; Rajabifard, A. *Land Administration for Sustainable Development*, 1st ed.; ESRI Press Academic: Redlands, CA, USA, 2010; ISBN 978-1-58948-041-4.
4. Williamson, I.P.; Rajabifard, A.; Holland, P. Spatially Enabled Society. In Proceedings of the Building the Capacity; FIG Congress 2010, Sydney, Australia, 11–16 April 2010.
5. Van der Eng, P. After 200 Years, Why is Indonesia's Cadastral System Still Incomplete? In *Land and Development in Indonesia*; ISEAS–Yusof Ishak Institute: Singapore, 2018; pp. 227–244.
6. Kaufmann, J.; Steudler, D. Cadastre 2014—A Vision for a Future Cadastral System. In Proceedings of the 1st Congress on Cadastre in the European Union, Granada, Spain, 15–17 May 2002.
7. Williamson, I.; Grant, D. The United Nations—International Federation of Surveyors declaration on land administration for sustainable development. *Z. Fur Vermess.* **2000**, *125*, 341–348.
8. United Nations; FIG. *The Bogor Declaration—United Nations Interregional Meeting of Experts on the Cadastre*; United Nations: Bogor, Indonesia, 1996.
9. Bennett, R.; Rajabifard, A.; Williamson, I.; Wallace, J. On The Need for National Land Administration Infrastructures. *Land Use Policy* **2012**, *29*, 208–219. [CrossRef]
10. Van Oosterom, P.; Lemmen, C. The Land Administration Domain Model (LADM): Motivation, standardisation, application and further development. *Land Use Policy* **2015**, *49*, 527–534. [CrossRef]
11. Aditya, T.; Maria-Unger, E.; vd Berg, C.; Bennett, R.; Saers, P.; Lukman Syahid, H.; Erwan, D.; Wits, T.; Widjajanti, N.; Budi Santosa, P.; et al. Participatory Land Administration in Indonesia: Quality and Usability Assessment. *Land* **2020**, *9*, 79. [CrossRef]
12. The Goverment of Indonesia. Peraturan Pemerintah Republik Indonesia Nomor 24 Tahun 1997 Tentang Pendaftaran Tanah. 1997. Available online: http://perundangan.pertanian.go.id/admin/file/PP-24-97.pdf (accessed on 16 November 2020).
13. International Federation of Surveyors (FIG); World Bank. *Fit for Purpose Land Administration*; International Federation of Surveyors (FIG): Copenhagen, Denmark, 2014.
14. UN Habitat; GLTN. *Fit for Purposes LAND Administration—Guiding Principles for Country Implementation*; UN Habitat: Nairobi, Kenya, 2016.
15. Zevenbergen, J.A. *Systems of Land Registration—Aspects and Effects*; Nederlandse Commissie Voor Geodesie, Netherlands Geodetic Commission: Delft, The Netherlands, 2002; ISBN 9061322774.
16. Lemmen, C.; van Oosterom, P.; Bennett, R. The Land Administration Domain Model. *Land Use Policy* **2015**. [CrossRef]
17. Oosterom, P.V.; Lemmen, C.; Uitermark, H. ISO 19152: 2012, Land Administration Domain Model published by ISO. In *FIG Work Week 2013*; ISO: Abuja, Nigeria, 2013.
18. Open Geospatial Consortium InfraGML. *LandInfra Core—Encoding Stand*; Open Geospatial Consortium InfraGML: Wayland, MA, USA, 2017.
19. Christiaan, L. *The Social Tenure Domain ModelA Pro-Poor Land Tool*; International Congress (24: 2010: Sydney); The International Federation of Surveyors (FIG): Copenhagen, Denmark, 2013.
20. Griffith-Charles, C. The application of the social tenure domain model (STDM) to family land in Trinidad and Tobago. *Land Use Policy* **2011**, *28*, 514–522. [CrossRef]
21. Paasch, J.M.; van Oosterom, P.; Lemmen, C.; Paulsson, J. Further modelling of LADM's rights, restrictions and responsibilities (RRRs). *Land Use Policy* **2015**, *49*. [CrossRef]
22. International Federation of Surveyors (FIG); UN Habitat; GLTN. *A Review of the Social Tenure Domain Model (STDM) Phase II*; International Federation of Surveyors (FIG): Copenhagen, Denmark, 2014.
23. Ramadhani, S.A.; Bennett, R.M.; Nex, F.C. Exploring UAV in Indonesian cadastral boundary data acquisition. *Earth Sci. Informatics* **2018**, *11*, 129–146. [CrossRef]
24. Rahmatizadeh, S.; Rajabifard, A.; Kalantari, M.; Ho, S. A framework for selecting a fit-for-purpose data collection method in land administration. *Land Use Policy* **2018**. [CrossRef]
25. Steudler, D.; Williamson, I.; Van Der Molen, P.; Kaufmann, J.; Adlington, G.; Jang, B.-B.; Koh, J.-H.; Lemmen, C.; Van Oosterom, P.; Germann, M.; et al. *CADASTRE 2014 and Beyond*; Hakapaino: Helsinki, Finland, 2014.
26. Zevenbergen, J.; Augustinus, C.; Antonio, D.; Bennett, R. Pro-poor land administration: Principles for recording the land rights of the underrepresented. *Land Use Policy* **2013**, *31*, 595–604. [CrossRef]
27. Hendriks, B.; Zevenbergen, J.; Bennett, R.; Antonio, D. Pro-poor land administration: Towards practical, coordinated, and scalable recording systems for all. *Land Use Policy* **2019**. [CrossRef]
28. Rohan, B.; Zevenbergen, J. The visible boundary: More than just a line between coordinates. In Proceedings of the GeoTech Rwanda 2015, Kigali, Rwanda, 18–20 November 2015; pp. 1–4.
29. Luo, X.; Bennett, R.M.; Koeva, M.; Lemmen, C. Investigating semi-automated cadastral boundaries extraction from airborne laser scanned data. *Land* **2017**, *6*, 60. [CrossRef]

30. Wassie, Y.A.; Koeva, M.N.; Bennett, R.M.; Lemmen, C.H.J. A procedure for semi-automated cadastral boundary feature extraction from high-resolution satellite imagery. *J. Spat. Sci.* **2018**, *63*, 75–92. [CrossRef]
31. Nyandwi, E.; Kohli, D.; Bennett, R.M.; Koeva, M. Comparing Human Versus Machine-Driven Cadastral Boundary Feature Extraction. *Remote Sens.* **2019**, *11*, 1662. [CrossRef]
32. Koeva, M.; Stöcker, C.; Crommelinck, S.; Ho, S.; Chipofya, M.; Sahib, J.; Bennett, R.; Zevenbergen, J.; Vosselman, G.; Lemmen, C.; et al. Innovative Remote Sensing Methodologies for Kenyan Land Tenure Mapping. *Remote Sens.* **2020**, *12*, 273. [CrossRef]
33. Arruñada, B. Evolving practice in land demarcation. *Land Use Policy* **2018**, *77*, 661–675. [CrossRef]
34. Yin, R.K. *Case Study Research and Applications*, 6th ed.; Sage Publication: Thousand Oaks, CA, USA, 2017; ISBN 9781506336169.
35. Çağdaş, V.; Stubkjær, E. Doctoral research on cadastral development. *Land Use Policy* **2009**. [CrossRef]
36. Steudler, D.D.; Williamson, I.P.I.; Rajabifard, A. The Development of a Cadastral Template. *J. Geospat. Eng.* **2003**, *5*, 39–47.
37. Rajabifard, A.; Williamson, I.; Steudler, D.; Binns, A.; King, M. Assessing the worldwide comparison of cadastral systems. *Land Use Policy* **2007**, *24*, 275–288. [CrossRef]
38. Rajabifard, A.; Steudler, D.; Aien, A.; Kalantari, M. The Cadastral Template 2.0, From Design to Implementation. In Proceedings of the FIG Congress 2014. In *Engaging the Challenges—Enhancing the Relevance;* International Federation of Surveyors (FIG): Kuala Lumpur, Malaysia, 2014; p. 25.
39. FIG. Cadastral Template. A Worldwide Comparison of Cadastral Systems Tersedia Pada. Available online: https://www.fig.net/organisation/comm/7/cadastraltemplate/index.htm (accessed on 16 November 2020).
40. The Goverment of Indonesia. *Undang-Undang Nomor 5 Tahun 1960 tentang Peraturan Dasar Pokok-pokok Agraria;* The Goverment of Indonesia: Jakarta, Indonesia, 1960.
41. The Ministery of Agrarian Affairs/Head of National Land Agency. *Peraturan Menteri Negara Agraria/Kepala Badan Pertanahan Nasional Nomor 3 Tahun 1997 tentang Ketentuan Pelaksanaan Peraturan Pemerintah Nomor 24 Tahun 1997 tentang Pendaftaran Tanah;* The Ministery of Agrarian Affairs/Head of National Land Agency: Jakarta, Indonesia, 1997.
42. Enemark, S.; Mclaren, R. Fit-for-Purpose Land Administration: Developing Country Specific Strategies for Implementation. In Proceedings of the 2017 World Bank Conference on Land and Poverty, Washington, DC, USA, 20–24 March 2017; pp. 1–18.
43. Muhadjir, N. *Metodologi Penelitian: Paradigma Positivisme Objektif Phenomenologi Interpretatif Logika Bahasa Platonis, Chomskyist, Hegelian & Hermeneutik Paradigma Studi Islam Matematik Recursion-, Set-Theory & Structural Equation Modeling Dan Mixed*, 6th ed.; Rake Sarasin: Yogyakarta, Indonesia, 2011; ISBN 9789798975196.
44. Supranto, J. *Teknik Sampling;* PT Rineka Cipta: Jakarta, Indonesia, 1998; ISBN 9795183141.
45. BPS—Tatistics Indonesia. *Statistical Yearbook of Indonesia 2020;* Badan Pusat Statistik: Jakarta, Indonesia, 2020.
46. Grant, D.; Enemark, S.; Zevenbergen, J.; Mitchell, D.; McCamley, G. The Cadastral triangular model. *Land Use Policy* **2020**, *97*. [CrossRef]
47. Enemark, S.; McLaren, R. Making FFP Land Administration Compelling and Work in Practice. In Proceedings of the FIG Commission 7 International Commission 7 2018 International Seminar, Bergen, Norway, 24–28 September 2018.
48. Bennett, R.; Tambuwala, N.; Rajabifard, A.; Wallace, J.; Williamson, I. On recognizing land administration as critical, public good infrastructure. *Land Use Policy* **2013**, *30*, 84–93. [CrossRef]
49. Mohammadi, H. The Integration of Multi-Source Spatial Datasets in the Context of SDI Initiatives. PhD. Thesis, The University of Melbourne, Melbourne, Australia, 2008.
50. Grant, D.B.; Mccamley, G.; Mitchell, D.; Enemark, S.; Zevenbergen, J. *Upgrading Spatial Cadastres in Australia and New Zealand: Functions, Benefits & Optimal Spatial Uncertainty;* The Cooperative Research Centre for Spatial Information: Melbourne, Australia, 2018.

Article

Securing Land Rights for All through Fit-for-Purpose Land Administration Approach: The Case of Nepal

Uma Shankar Panday [1,*], Raja Ram Chhatkuli [2], Janak Raj Joshi [3], Jagat Deuja [4,5], Danilo Antonio [6] and Stig Enemark [7]

1 Department of Geomatics Engineering, Kathmandu University, Dhulikhel 45200, Nepal
2 UN-Habitat, P.O. Box 107, Kathmandu 44600, Nepal; rr.chhatkuli@unhabitat.org.np
3 Ministry of Land Management, Cooperatives and Poverty Alleviation, Kathmandu 44600, Nepal; janakraj.joshi@nepal.gov.np
4 Community Self Reliance Centre (CSRC), Kathmandu 44700, Nepal; deujaj@csrcnepal.org
5 Land Issues Resolving Commission (LIRC), Kathmandu 44620, Nepal
6 UN-Habitat/LHSS, P.O. Box 30030, Nairobi 00100, Kenya; danilo.antonio@un.org
7 Department of Planning, The Technical Faculty of IT and Design, Aalborg University, 9000 Aalborg, Denmark; enemark@plan.aau.dk
* Correspondence: uspanday@ku.edu.np; Tel.: +977-11-415100

Abstract: After the political change in Nepal of 1951, leapfrog land policy improvements have been recorded, however, the land reform initiatives have been short of full success. Despite a land administration system based on cadaster and land registries in place, 25% of the arable land with an estimated 10 million spatial units on the ground are informally occupied and are off-register. Recently, a strong political will has emerged to ensure land rights for all. Providing tenure security to all these occupants using the conventional surveying and land administration approach demands a large amount of skilled human resources, a long timeframe and a huge budget. To assess the suitability of the fit-for-purpose land administration (FFPLA) approach for nationwide mapping and registration of informality in the Nepalese context, the identification, verification and recordation (IVR) of the people-to-land relationship was conducted through two pilot studies using a participatory approach covering around 1500 and 3400 parcels, respectively, in an urban and a rural setting. The pilot studies were based on the FFPLA National Strategy and utilized satellite imageries and smartphones for identification and verification of land boundaries. Data collection to verification tasks were completed within seven months in the urban settlements and for an average cost of 7.5 USD per parcel; within the rural setting, the pilot study was also completed within 7 months and for an average cost of just over 3 USD per parcel. The studies also informed the discussions on building the legislative and institutional frameworks, which are now in place. With locally trained 'grassroots surveyors', the studies have provided a promising alternative to the conventional surveying technologies by providing a fast, inexpensive and acceptable solution. The tested approach may fulfill the commitment to resolve the countrywide mapping of informality. The use of consistent data model and mapping standards are recommended.

Keywords: tenure security; land tenure; land policy; fit-for-purpose land administration; pilot study

Citation: Panday, U.S.; Chhatkuli, R.R.; Joshi, J.R.; Deuja, J.; Antonio, D.; Enemark, S. Securing Land Rights for All through Fit-for-Purpose Land Administration Approach: The Case of Nepal. *Land* **2021**, *10*, 744. https://doi.org/10.3390/land10070744

Academic Editor: Volker Beckmann

Received: 7 March 2021
Accepted: 13 July 2021
Published: 16 July 2021

1. Introduction

The land tenure system in Nepal has a history of orientation to feudal landlordism. Historically, until the political change of 1951 when the country was turned into a democratic set-up, most of the fertile land of the state was granted by the aristocratic rulers to various elites who turned to be landlords. Peasants, the real tillers on land, were left behind with limited access and poorly recognized tenure rights on the land they operate. Even today, a large proportion of the arable land is unregistered in the national record [1–3]. Several areas and settlements cultivated for decades are officially unrecognized and undocumented. The rights and legal ownership of the occupants being unrecognized, land transactions are often conducted in the informal market and the state is losing large revenue that could have been collected as land taxation and registration fees when land tenure is formally recognized. More alarmingly, being legally unrecognized despite decades of use, the poor landless slum dwellers and peasants in the informal settlements who come generally from different socio-economically vulnerable groups like Dalits and Sukumbasi (see definitions listed at the end of the article) face constant fear of eviction. This uneven allocation, poor access and unsecured tenure of vulnerable groups over the land they hold triggered various political unrest [4], civil movement [5,6] and even an armed conflict [7–9].

With time elapsed and several restructuring of the political system, the land administration system has undergone various changes. Different commitments were made by key political parties and the governments in office to address these issues leading to some leapfrog policy improvements in 1964, 1989 and 2007 [10]. Three high-level land commissions and at least 12 landless problem resolving commissions were constituted to provide secured land rights to the landless. However, the problems remain [11]. The latest estimate shows that about 25% of the arable land with an estimated 10 million spatial units on the ground are off-register and the related land rights are legally unrecognized [1–3]. Around 1.34 million households consisting of around 25% landless peasants and squatters called Sukumbasi, around 17% landless Dalits and 57% informal land-holders are living under informality and are under constant fear of eviction [12]. More than one-fourth of the households, mainly ethnic minorities, indigenous people and Dalits do not have their land to farm on [13]. These families cultivate land owned by the landlords under different tenure arrangements for a living which are mostly exploitative in nature. One such examples is share-cropping system wherein the peasant pays half of the farm production to the legally registered land owner [14–16].

The demand for access to land to the landless and tenure security for all was one of the triggers of the armed conflict (1996–2006) as reflected in the 40-points demand of the United People's Front [17]. After the Comprehensive Peace Agreement (CPA 2006) [18], a national consensus and a strong political will have evolved to resolve the historical injustice and the outstanding issues on land governance arrangements. The new constitution promulgated in 2015 [19] and the Local Government Operation Act 2017 [20] have guaranteed various rights to the municipalities including issues such as levying local taxes, providing local services and land development activities, management of local records and land management activities. The act has ensured the autonomy of local governments to devise their own rules and regulations related to governance, protection and utilization of local resources under their jurisdiction including the governance of rights on the land. They are also empowered to prepare long-term development plans, formulate policies and implement them. Land is a prerequisite for any kind of infrastructure development. Land tenure security is a must for not only increasing household income, food security and equity [21] but is also important for generating revenues by local governments that are crucial for financing their local development projects. However, the local government bodies lack sufficient technical know-how for land administration and management and related skills such as land surveying and mapping. Additionally, given the volume of work, surveying and mapping with conventional techniques would require thousands of highly skilled human resources, enormous budget and decades to complete the work [22]. On contrary, a fit-for-purpose land administration (FFPLA) system can record the existing situation of informal land tenure relatively quickly, within affordable costs and using a participatory approach without requiring highly skilled professional personnel and a costly infrastructure [3,23]. The FFPLA approach is well considered as an overarching tool in the Framework for Effective Land Administration (FELA) [24]. However, the actual implementation is contextual which differs from country to country. In this regard, the experiences and lessons learned from an earlier application of FFPLA for rehabilitation of displaced people in the post-earthquake context [25–28] catalyzed to test the approach in the context of providing land tenure security for the landless Dalits, Sukumbasi and informal settlers.

Based on the experience and lessons learned from the two pilot studies in urban and rural settings in Nepal, a FFPLA implementation approach is discussed that is estimated to provide secure land rights at the national level, through the local level initiatives. A roadmap for nationwide scale-up is provided. The objectives of this study are:

1. To assess the suitability of the Fit-For-Purpose Land Administration (FFPLA) approach for identification, verification and recordation (IVR) of informal land rights in the Nepalese context; and
2. To validate the acceptability of locally initiated mapping results at the national level.

The remainder of the article is organized as follows: Section 2 describes the historical social discrimination, inequality in access to land and recent government initiatives to fill the gap. Section 3 presents the rationale for conducting this research, Section 4 highlights the methodology adapted while Section 5 elaborates the major results. Section 6 discusses the results, compare them in international scenarios and analyses estimated timeframe for nationwide scale-up as well as the potential policy implications and challenges. Finally, the conclusion of this research and future outlook are presented in Section 7.

2. Addressing Social Discrimination and Inequality in Access to Land

Nepal is an ethnically diverse country [29,30] where discrimination on the basis of caste, ethnicity and gender has been prevalent for centuries [31]. The people of higher caste enjoyed social and political power while those from lower caste were discriminated. For example, the Civil Code of 1854 provided harsher punishments for lower caste people for similar offences compared to people of higher caste [32]. Moreover, it also institutionalized gender discrimination, especially providing inheritance rights on properties to only male members of the family. Ownership of land as a source of power was generally held by higher caste elites; peasants and low caste, especially Dalits, turned into indenture laborer and artisans. With the advance of democracy in 1951, Nepal started into the path of progressive provisions and to reduce the discrimination on access to land [10]. The constitution enacted in 1990 [33], the Interim Constitution 2007 [34] and the present Constitution of Nepal [35] have gradually become more inclusive and thus discriminations on the basis of caste and gender have been gradually diminishing. The Constitution of Nepal enacted in 2015 [35] prohibits discrimination on the basis of religion, race, gender, caste, tribe, origin, language or ideological conviction and proactive provisions on protecting rights of women and socially excluded groups do exist.

Land is the most significant asset in the Nepalese rural-agrarian economy. Nearly 60% of Nepalese population depend on agriculture for their livelihood [36]. However, land is concentrated in the hands of the few. The richest 7% of households own about 31% of agricultural land [37]. Size of holdings are marginal. In 2001/02 some 47% of the Nepalese farmers owned less than 0.5 hectare of land while in 2011/12 this was enlarged to 54% [38]. Additionally, 29% of the rural households are landless [37] and up to 85% of the rural households are described as "land poor" [31]. Five to seven million urban poor are landless squatters and have even lost their daily livelihoods in 2020 due to the COVID-19 pandemic [37]. Skewed patterns of land ownership have also been compounded by a deeply discriminatory and strictly hierarchical society that has excluded women, ethnic minorities and especially people of 'low caste' (particularly Dalits). Dalits comprising over 20% of the country's population possess only 1% of arable land [31], 75% of them are functionally landless [39]. Gender inequality in access to land is also prevalent in Nepal, though the female ownership is gradually increasing. To increase women's access to land, the Government of Nepal, through the Eleventh Periodic Three Year Plan (2007–2010), introduced tax exemption for registering land in women's name and under joint ownership of husband and wife [40]. Female ownership of land reached to 19.71% [40] in 2011 with an increase of nearly 10% from 2001 [41]. Land access has also significant impacts on the lives of indigenous people. Establishment of wildlife reserves and national parks in the territories of the indigenous people (especially of Tharu tribe) have displaced them from the land upon which they had relied on for generations. Many of them have turned into squatters after losing the ancestral land and traditional livelihood. Approximately 80% of Nepal's indigenous population are marginal landowners, meaning they have less than 0.4 hectare of landholding [14]. Recently, the Government has taken major steps to ensure land rights for all. It has adopted the National Land Policy 2019 [42] and has developed a FFPLA National Strategy for implementing secure land rights for all [2]. These two documents are the landmark provisions for addressing unresolved issues of security of tenure and inequitable access to land for landless, smallholders and informal settlers in the country. The FFPLA approach is considered to minimize the time involved in recording and registering the land rights at scale using locally trained "grassroots surveyors". The FFPLA National Strategy was mainly aimed at being implemented as a national-level program to deliver security of tenure for all by recognizing and legalizing all legitimate land rights across the country. It recommended to conduct two pilot studies, one in a rural and another in a peri-urban informal setting, in order to test the suitability of data acquisition and processing using the FFPLA approach [2]. This article provides the design and results of the two pilot studies and discusses the implications and recommendations for implementation at a national scale.

3. Rationale of the Research

The FFPLA approach is widely tested and discussed in the literature [23,25,27,43]. However, it is not extensively implemented on the ground at scale [22] except for a few African and Asian countries [44,45] where the majority of land are not covered by the formal land administration system [45]. While the Nepalese formal cadastre covers about 75% of the arable lands and settlements, the rest are informally occupied. Providing secured legal tenure to these informal occupants through the conventional surveying approach and land administration is estimated to require a large number of professionally skilled personnel, a long timeframe and a huge budget [46], while it may alternatively be achieved quickly and affordably with the FFPLA approach. Nonetheless, the approach has not been widely tested in a country with the diverse culture, ethnic setup, varying topography and more importantly, having a hierarchical institutional setup as well as complex legal arrangements. Tenure security improvement is context dependent where an action-oriented research would be necessary to validate an alternative approach [47]. This article attempts to assess the application of the FFPLA country implementation strategy in Nepal [2] for IVR of informal lands through the local level initiatives and to confirm the suitability of the results obtained through the local level initiatives at the national level. The knowledge and experience gained from this research may encourage land academics, land professionals, land agencies, politicians and government decision makers to apply the approach in similar scenario. Furthermore, the methodology and procedure recommended in this article may be instrumental especially in low-income countries where managing resources for land administration with large amount of informally occupied lands and unsecured tenureships are an issue.

4. Materials and Methods

4.1. Study Area

The two pilot study sites were chosen representing two different characteristics: a dense urban and an open rural setting. While Ratnanagar Municipality has compact residential areas, Belaka Municipality Ward No. 3 is dominated by relatively open and large agriculture fields. In addition, Sukumbasi are dominant in Ratnanagar while informal land occupancies exist in Belaka. The diverse sites were chosen such that the success of the approach can provide lessons for nationwide replication. The locations of study sites are depicted in Figure 1.

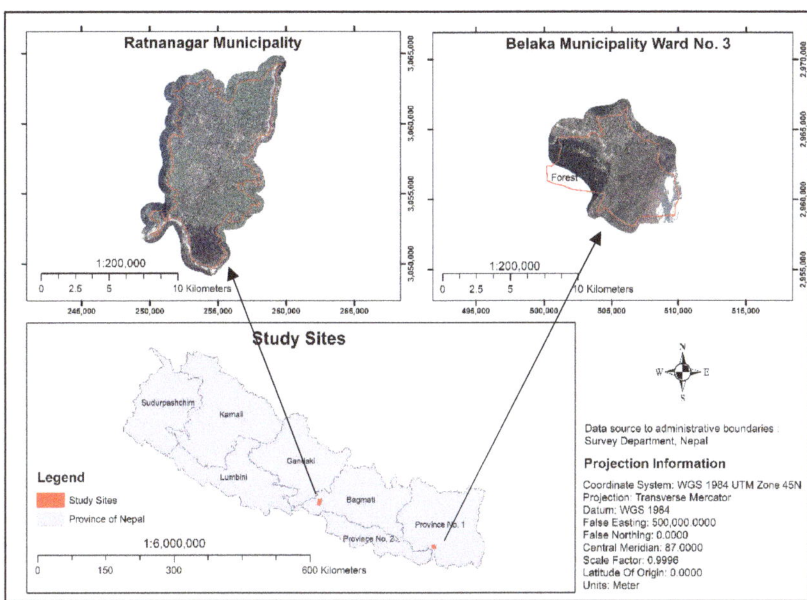

Figure 1. Study sites.

Ratnanagar Municipality lies in Chitwan District of Bagmati Province. Official Census is carried out every 10 years in Nepal. The next census is scheduled in 2021. In 2011, a population of 69,848 inhabitants was recorded which is spread over 16 wards (smallest administrative unit) of the municipality comprising a total of 68.68 sq. km. area. No other consistent reliable data for both the municipalities are available. Though traditionally tribal Tharu communities were the majority of residents in the municipality, people of various ethnic groups from different parts of the country have migrated to the area due to its suitable topographic conditions, better livelihood opportunities and ease of accessibility through the East–West highway which has contributed to mushrooming of informal settlements. There are a total of 58 large and small informal settlements which host more than 1500 households of Sukumbasi and some seven thousand people scattered in an area of 0.41 sq. km. Most of the households are residing at the government or public lands, some at the private lands legally owned by others and few at the lands owned by schools and religious entities like temples. A major proportion of the people have resided at the government or public lands for generations without having any land tenure documents. Only a few families have been identified as landless or informal settlers by different land-related commissions formed earlier by the Government of Nepal. They have been issued with documents recognizing them as such, however, the land itself is not formally registered. Some people have settled there during different political transitions since 1950s with the consent of local leaders associated with different political parties. Nevertheless, they have received no legal tenureship documents. Few families from indigenous communities like Mushahar, Tharu and others have been living on community land with common land tenureship, after being evacuated by natural disasters. This study covered all 58 informal settlements in the municipality.

Belaka Municipality lies in Udayapur District of Province No. 1. It encompasses 344.73 sq. km. area which is divided into 9 wards with a total population of 42,386 as per the 2011 Census, out of which the selected pilot study area lies in Ward No 3. The total area covered in the pilot study is 10.33 sq. km. Recently declared as an urban municipality, it is slowly emerging and still holds by and large rural character with most of the people engaged in agriculture. It is situated in the foothills of the Churia mountains and is mostly inhabited by migrants from the mountain districts in the north who have settled in search of fertile farming land in the past few decades. Few small settlements are inhabited by the traditional Tharu tribe but the rest of the area is inhabited by other different migrant ethnicities. As such, most of the land are still not registered and are under informal occupancy.

4.2. Methods

The FFPLA National Strategy (refer for Figure 12 of the National Strategy) [2] provided the basis for IVR of informal land rights through local level initiatives, which can later be recorded in the national register and reviewed for land right reforms. The authors of this article designed the methodology, supervised and closely monitored the process and initiated policy dialogues at the municipal, provincial and the national levels. The results from the two pilot studies jointly conducted by the Federal Ministry of Land Management, Cooperatives and Poverty Alleviation (MOLMCPA), UN-Habitat Nepal and the respective municipality, validated the proposed methodological approach and the FFP process as included in the strategy. IVR of informal lands is conducted through local pro-poor initiatives. The approach identifies and records the actual legitimate land rights through a participatory approach. More information on the FFPLA process for recognizing, recording and reviewing land rights by local-level initiatives in Nepal are well documented in [2]. This research considers all three frameworks of the FFPLA National strategy: The Spatial, Legal and Institutional Framework. It assesses a technological option of developing the spatial framework and informs on the development of legal and institutional frameworks to regularize informality for secured land tenure. The methodological workflow is presented in Figure 2.

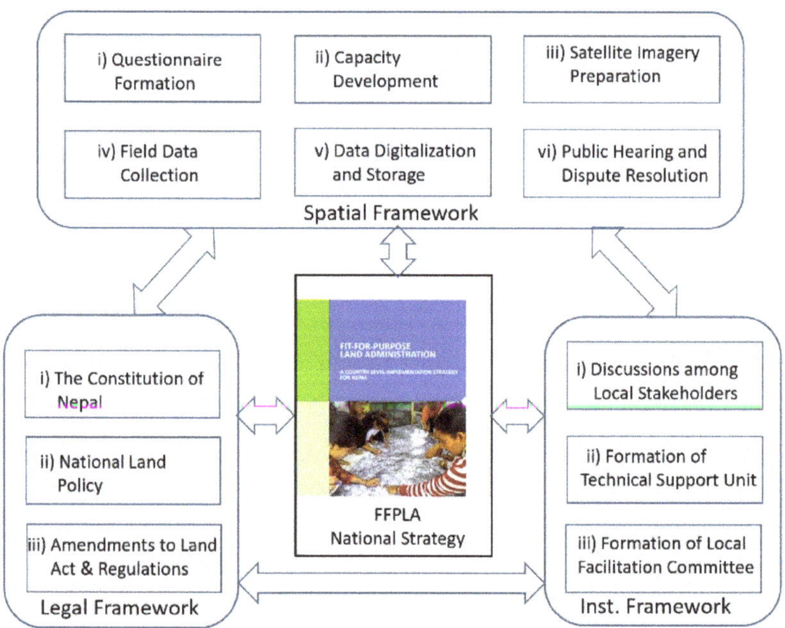

Figure 2. Overall methodological workflow of the research.

4.2.1. Spatial Framework

Spatial data acquisition on informality was divided into six main components as listed below:

(i) Questionnaire Formation

After discussion with local stakeholders, a questionnaire was prepared for the purpose of identifying land occupancy, recording people-to-land relationship, time span and circumstances regarding their holding on the land, land use and information on the socio-economic status of the family under informality. The questionnaire was required to identify the Sukumbasi as well as the status of the informal settlers. A consultation was held on the draft questionnaire with the Ministry and UN-Habitat Nepal. It was finalized after discussion with the stakeholders such as Community-Based Organizations (CBOs), civil society members and locally elected representatives of the municipality and finally, approved by the concerned municipality. The original questionnaire is in Nepali language and key points incorporated are provided in English in a simplified form the Appendix A, Table A2.

The final questionnaire contained questions related to family and family members details, occupation, income, land tenure documents, associated land rights, duration of occupancy, occupancy of land elsewhere in the country and present/potential risk on the land they utilize. The questions related to the socio-economic status were included to identify the Sukumbasi and the informal settlers' status.

(ii) Capacity Development

The training was conducted on Social Domain Tenure Model (STDM) tools and survey procedures for local enumerators identified by the municipalities, community leaders, CBOs and municipal/ward representatives. The training covered the introduction to the FFPLA but mainly focused on technical aspects—using imageries for identification of visual boundaries and their delineation, collecting locational data using Global Navigation Satellite System (GNSS)-enabled smartphone, documents digitalization and operations of STDM tools.

(iii) Satellite Imagery Preparation

Four-band natural color pan-sharpened ortho-rectified Pleiades [48] high-resolution satellite imageries of 0.5 m spatial resolution from the official vendor were used. Alternatively, satellite imageries from Google Earth may also be considered if their available resolution in the area fits the purpose. The imageries were mosaiced together in tiff format. True color composite orthophoto maps were printed at a scale of 1:1250. The scale is comparable with that of local cadastral maps. These printed satellite imageries were used by the enumerators, community members and the land-holders for identification, recognition and delineation of visible boundaries on the imagery through the participatory process.

Satellite imageries of 9th January 2018 were used for mapping informal settlements in Ratnanagar Municipality while those of 24th November 2017 were utilized for mapping agricultural fields having informal land tenure in Belaka Municipality Ward No. 3.

(iv) Field Data Collection

The municipal Ward Office informed/notified the landholders in advance regarding the date, time and venue for data enumeration and land survey. A survey team comprising of two enumerators trained as 'grassroots surveyor', a representative of the informal tenure holder community and the elected municipality ward leader visited the area with the questionnaire/form and the printed imagery of the area. The respective land/house-holder and the neighbors joined the survey team. First, of all, the team collected the information of the individual families, their land holdings, socio-economic parameters as included in the questionnaire and checked that it was all filled in appropriately. Photo images of the household head along with his/her spouse were acquired with a smartphone. The supporting documents were also photographed. The image IDs were recorded on the form so that they can be appropriately linked with the other data in the STDM software. The form was read aloud by a member of the survey team. When the participants agreed on the information, the head of the household/informant was asked to sign the form along with his/her thumbprints. This was attested by the community representative and the municipality/ward leader.

After a questionnaire survey for a group of households was completed, the participants/informants and the locals were asked to identify the houses and the land boundaries.

The identified visible boundaries were delineated on the printed satellite imagery, given a unique ID and recorded on the questionnaire form. The form was verified by the representative of the ward office and the community representative. Finally, it was approved by the ward chairperson and stamped to get officially recognized by the municipality.

Due to the very small size of spatial units and congested settlements in Ratnanagar Municipality, visible boundaries were non-existent or could not be recognized on the satellite imageries and, instead, Global Positioning System (GPS) coordinates of individual houses were collected through a smartphone. Further, the surveyors visited the doorsteps and informants were asked to identify their house in relation to the roads and neighboring houses which was marked on the printed satellite imagery. It was done this way in order to cope with the inaccuracy of the point coordinates obtained through the smartphone's GNSS.

Data collection and boundaries survey work for Ratnanagar and Belaka were respectively conducted between March–September 2019 and January–July 2019.

(v) Data Digitization and Storage

STDM tool was customized to incorporate data as per the respective questionnaires. Local data entry personnel, trained on data entry and digitization using STDM, undertook data entry from the questionnaire and referred to the satellite imagery with delineated boundaries for digitization on QGIS. The data collected through the questionnaire survey was digitized and imported into the STDM (see Appendix A Figures A1 and A2). The images of the household head together with the spouse when applicable, as well as images of the house, citizenship certificate and other supporting documents, if available were linked with the respective data. The pdf copies of the filled-in questionnaire forms were also inputted as evidence of field enumerated attested/certified information.

(vi) Public Hearing and Dispute Resolution

A public hearing was held at each of the communities to validate the information collected through the household's survey and participatory mapping of informally occupied lands. In addition, residents' active participation, data entry staffs, field enumerators, municipal/ward representatives and community leaders contributed to the validation process. The disputes, if any, were sorted out by a local judicial committee. Necessary changes recommended by the committee were updated on the database.

Finally, the mapping cost was estimated considering the direct cost for satellite imageries procurement, human resources costs for field data collection and verification. However, the mapping costs do not include pro bono costs incurred by the implementing partners in terms of computer hardware, customization of STDM software, providing training and technical guidance and monitoring.

4.2.2. Legal Framework

The FFPLA country-level implementation strategy [2] is the key concept which is kept in the center of this research. The legal framework in this strategy envisages flexible and implementable provisions in existing policy and legal frameworks and recommends such amendment to existing Lands Act and Lands Regulation, which can be applied further for addressing informal tenure on land. In this context, some other soft policies such as Voluntary Guidelines on the Governance of Tenure (VGGT) by FAO, which is also referred by the FFPLA strategy, were also taken into account. The VGGT explicitly suggests to identify, record and respect legitimate tenure rights on land [49]. Addressing the legal framework requirements was divided into three main components as listed below.

(i) The Constitution of Nepal

Fundamental concept and basic guidelines for this research is taken from the Constitution of Nepal [35], which made a provision for addressing the issue of land-lessness by providing land to the landless farmers and shelter/house to the homeless. The Constitution mandates for one time appropriation of land to the landless Dalits from the state and addressing the issue of informal settlers.

(ii) National Land Policy

This research was designed to achieve the aspiration of National Land Policy 2019 [42], which advocates defining informal tenure, recognizing it and addressing the problem of informality by regularizing the rights of settlers on that land. The policy also assures the equitable access on land to the marginalized people.

(iii) Land Act and Land Regulation Amendments

The research activities were conducted vis-à-vis ongoing discussions on the legal framework and informed on appropriate legal amendments in terms of the 8th Amendment to the Land Act [50] and the 18th Amendment to the Land Regulations [51] which have now come into place. Policy dialogues among stakeholders held (see dates in Appendix A, Table A1) at the municipal, provincial and the national levels provided inputs for the necessary amendments. The legal amendments were proposed to accommodate the process of spatial data acquisition and institutional arrangement as the former land legislation did not allow regularizing and formalizing the informal land rights. This kind of landholding was then legally considered as an encroachment on the Government/public land and the people settled therein were, in principle, subject to eviction.

4.2.3. Institutional Framework

The institutional arrangement was tested by ensuring the involvement of local government authorities. The Ministry led the process on the policy and management front. The concerned municipality planned and operated the whole IVR process of the people-to-land relationship at the local level and submitted the records, approved by the municipal council, to the Ministry. Addressing the institutional framework was divided into three main components as listed below.

(i) Discussions among Local Stakeholders

Discussions were held with stakeholders before the questionnaire formation, to seek their suggestion on the questionnaire as well as during the survey works. The municipalities leader and officials, CBOs, NGOs working in the field of land rights, civil society members and land occupants took part in the local level discussion and provided important feedback.

(ii) Formation of Technical Support Unit

A unit of locally trained surveyors/enumerators was constituted for data collection in the municipality, digitization and linking into the QGIS plug-in STDM. The co-authors of this article and the local UN-Habitat staff guided the work and were occasionally present on the site to sort out any technical issues during the data collection, form digitization and to provide support of using the STDM.

(iii) Formation of Local Facilitation Committee

Ward level local facilitation committee was formed to support the technical support unit to collect and verify the data in the village blocks. This local committee was aimed to inform, aware, mobilize and motivate local people to cooperate with the survey team. The committee was also involved in the adjudication process of land rights. The validation process was inbuilt at different steps. In case of any dispute regarding rights and occupancy on the land among the neighbors, the committee was responsible to facilitate the dispute resolution process in the local judicial committee led by the Deputy Mayor of the municipality. A supervisory team consisting of municipal representatives and local staffs of supporting Non-Governmental Organization (NGO) was present in order to guide, coordinate and help manage the data collection. The Municipal Council, after the resolution of disputes, if any, finally approved the data and inventory of the people-to-land relationship collected by the team in the field.

5. Results

The two study areas selected for this research—whole of the area under informality in Ratnanagar Municipality of Bagmati Province and Ward No. 3 of Belaka Municipality of Province No. 1—are diverse in characteristics. Agriculture land use is dominant in the Belaka Municipality where physical boundaries are visible and identifiable on the printed satellite imagery. In Ratnanagar Municipality, dense residential areas with tiny spatial units hardly identifiable on the available satellite imageries are prominent.

5.1. Spatial Framework

5.1.1. A Pilot Study in Dense Urban Setting (Ratnanagar Municipality)

A total of 1504 households consisting of 1517 spatial units from 58 communities distributed across 16 wards of the Ratnanagar Municipality were mapped by 6 groups of enumerators in 7 months. The STDM database was configured to store all the information collected from the field. The information obtained through the questionnaire survey was digitalized and first stored in spreadsheets. The questionnaires provisioned to collect documents such as certificate of citizenship, proof of the people living at that place such as voters' list or voters' card, local government's recommendations, certificates from community clubs or institutions and/or facilities provided by the municipality such as certificates of electricity connection, telephone connection, drinking water supplies or any other developments on the particular land. Photo images of these documents were captured by locally deployed grassroots surveyors using a smartphone which was later linked with the people-to-land relationship maintained in the STDM software. Where there are no such existing documents, a validation through participatory enumeration and certification by the ward representatives provided the evidence.

Spatial units were digitized on the QGIS as point features (Figure 3), with reference to GPS coordinates and the locations marked on the printed satellite imagery used for the land surveying. The same satellite imagery was used as a backdrop for the designation of the spatial units on the QGIS. Attribute information was imported into the STDM and linked with the respective spatial units. Relevant images of household head and the spouse, houses, filled-in and officially certified questionnaire forms and the supporting documents were uploaded and linked to the respective data. The inclusion and linking of these data and documents in the GIS platform enabled to query information based on parameters of the people-to-land relationship and perform location-specific analyses, for example. Likewise, linking documents and scanned questionnaire forms provide proof for legal validation.

The data collection to verification tasks were completed with an average of 7 parcels per day. The average costs for satellite imagery, field data collection, result verification and data entry were reported to be 7.40USD per parcel.

5.1.2. A Pilot Study in Rural Setting (Belaka Municipality Ward No. 3)

A total of 1783 households consisting of 3373 spatial units were surveyed in the Belaka Municipality Ward No. 3, also in seven months. Similar to the method applied in Ratnanagar Municipality, data were converted to digital form. Since the informal lands in the ward are dominated by agricultural land use, they were generally easy to delineate on the printed satellite imagery (Figure 4). Hence, the field boundaries were digitized (see Appendix A, Figure A2) using the imagery on the QGIS. All the data were imported to the STDM software and supporting documents including the pictures of the household head and the spouse were linked, as in the case of Ratnanagar Municipality. The inclusion of such data and documents in the QGIS enabled performing different types of spatial and non-spatial analyses, as has been described for the Ratnanagar case. Further, linking of supporting documents and the scanned questionnaire forms supported evidence for legal validation.

Figure 3. Identification of house/property in one of the compact settlements of Ratnanagar Municipality. The individual house location is shown with red-colored dots with satellite imagery in the background.

Figure 4. Delineating and coding spatial units on the printed satellite image map.

The data collection to verification tasks were completed with an average of 16 parcels per day. Likewise, the direct costs for satellite image procurement, data collection, verification and data entry were 3.11USD per parcel.

5.2. Legal Framework

Various new legal provisions were made in the 8th amendment of Lands Act and the 18th amendment of Land Regulations by incorporating recommendations and feedback from the municipal, provincial and national-levels stakeholders discussions, which were conducted during the pilot studies. The informal occupancy on the land that was considered an encroachment, can now be awarded a tenure certificate after necessary IVR process. Similarly, the study did inform on the typology of land areas under informal tenure which reflected as ceiling on the land area to be granted to the landless, landless Dalits and informal settlers. A different ceiling amount for rural agricultural land and urban residential area, as recommended, is now incorporated in the new Land Regulations.

Provision of regularization fee was recommended by the stakeholders and municipal corporations during field study and discussions. The regularization fee should be based on various factor such as location of land, land use class, land value and infrastructure and associated services in the surrounding. Similarly, it was recommended that a higher fee should be charged to the settlers with good socio-economic status whereas less fee should be charged to socially and economically disadvantaged and marginalized groups. The landless and landless Dalits should be provided land without charging any fee. The Land Act amendment incorporated these recommendations and such provisions are now incorporated in the land legislations.

A new institutional setup was also recommended by the study which is now materialized in the legislation. The Land Issues Resolving Commission (LIRC) is setup at the center. Along with this legislative provision for institutional setup, an executive order is now issued for its smooth operation and various procedure and guidelines have been developed for execution of the task at the local municipality level in accordance with the lessons learnt from the pilot studies.

5.3. Institutional Framework

The landholders received social and local government's recognition after mapping and local-level validation. The record was maintained by the municipality after the validation by the community and provided a document to the occupant mentioning that his/her application was recorded. However, they will eventually get a land ownership certificate once the Ministry approves the legitimacy of the land rights in line with the recently amended legal framework.

Discussion with the stakeholders helped formulate more relevant questions and improvise them. More importantly, collecting information as well as land survey was smooth because of inclusion and active participations of the occupants, CBOs, NGOs and municipalities officials. Provision of local facilitation committee helped verify information and correct any falsified claims of the occupants. Dedicated involvement of local government leaders and authorities was instrumental for planning and collection of the people-to-land relationships.

In Ratnanagar, there were issues related to land informally occupied in the private land or land owned by entities like School, Temple etc. The municipality board and other stakeholders had agreed before the start of the participatory mapping exercise to exclude them. Three family disputes surfaced during the enumeration. Though female siblings were occupying the lands and houses, they were claimed by their male siblings stating that the lands were temporarily allocated to the sisters. The local facilitation committee resolved these disputes and the lands were enumerated in the names of the female siblings. All disputes as evolved did resolved at the facilitation committee and no dispute was reported to the municipal judicial committee in Ratnanagar. In Belaka, three dispute cases were resolved through the local facilitation committee and enumerated as such. In another dispute, a spatial unit was claimed by a maternal family of the present holder stating that it was handed over for temporary use only. After not being resolved by the local facilitation committee, the dispute was referred to the judicial committee which decided to enumerate the land in the name of present holder. During enumeration stages, there used to occur debates or discussions regarding the land boundaries and their identification/delineation on the imagery. However, they were all resolved with participatory approach. This is a strength of the approach where chances of resolving disputes by the stakeholders themselves are high.

6. Discussion and Recommendations

The results obtained and the lessons learned from these two pilot studies were instrumental and have contributed to developing the spatial, legal and institutional setups for IVR of landless Dalits, Sukumbasi and informal settlers. The methodological development provided a mechanism to IVR of the people-to-land relationship. The two pilot studies provided information on the socio-economic conditions of the Sukumbasi and informal settlers and the type of holdings under informal tenure. Several cases were reported that need to be vacated due to cultural conflicts, environmental hazard risk and restrictions on land use. These cases were brought to the attention of the government legislation drafting team for necessary amendment in the legislation for incorporating issues related to the relocation.

Data collection, verification, storage in the STDM software and approval from the municipal office were completed in 7 months in both pilot areas with an average mapping cost of 3USD and 7.5USD respectively in rural and urban settings. The costs are context-specific which may differ between locations. Furthermore, time and cost may be reduced when the work is conducted at scale. The higher mapping cost in the Ratnanagar may be partly attributed to scattered settlements, comparatively expensive human resources due to the area being more urbanized and complexity in identifying compact houses on the satellite imagery.

The FFPLA tools have been applied in other countries too [44,45]. Rwanda has completed land registration of more than 11.4 million parcels with four years and at an average cost of around 6USD per parcel [44,45]. Similarly, Ethiopia and Tanzania have reported average costs of less than 10USD per parcel [44,45] when the mapping were completed at scale. Further, these studies suggest that the data collected and the system administered at the local level could be migrated to national systems provided there is due consideration for the underlying data model and mapping standards.

The experiences from India having social and caste-based structure similar to Nepal confirm that the implementation of FFPLA can be achieved at scale through local level initiative [52]. Similar to Nepal, India has 31% landless households, majority of whom belong to Dalits, tribal and lower caste communities [53]. As is evident like in India and other countries, these most vulnerable sector of the population will benefit from implementation of the FFPLA approach.

The experience of mapping informality in the pilot study sites infers that the FFPLA approach could be used to provide secure land tenure for all through local level initiatives, both faster and cheaper and support results as expressed in the political will. Further discussions have been divided into three themes corresponding to each of the FFPLA frameworks: The Spatial, Legal and Institutional Framework. Likewise, potential policy implications and time estimations for nationwide mapping of informality are also presented.

6.1. Spatial Framework

Spatial units' size in Belaka Municipality were generally rather large and the physical boundaries were identifiable on the satellite imagery that was then sufficient to map the area. The grassroots surveyors and occupants were able to recognize and demarcate the field boundaries. From field visits of few areas in other wards of the municipality, it was observed that the field boundaries were sometimes obscured by vegetation. Land boundaries in dense settlements in the other areas of the municipality were also difficult to be identified on 0.5 m resolution satellite imagery.

In the case of Ratnanagar Municipality, the spatial resolution of available satellite imagery did not support the identification of actual land boundaries (polygon) due to the presence of compact houses. Thus, the use of the smartphone's GNSS was tested for collecting the location of each house. However, horizontal position obtained from the smartphones contained a random error of up to 10m [54] which overlaid on satellite imagery showed a shift of spatial unit's location. Some adjustment with reference to the location and physical boundaries on the ground itself was necessary. Thus, houses were marked as point features utilizing neighborhood information, e.g., roads and adjacent houses. A similar approach has been reported in other research as well [25].

Hence, in line with recommendations of the FFPLA approach [23] and preliminary unpublished results obtained using drone survey from the dense area of Belaka Municipality, the authors recommend applying a hybrid method of boundary mapping using drone imageries in densely populated areas. The use of ultra-high-resolution drone imageries have shown promising results for boundary mapping in compacted areas as reported elsewhere as well [55]. Furthermore, conventional ground survey methods should be applied where the number of spatial units is low and the size of the plot under consideration is very small. This is recommended to cut down the cost of drone surveys, discard extra work of securing flight permissions as well as to reduce the technical complexity of drone survey and data processing and arrange additional computing resources, for smaller areas. Such conventional surveying methods or drone imageries should be restricted to locations where delineation of land boundaries and area computation is mandatary and the price of land is very high and therefore—demanding more accurate mapping of the boundaries.

As an alternative, point cadastre [1,56] may be used in the case of very dense settlements as tested in Ratnanagar Municipality when this is supported by the legislative framework. To improve the accuracy of point locations, handheld GPS can replace smartphone's GNSS which can provide an accuracy of approximately 3m when the Wide Area Augmentation System (WAAS) signal is available [57], but this may as well be considered when the size of spatial units is adequately large. This would be a much cheaper and faster way of data collection [1,56]. If the WASS signal is problematic, mapping grade GNSS equipment in deferential mode can be used which can provide an accuracy of 1–3 m [57]. Nonetheless, if boundaries are not demarcated, conflicts may arise in the future where the price of land escalates and as such should be supported by the Alternative Dispute Resolution (ADR) mechanism within the legislative and institutional framework. As is evident, surveying is a tool to make a "good estimate" of the true value and all survey estimates are "good enough" when they serve the purpose. Accuracy assessment was not the purpose of this study; however, several studies have indicated that the FFPLA approach provides good enough results [58] while the accuracy of conventional techniques like plane-tabling applied in many areas in Nepal [59,60] could be even more challenging.

6.2. Legal Framework

Despite the strong political will to resolve the issues of informal landholding and securing tenure for all after the proclamation of the new Constitution in 2015, the lack of an appropriate legal framework was considered as one of the barriers. The existing institutions managing the land administration, dealt only with the formal land tenure as per the (then) legal system. The policy discussions and testing of appropriate tools were conducted simultaneously. The National Land Policy [42] adopted in 2019 accepts and recognizes the existence of informal tenure and realizes the need for its formalization. Simultaneously, the FFPLA National Strategy [2] was developed to conduct the formalization process and an implementation framework was proposed. Piloting the strategy at two sites as pilot studies, representing a rural and a semi-urban with dense settlements, suggested the need for an appropriate legal mechanism to carry out the formalization process smoothly. Thus, the 8th Amendment to the Land Act 1964 [50] adopted in 2020 accommodates the substantial issues and the 18th Amendment of Land Regulations [51] adopted in 2020 complements to define procedures of the formalization process. The specific dates for major activities are tabulated in Appendix A, Table A1. Necessary executive orders, procedural guidelines and directives have been issued to speed up the process.

This framework provides a legal basis for identifying the various kinds of land rights of the settlers, recognizes and verifies these rights and records them properly in the system. The Act defines the settlers as either landless or informal settlers. As per the Act, the landless (landless Dalits and Sukumbasi included) will be able to get land and land title (ownership certificate with due restriction) over the specified land without paying a regularization fee whereas the informal settler having some registered land elsewhere should pay a specified regularization fee to get it formalized up to a specified area. The informal settlers must give-up the extra land exceeding the defined area threshold to be taken over by the Government. Along with the rights established on the land, the law also imposes some restrictions for the settlers on the land. The title will be issued with the restriction on the right to dispose of the property for 10 years except for inheritance by legal heirs. The restriction also ensures that the settler would not dispose of the land rendering him/herself landless again. The restriction on sale for 10 years is marked on the land record (Shresta—a copy of the ownership certificate that stays at the District Land Revenue Office) and the ownership certificate itself. It is believed that the family may become financially independent within this restricted period after which they can enjoy absolute rights over the land. The Act has a strong gender focus with a provision that the land must be registered jointly with the husband and wife when both the spouses are existing. The Act provisions evacuation of illegal occupancy and relocation of landless residing in unsecured areas and area designated for protection due to cultural reasons and land use restrictions.

The new legal framework also encompasses the technological (spatial) and institutional arrangements for carrying out the formalization process. As per the law, the Land Issues Resolving Commission (LIRC) and the municipality can adopt any technical tools and methodology appropriate to collect spatial information on the settlers and their land rights including spatial boundaries and preparation of cadastral maps. This is evidently based on the FFPLA National Strategy and supports the application of appropriate, hybrid and non-conventional technologies which is faster, cheaper and good enough to address the current needs and is upgradable and updatable to the needs of specific situations. Similarly, the legal framework makes a provision for the participation of local governments and other stakeholders in the formalization process. The need for these kinds of mechanisms in the new amendment of the law was realized as informed by the pilot studies whose outcomes are discussed in this article. Given the volume of work and the available resources—human, financial and time—the FFPLA approach is considered as the viable and scalable solution. Hence, the legal provisions as amended and now existing provide a path for mapping informality initiated by local governments, which would allow to provide legal land rights over the lands to the legitimate holders.

6.3. Institutional Framework

Whilst the pilot projects were conducted, the legislative and the institutional frameworks were under discussion and were not fully in place. As the legitimacy of the survey could not be fully assured, collecting authentic information from the household with the relevant supporting documents was a big challenge for the enumerators. The proper training provided helped them in getting reliable information. A similar approach is reported to be used elsewhere [61–63]. Further, the active participation of the local representatives and representatives from the organization of local informal settlements was instrumental. As they knew every detail of the respective households, it helped in verifying the information provided by the households to confirm that the information is true. In addition to inclusive and flexible community awareness programs, the role of committed representatives from the community and local offices supported obtaining more reliable and quality datasets. Such experience has also been observed in Indonesia, for example [22]. It is understood that the effective mobilization of these representatives and active participation of local elected representatives throughout the household survey was crucial for successful collection of authentic information with proper evidence.

Considering the importance of the participatory approach, the needs of local government initiatives and in line with the recent provisions of the legislative and institutional frameworks, a process workflow (Figure 5) has been devised. It is based on the experience and lesson learned from the pilot studies and after broader discussions with the stakeholders. It encompasses the identification of informal land tenure, their survey and mapping, verification and the distribution of land ownership certificates. The municipalities should play an active role in the identification and mapping of existing scenarios and planning land distribution/formalization/relocation/resettlement in line with the legislation and as foreseen as appropriate in a given municipal context. This will provide a technical framework for nationwide scale-up of the FFPLA approach from data collection on the people-to-land relationship and adjudication and distribution of land title to the landless and informal tenure-holders in Nepal. The district committees should support the municipalities in mapping and planning activities while the provincial government should facilitate inter municipalities relocations/resettlements activities, together with the identification of lands under hazard risks as mandated by the legislation. The LIRC [64], as constituted, is now mandated to give final approval of the legitimate land rights recorded at the local level and award ownership certificate to the eligible person/family. Additionally, the LIRC should design and develop a centralized system following specific data model, prepare uniform templates, guidelines mapping standards and standard operating procedures (SOPs), guide with specific methodology to the municipalities.

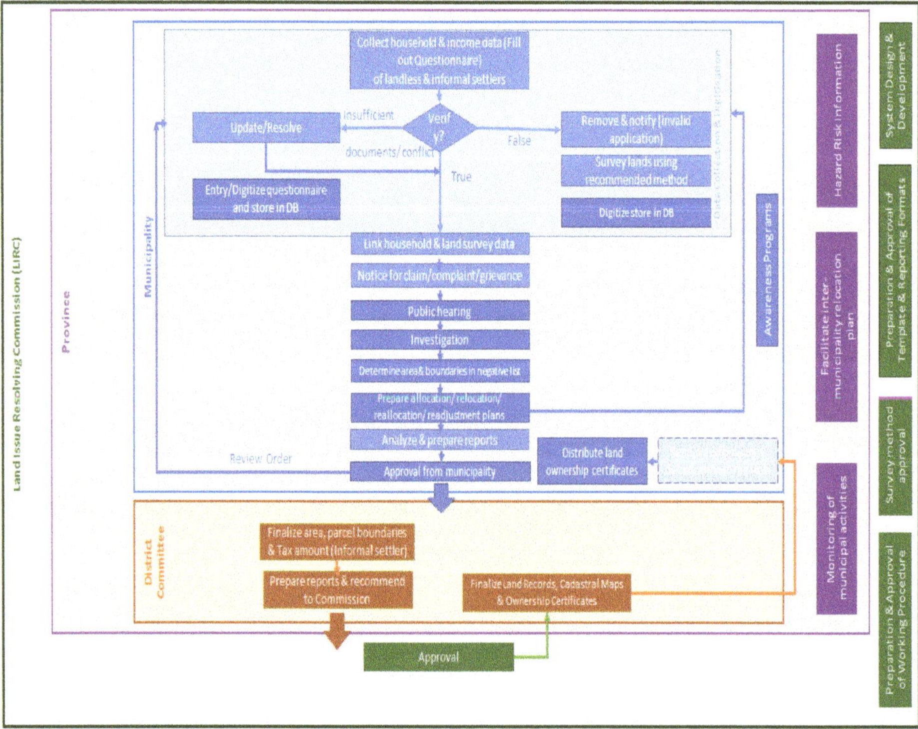

Figure 5. Process workflow depicting roles and responsibilities of different institutions (The box with dashed boundary is not mandatory).

Data management and institutional arrangement relating to integration with the existing formal land administration system need due consideration. It is recommended that a central web-based system be developed which should host nationwide data of people under informality, spatial units and people-to-land relationships. Further, the system should host satellite/drone imageries, administrative boundaries and other relevant data that would be used for mapping informality. The system should be maintained and governed by the LIRC where appropriate access would be given to the municipalities, the district committees and the provincial governments. When the formalization of informal spatial units is complete by delineation of the land boundaries and distribution of land ownership certificate to the holders, all relevant records should be handed over to the relevant land administration entities. While spatial data should be transferred to the Survey Department, non-spatial information should go to the Department of Land Management and Archives (DOLMA) who will maintain the records at the center providing necessary use and update rights to the municipalities for day-to-day service delivery. Since a large number of land parcels would be added to the formal system after the formalization of informal tenureship, it is expected that the service demand will increase after the release of restrictions in 10 years. Hence, necessary human resources and infrastructure should gradually be developed, mainly at the municipalities for improved service delivery.

6.4. Potential Policy Implications

The implementation of the FFPLA National Strategy is expected as a huge policy shift considering the existing traditional and conventional surveying and mapping approach, complex legal and procedural phenomena and overlapping institutional silos. The pilot studies show that new and out-of-the-box technologies in surveying are suitable enough to acquire the data on the people-to-land relationship and the physical boundaries in comparatively less time and with limited financial and semi-skilled human resources. Similarly, newly enacted amendments in the land act and land regulations are expected to fill the past gaps and address the legal issues experienced in the past. This will pave the way to formalize and legalize the legitimate occupancy of informal settlers and provide land and title to the landless. Further, the newly developed institutional framework will ensure participation of beneficiaries and/or stakeholders which would facilitate hassle-free identification, verification and recordation of the landless and informal settlers for ensuring land rights quickly.

6.5. Time Estimation for Nationwide Scale-Up

With the 10 million spatial units reported under informality, IVR of these lands is estimated to be completed in 4 to 5 years if the implementation of the FFPLA approach is simultaneously replicated, respectively, at 150 to 100 municipalities with implementation at an average of 4 wards in parallel (refer to Table 1 for details). Nepal has 753 municipalities/rural municipalities, which are divided into 6743 smallest administrative units called wards (https://www.sthaniya.gov.np/gis/, accessed on 2 July 2021) A municipality/rural municipality has a minimum of 5 to a maximum of 33 wards, with an average of about 9 wards. The most time-consuming field work is conducted at the ward/municipality level and managed by the respective offices. While the district committees support the municipalities/wards planning and mapping tasks, the LIRC ensures mapping standards and consistency of the work. Thus, the FFPLA approach enables to work in parallel throughout the country. Furthermore, the semi-skilled grassroots surveyors can be trained at the local level. However, slightly uneven distribution of informality across ecological reasons and among municipalities may lead to slightly higher mapping time. Likewise, the management time for district level validation, revenue collection and final approval from the LIRC are not included in the time estimation which may further delay the final approval and providing land ownership certificates to the legitimate holders.

Table 1. Time estimation for nationwide mapping of informal lands using the FFLA approach.

Description	Estimates	Remarks
Spatial units under informality	10,000,000	[12]
Number of Spatial Units Mapped per month (Urban Scenario)	217	as per the results of the two pilot studies.
Number of Spatial Units Mapped per month (Rural Scenario)	482	
Average Number of Spatial Units that can be mapped per month	402	with an estimated Urban-Rural spatial units under informality of 30:70
Estimated Mapping Time in Years (Scenario 1)	3.5	Simultaneous Implementation at 150 municipalities with an average of 4 wards in parallel
Estimated Mapping Time in Years (Scenario 2)	5.2	Simultaneous Implementation at 100 municipalities with an average of 4 wards in parallel

6.6. *Future Challenges*

Despite all these arrangements, the implementation is not without challenges. Hesitation of the traditionally trained surveyors and the beneficiaries who have seen instrument survey as a legitimate method in the past, in adopting new technologies such as satellite/drone imageries for boundaries demarcation may be one of the challenges. Similarly, conflicting interests of various stakeholders on the state land is another challenge, which might delay the process. Further, some of the newly formed local institutions in rural remote area may find difficulties in managing technological, human and financial resources to deal with the identification and mapping of landless, informal settlers and their land rights. In addition, coordination and supervision of work simultaneously in 100 to 150 municipalities at a time is a challenge. Additionally, concept of point cadastre is currently not supported by the existing legal provision which encourages increased application of drone imageries or ground-based conventional survey methods in dense settlements. Consequently, it may put more pressure on municipal resources and may cause delay in securing land rights for all.

7. Conclusions

Informally occupied lands were mapped using participatory methods and the Fit-For-Purpose Land Administration (FFPLA) approach. Satellite images, GNSS-enabled smartphones and free and open-source software tools were used for mapping in two different scenarios—dense settlements and areas dominated by relatively large agricultural fields. The pilot studies conclude that the high-resolution satellite imageries and freely available open-source software like STDM could be used in a participatory approach by locally trained 'grassroots surveyors' to quickly and inexpensively map large areas which would otherwise require highly-skilled human resources, large infrastructure and decades of time for mapping while using the conventional approaches. Point cadastre can be used in dense settlements for quick mapping where accuracy is not an issue. Nonetheless, it should be supported by necessary legislation which is so far not in place in Nepal.

The two pilot studies recommend that the FFPLA approach, supplemented by hybrid surveying technologies with the use of satellite imageries, drone imageries and ground survey techniques when appropriate can be applied. The technologies, tools and methodology for a spatial framework would meet the demand of high political will for providing land to the landless Dalits and Sukumbasi and tenure security for all within a limited timeframe. The study did inform on the suitable options for the adoption of the legislative and institutional framework in Nepal. However, scaling the FFPLA approach nationwide and integrating with the existing land administration system is not without further challenges.

The use of consistent data model, guidelines, uniform templates and mapping standards are recommended for nationwide scale-up.

Definition 1. *Sukumbasi is a landless person/family without land country-wide and a financial status incapable of procuring a piece of land on their own.*

Definition 2. *Informal Settlers are persons or families utilizing some of land(s) without having a formal land tenureship, however, possesses land(s) somewhere in the country that is(are) legally registered in their (or their family) name.*

Definition 3. *Dalits were previously considered to be in the lowest caste and part of the so-called "untouchables". Nepal legally abolished the caste-system and criminalized caste-based discrimination, including "untouchability" in 1963.*

Author Contributions: Conceptualization, U.S.P., R.R.C. and J.R.J.; methodology, R.R.C., J.R.J., J.D. and U.S.P.; software, U.S.P. and R.R.C.; validation, J.D., R.R.C. and J.R.J.; formal analysis, U.S.P., R.R.C. and J.R.J.; resources, R.R.C., J.D. and J.R.J.; data curation, U.S.P. and R.R.C.; writing—original draft preparation, U.S.P.; writing—review and editing, ALL; visualization, U.S.P.; project administration, R.R.C. and J.D.; funding acquisition, D.A., R.R.C., J.D. and J.R.J. All authors have read and agreed to the published version of the manuscript.

Funding: Funding of the publication costs for this article has kindly been provided by the School of Land Administration Studies, University of Twente, in combination with Kadaster International, The Netherlands.

Data Availability Statement: The data presented in this research are available on request from the authors except from the personal's data of the occupants.

Acknowledgments: The study is based on the findings of the UN-Habitat/GLTN funded project (Contract No. SAoC18-007 and SAoC-008) in Nepal. The authors would like to acknowledge the contribution of the Government of Nepal, Ratnanagar and Belaka municipalities, and National Land Rights Forum (NLRF) for partial co-funding. We would like to thank Lumanti and CSRC for their support in data collection.

Conflicts of Interest: The authors declare no conflict of interest.

Appendix A

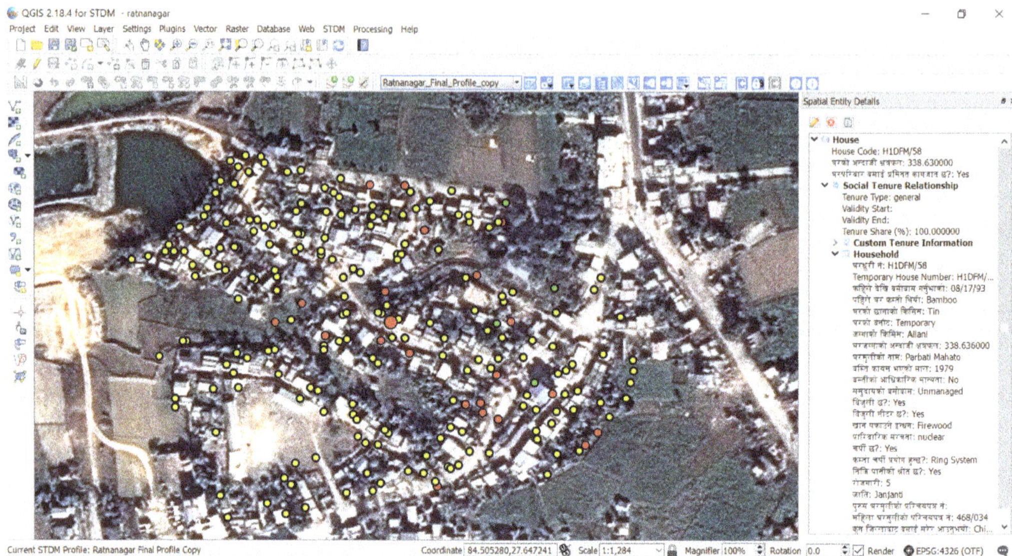

Figure A1. Identification of house/property in one of the compact settlements of Ratnanagar Municipality (shown with red colored dots with satellite imagery in the background; Red, Yellow and Green colored dots respectively represent temporary, semi-permanent and permanent houses) in compact settlements of Ratnanagar Municipality. The right most portion of the picture represents people-to-land relationship's parameters (in Nepali language) of a select spatial unit (depicted by the enlarged Red colored dot in the center).

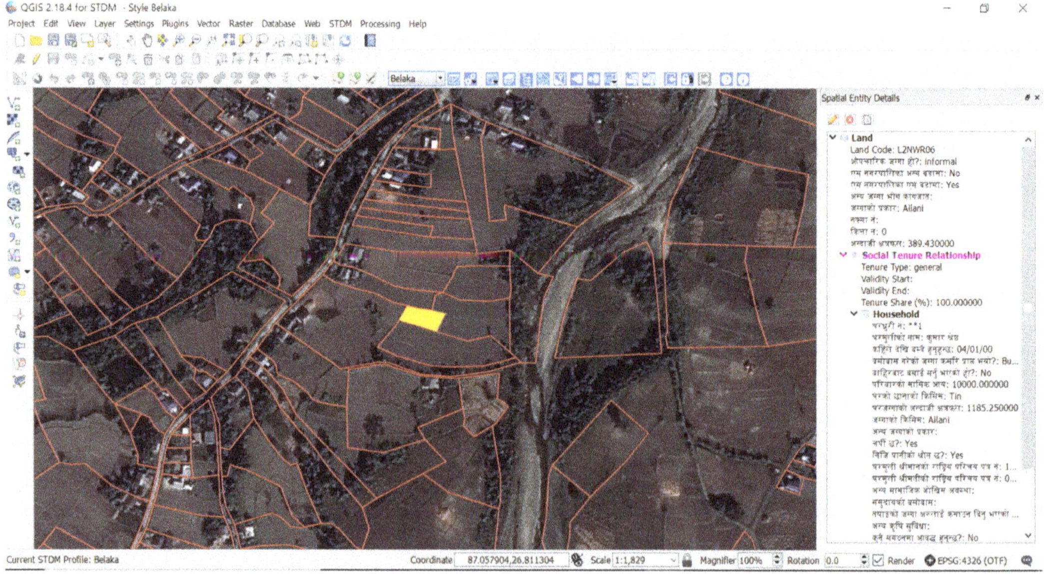

Figure A2. Delineated spatial units overlaid on satellite image in QGIS. The right most portion of the picture represents people-to-land relationship's parameters (in Nepali language) of a select spatial unit (shown in Yellow color).

Table A1. Important Dates.

Framework Issues	Date
Spatial Framework	
Image Date (Belaka/Ratnanagar)	24 November 2017–9 January 2018
Orientation on STDM tools and field-based training (Belaka/Ratnanagar)	January–March 2019
Survey Start (Belaka/Ratnanagar)	January–March 2019
Survey End (Belaka/Ratnanagar)	July–September 2019
Legal Framework	
Consultation and Drafting Land Act Amendment	September 2018–May 2019
Land Act Bill passed by the Parliament	August 2019
Land Act approved by the President	February 2020
Consultation and Drafting Land Regulation Amendment	March–May 2019
Land Regulation approved by the Cabinet	December 2020
Land Regulation Act Published in Gazette	December 2020
Institutional Framework	
Consultation meeting, to finalize the survey questionnaires (Belaka/Ratnanagar)	January 2019/January 2019
Provincial dialogue on land issues and possible solution	February–April 2019
Formation Order of Land Issues Resolving Commission (LIRC)	April 2020

Table A2. Simplified Questionnaire (translated from Nepali language).

Form and Basic Information	
1. Questionnaire number 2. Date of household Survey 3. Ward No. 4. GPS-ID number 5. Camera ID number 6. Household (HH) number 7. Caste (*Dalit*/Indigenous/*Brahmin-Chhetri*/Others) 8. Household head 9. Citizenship number of HH head 10. Father/Husband Name 11. Grandfather/Father-in-law Name 12. Respondent Name and Relation to HH Head (If different than house head)	
Section A Family details for each adult member of the family	**Section B** Land occupancy
1. Name 2. Gender 3. Date of birth 4. Marital status 5. Education status 6. Occupation 7. National ID type and number 8. Physical and Health Condition 9. Relation to HH head 10. For how long you have been cultivated or residing on this land (in years)? 11. How did you receive this land (i. Inheritance ii. From relatives iii. Purchased iv. Started cultivation by Self v. Allocated by someone vi. Others-mention source)? 12. The purpose of your migration to this location (i. Opportunity for better livelihood ii. Better education for kids iii. Displaced due to natural disaster iv. Displaced due to war v. Ruined by development of physical infrastructure vi. Opportunity for land access vii. Rehabilitation viii. Others-specify)? 13. What is your family monthly income and expenses (in Nepalese rupees)?	1. Do your family have own house? What's your house structure (Permanent/Temporary/Other-specify)? 2. What is the type of land where the house is positioned (i. Private ii. Public iii. Guthi iv. Government v. Forest vi. Other-specify)? 3. Citizenship certificate/National ID and photo number (of spouse) 4. Location code of land 5. Approximate area 6. GPS File number (of land boundaries/point coordinate) 7. Photo number (House) 8. Evidence document on land occupancy (i. Temporary land ownership certificate ii. Revenue bill against occupied land iii. House migration document iv. Recommendation document v. No evidence vi. Other-specify) 9. Four corners of the land (E/W/N/S)

References

1. Joshi, J.R.; Panday, U.S.; Chhatkuli, R.R.; Enemark, S.; Antonio, D.; Deuja, J.; Sylla, O. Fit-For-Purpose Land Administration Strategy: An Innovative Approach to Implement Land Policies in Nepal. In Proceedings of the Land and Poverty Conference 2019: Catalyzing Innovation, Washington, DC, USA, 25–29 March 2019; p. 25.
2. MOALMC; GLTN; UNHABITAT; CSRC. Fit-For-Purpose Land Administration: A Country Level Implementation Strategy for Nepal. 2018. Available online: https://gltn.net/download/full-report-fit-for-purpose-land-administration-a-country-level-implementation-strategy-for-nepal/?wpdmdl=12829&ind=0 (accessed on 2 July 2021).
3. Panday, U.S.; Joshi, J.R.; Chhatkuli, R.R.; Enemark, S.; Antonio, D.; Deuja, J. Development of Fit-For-Purpose Land Administration Country Strategy: Experience from Nepal. In Proceedings of the FIG Working Week 2019, Hanoi, Vietnam, 22–26 April 2019; p. 15.
4. Cox, T. Land Rights and Ethnic Conflict in Nepal. *Econ. Polit. Wkly.* **1990**, *25*, 1318–1320.
5. Sapkota, M. Public land movement in Nepal: Expanding coverage and diminishing achievements. *Nepal J. Soc. Sci. Public Policy* **2016**, *4*, 19.
6. Tamang, M. *Accord-An International Review of Peace Initiatives*; Conciliation Resources: London, UK, 2017; p. 10.
7. The Asia Foundation. *The State of Conflict and Violence in Asia*; The Asia Foundation: San Francisco, CA, USA, 2017.
8. Upreti, B.R. *Nepal's Armed Conflict: Security Implications for Development and Resource Governance*; The International Centre for Integrated Mountain Development: Kathmandu, Nepal, 2006.
9. Macours, K. Increasing inequality and civil conflict in Nepal. *Oxf. Econ. Pap.* **2011**, *63*, 26. [CrossRef]
10. Chhatkuli, R.R.; Dhakal, S.; Antonio, D.; Singh, S. Statutory Versus Locally Existing Land Tenure Typology: A Dilemma for Good Land Governance in Nepal. In Proceedings of the FIG Working Week 2019 Geospatial Information for A Smarter Life and Environmental Resilience, Hanoi, Vietnam, 22–26 April 2019; p. 10.
11. Wily, L.A.; Chapagain, D.; Sharma, S. *Land Reform in Nepal Where Is It Coming from and Where Is It Going? The Findings of a Scoping Study on Land Reform for DFID Nepal*; Tribhuvan University: Kirtipur, Nepal, 2009; ISBN 978-99937-213-90-5.
12. *LIRC Preliminary Data Collected by Land Issues Resolving Commission (LIRC)*; LIRC: Kathmandu, Nepal, 2020.
13. AI-Nepal; CSRC; JURI-Nepal. *Nepal: Land for Landless Peasants-Comments and Recommendations on Amendment to the Lands Act 1964*; Amnesty International Nepal: Kathmandu, Nepal, 2019. Available online: https://www.amnesty.org/download/Documents/ASA3112212019ENGLISH.pdf (accessed on 14 July 2021).
14. Adhikari, J. *Land Reform in Nepal: Problems and Prospects*; Nepal Institute of Development Studies: Kathmandu, Nepal, 2008; p. 162.
15. Government of Nepal. *The Lands Act, 2021 (1964 BS)*; Nepal Law Commission: Kathmandu, Nepal, 1964.
16. CSRC. *Locally Present Land Tenure Typology in Nepal: A Study Report*; GLTN Publications: Kathmandu, Nepal, 2018.
17. *United People's Front Nepal Translation of 40 Point Demand to the Prime Minister of Nepal by the United People's Front*; Nepal, 1996. Available online: https://www.satp.org/satporgtp/countries/nepal/document/papers/40points.htm (accessed on 2 July 2021).
18. *Government of Nepal Comprehensive Peace Agreement between the Government of Nepal and the Communist Party of Nepal (Maoist)*; 2006; p. 13. Available online: https://peacemaker.un.org/nepal-comprehensiveagreement2006 (accessed on 14 July 2021).
19. Government of Nepal. *The Constitution of Nepal, 2015 (2072 BS)*; Nepal Law Commission: Kathmandu, Nepal, 2015.
20. Government of Nepal. *Local Government Operation Act, 2017 (2074 BS)*; Nepal Law Commission: Kathmandu, Nepal, 2017; p. 86.
21. World Bank Land. Available online: https://www.worldbank.org/en/topic/land (accessed on 23 November 2020).
22. Aditya, T.; Unger, E.-M.; vd Berg, C.; Bennett, R.; Saers, P.; Syahid, H.L.; Erwan, D.; Wits, T.; Widjajanti, N.; Santosa, P.B.; et al. Participatory land administration in Indonesia: Quality and usability assessment. *Land* **2020**, *9*, 79. [CrossRef]
23. Enemark, S.; Mclaren, R.; Lemmen, C. *Fit-For-Purpose Land Administration: Guiding Principles for Country Implementation*; United Nations Human Settlements Programme: Nairobi, Kenya, 2016.
24. UNGGIM. *Framework for Effective Land Administration: A Reference for Developing, Reforming, Renewing, Strengthening, Modernizing and Monitoring Land Administration*; UNGGIM: New York, NY, USA, 2020. Available online: https://ggim.un.org/meetings/GGIM-committee/10th-Session/documents/E-C.20-2020-29-Add_2-Framework-for-Effective-Land-Administration.pdf (accessed on 14 July 2021).
25. Unger, E.-M.; Chhatkuli, R.R. *Fit-For-Purpose Land Administration in a Post Disaster Context: Lessons and Applications from Nepal*; United Nations Human Settlements Programme: Nairobi, Kenya, 2019.
26. Unger, E.M.; Chhatkuli, R.R.; Antonio, D.; Lemmen, C.H.J.; Zevenbergen, J.A.; Bennett, R.M.; Dijkstra, P. Creating Resilience to Natural Disasters through FFP Land Administration—An Application in Nepal. In Proceedings of the 20th Annual World Bank Conference on Land and Poverty 2019: Catalyzing Innovation, Washington, DC, USA, 25–29 March 2019; p. 20.
27. Dijkstra, P.; Antonio, D.; Chhatkuli, R.R.; Beshah, W.T.; Shrestha, S.S.; Lemmen, C. Implementation of Fit for Purpose Land Administration Approaches in Nepal. In Proceedings of the FIG Congress 2018: Embracing Our Smart World Where the Continents Connect-Enhancing Geospatial Maturity of Societies, Istanbul, Turkey, 6–11 May 2018.
28. SILTIP; HURADEC-Nepal. *Support for Land Reform in Nepal and Land Tenure Initiative in Asia-Pacific*; Charikot: Dolakha, Nepal, 2017.
29. Giri, H. An Exploration of Ethnic Dynamics in Nepal. *J. Popul. Dev.* **2020**, *1*, 71–78. [CrossRef]
30. MOFA Nepal Profile. Available online: https://mofa.gov.np/about-nepal/nepal-profile/ (accessed on 10 May 2021).
31. Leitner Center for International Law and Justice. *Land Is Life, Land Is Power: Landlessness, Exclusion, and Deprivation in Nepal*; Leitner Center for International Law and Justice: New York, NY, USA, 2011.
32. Höfer, A. *The Caste Hierarchy and the State in Nepal: A Study of the Muluki Ain of 1854*, 2nd ed.; Himal Books: Kathmandu, Nepal, 2004; ISBN 9789993343585.

33. Macalester College. Constitution of the Kingdom of Nepal VS 2047 (1990). *Himalaya J. Assoc. Nepal Himal. Stud.* **1991**, *11*, 51.
34. GoN. *The Interim Constitution of Nepal 2007 (2063 BS)*; 2007; p. 167. Available online: http://www.ilo.org/wcmsp5/groups/public/---ed_protect/---protrav/---ilo_aids/documents/legaldocument/wcms_126113.pdf (accessed on 14 July 2021).
35. GoN. *The Constitution of Nepal, 2015 (2072 BS)*; 2015; p. 240. Available online: https://www.mohp.gov.np/downloads/Constitution%20of%20Nepal%202072_full_english.pdf (accessed on 14 July 2021).
36. NPC. *The Fifteenth Plan (Fiscal Year 2019/20–2023/24)*; The National Planning Commission: Kathmandu, Nepal, 2020.
37. NPC; UNDP. *Beyond Graduation: Productive Transformation and Prosperity*; UNDP: Kathmandu, Nepal, 2020.
38. MOAD. *Agriculture Development Strategy (ADS) 2015 to 2035*; Ministry of Agricultural Development: Kathmandu, Nepal, 2015.
39. Aryal, J.P.; Holden, S.T. Land reforms, caste discrimination and land market performance in Nepal. In *Land Tenure Reform in Asia and Africa*; Holden, S.T., Otsuka, K.D.K., Eds.; Palgrave Macmillan, London: London, UK, 2013; pp. 29–53. ISBN 978-1-349-46586-6.
40. IOM. *Barriers to Women's Land and Property Access and Ownership in Nepal*; International Organization for Migration (IOM): Kathmandu, Nepal, 2016. Available online: https://www.iom.int/sites/default/files/our_work/DOE/LPR/Barriers-to-Womens-Land-Property-Access-Ownership-in-Nepal.pdf (accessed on 14 July 2021).
41. FAO. *Country Gender Assessment of Agriculture and the Rural Sector in Nepal*; Food and Agriculture Organization of the United Nations: Kathmandu, Nepal, 2019. Available online: http://www.fao.org/3/CA3128EN/ca3128en.pdf (accessed on 14 July 2021).
42. *MoLMCPA National Land Policy*; The Ministry of Land Management, Cooperatives and Poverty Alleviation (MoLMCPA): Kathmandu, Nepal, 2019; p. 24. Available online: https://molcpa.gov.np/storage/uploads/land%20use%20policy__2015_ENGLISH_1514355245_1542355707.pdf (accessed on 2 July 2021).
43. Unger, E.-M.; Bennett, R. *Fit-For-Purpose Land Administration for All: A Guide for Surveyors on Adoption and Adaptation of Fit-For-Purpose Land Administration*; United Nations Human Settlements Programme: Nairobi, Kenya, 2019.
44. Baldwin, R.; English, C.; Lemmen, C.; Rose, I.; Smith, A.; Solovov, A.; Sullivan, T. Are Local Registers the Solution? In Proceedings of the 19th Annual World Bank Conference on Land and Poverty, Washington, DC, USA, 19–23 March 2018; p. 27.
45. Enemark, S.; Bell, K.C.; Lemmen, C.; McLaren, R. *Fit-For-Purpose Land Administration*; FIG Publication: Copenhagen, Denmark, 2014.
46. Rajdevi Engineering Consultant. *Carrying Out Study on Comparison of Various Cadastral Data Acquisition Technology*; Rajdevi Engineering Consultant (P) Ltd.: Kathmandu, Nepal, 2016.
47. Coughlan, P.; Coghlan, D. Action Research: Action Research for Operations Management. *Int. J. Oper. Prod. Manag.* **2002**, *22*, 220–240. [CrossRef]
48. Apollo Mapping PLÉIADES 1. Available online: https://apollomapping.com/pleiades-1-satellite-imagery (accessed on 1 April 2021).
49. FAO. *Voluntary Guidelines on the Responsible Governance of Tenure of Land, Fisheries and Forests in the Context of National Food Security*; Food and Agriculture Organization: Rome, Italy, 2012.
50. Government of Nepal. *The 8th Amendment to the Land Act 1964*; The Ministry of Law, Justice and Parliametry Affairs: Kathmandu, Nepal, 2020; p. 16.
51. Government of Nepal. The 18th Amendment of the Land Regulations. Available online: http://lirc.gov.np/storage/uploads/-1881417450.pdf (accessed on 2 July 2021).
52. Ho, S.; Choudhury, P.R.; Haran, N.; Leshinsky, R. Decentralization as a Strategy to Scale Fit-for-Purpose Land Administration: An Indian Perspective on Institutional Challenges. *Land* **2021**, *10*, 199. [CrossRef]
53. Mint How Many Indians are landless? Available online: https://www.livemint.com/Opinion/PUzqHSs3xejXk4hm2djTPM/How-many-Indians-are-landless.html (accessed on 12 May 2021).
54. Merry, K.; Bettinger, P. Smartphone GPS accuracy study in an urban environment. *PLoS ONE* **2019**, *14*, e0219890. [CrossRef] [PubMed]
55. Chio, S.H.; Chiang, C.C. Feasibility study using UAV aerial photogrammetry for a boundary verification survey of a digitalized cadastral area in an Urban City of Taiwan. *Remote Sens.* **2020**, *12*, 1682. [CrossRef]
56. Hackman-Antwi, R.; Bennett, R.M.; de Vries, W.T.; Lemmen, C.H.J.; Meijer, C. The point cadastre requirement revisited. *Surv. Rev.* **2013**, *45*, 239–247. [CrossRef]
57. USGS. USGS Global Positioning Application and Practice. Available online: https://water.usgs.gov/osw/gps/ (accessed on 1 April 2021).
58. Yagol, P.; Shrestha, E.; Thapa, L.; Poudel, M.; Bhatta, G.P. Comparative Study on Cadastral Surveying Ising Total Station and High Resolution Satellite Image. In Proceedings of the FIG-ISRPS, 2015: International Workshop on Role of Land Professionals and SDI in Disaster Risk Reduction: In the Context of Post 2015 Nepal Earthquake, Kathmandu, Nepal, 25–27 November 2015; p. 10.
59. Kohli, D.; Unger, E.; Lemmen, C. Validation of A Cadastral Map Created Using Satellite Imagery and Automated Feature Extraction Techniques: A Case of Nepal. In Proceedings of the FIG Congress 2018: Embracing Our Smart World Where the Continents Connect: Enhancing the Geospatial Maturity of Societies, Istanbul, Turkey, 6–11 May 2018; p. 12.
60. Koirala, M.P. Provided Cadastral Survey Maps Commencing Risks to Identifying and Valuing the Land in Nepal. *World J. Eng. Res. Technol.* **2019**, *5*, 263–273.
61. Vuksanović-Macura, Z. The mapping and enumeration of informal Roma settlements in Serbia. *Environ. Urban.* **2012**, *24*, 685–705. [CrossRef]

62. Bennett, R.M.; Alemie, B.K. Fit-for-purpose land administration: Lessons from urban and rural Ethiopia. *Surv. Rev.* **2016**, *48*, 11–20. [CrossRef]
63. Balas, M.; Carrilho, J.; Joaquim, S.; Murta, J.; Lemmen, C.; Matlava, L.; Marques, M.R. Mozambique Participatory Fit for Purpose Massive Land Registration. In Proceedings of the Responsible Land Governance: Towards and Evidence Based Approach, Annual World Bank Conference on Land and Poverty, Washington, DC, USA, 20–24 March 2017.
64. GoN. *Formation Order: Land Issues Resolving Commission (in Nepali Language)*; Nepal Law Commission: Kathmandu, Nepal, 2020; p. 22.

Article

Fit for Purpose Land Administration: Country Implementation Strategy for Addressing Uganda's Land Tenure Security Problems

Moses Musinguzi [1,*], Stig Enemark [2] and Simon Peter Mwesigye [3]

1 Department of Geomatics and Land Management, Makerere University, Kampala P.O. Box 7062, Uganda
2 Department of Planning, The Technical Faculty of IT and Design, Aalborg University, 9000 Aalborg, Denmark; enemark@plan.aau.dk
3 Land, Housing and Shelter Section UN-Habitat/Global Land Tool Network, 13-15 Parliament Avenue, Kampala P.O. Box 7062, Uganda; simon.mwesigye@un.org
* Correspondence: moses.musinguzi@mak.ac.ug; Tel.: +256-772-511-119

Abstract: The Republic of Uganda is one of the five countries within the East African region. Uganda's efforts to increase land productivity are hampered by land tenure insecurity related problems. For more than ten years, Fit for Purpose Land Administration (FFPLA) pilot projects have been implemented in various parts of the country. Uganda is now in advanced stages of developing a country strategy for implementing a fit for purpose approach to land administration, to define the interventions, time and cost required to transform the existing formal (western type) land administration system into an administration system that is based on FFPLA principles. This paper reviews three case studies to investigate how lessons learnt from pilot projects informed a FFPLA country implementation strategy. The review is based on data collected during the development of the FFPLA strategy, in which the authors directly participated. The data collection methods included document review, field visits and interviews with purposively selected respondents from the pilot sites and institutions that had piloted FFPLA in Uganda. The study identified that pilot projects are beneficial in highlighting specific gaps in spatial, legal and institutional frameworks, that have potential to constrain FFPLA implementation. Pilot projects provided specific data for informed planning, programing and costing key interventions in the FFPLA country implementation strategy. The lessons learnt from the pilot projects, informed the various steps and issues considered while developing the national strategy for implementing a FFPLA approach in Uganda. On the other hand, the study identified that uncoordinated pilot projects are potential sources of inconsistencies in data and products, which may be cumbersome to harmonize at a national level. In order to implement a fit for purpose approach for land administration at a national level, it is necessary to consolidate the lessons leant from pilots into a unified country implementation strategy.

Keywords: fit for purpose; land administration; case studies; Uganda; customary tenure

Citation: Musinguzi, M.; Enemark, S.; Mwesigye, S.P. Fit for Purpose Land Administration: Country Implementation Strategy for Addressing Uganda's Land Tenure Security Problems. *Land* **2021**, *10*, 629. https://doi.org/10.3390/land10060629

Academic Editor: Hossein Azadi

Received: 14 May 2021
Accepted: 6 June 2021
Published: 11 June 2021

Publisher's Note: MDPI stays neutral with regard to jurisdictional claims in published maps and institutional affiliations.

1. Introduction

Like many developing countries, Uganda is faced with challenges of making the best use of its land and natural resources to support a large proportion of the population living in rural areas. It is estimated that 77% of Uganda's population lives in rural areas practicing subsistence agriculture as the major source of livelihood [1]. Uganda aspires to transform the agriculture sector from subsistence farming to commercial agriculture in order to make agriculture profitable, competitive and sustainable, so as to provide food and income security to the people [2], p. 45. Uganda, therefore, aspires to adjudicate and document land rights, and issue legal documents as means to provide security of tenure to the land rights holders.

Land tenure is the relationship, whether legally or customarily defined, among people, as individuals or groups, with respect to land [3]. The concept of tenure security has largely evolved in response to clarify investment incentives for property holders [4,5]. Security of tenure is the certainty that a person's rights to land will be recognized by others and protected in cases of specific challenges [3]. People with insecure tenure face the risk that their rights to land will be threatened by competing claims, and even lost as a result of eviction. Without security of tenure, households are significantly impaired in their ability to secure sufficient food and to enjoy and improve sustainable rural and urban livelihoods. Research effort to link land registration and tenure security is not very conclusive [6,7], but there is international recognition that issuance of legal land rights documents such as land titles enhances land tenure security, investment and environmental management [8,9]. In Uganda, recent research [10] has identified that tenure security is a pre-requisite for introducing successful commercial agriculture programs. It is along the same lines that Uganda's land policy (2013) prioritized issuance of legal documents to land owners as a means to secure their land rights [11]. However, Uganda's effort to provide tenure security for all land rights holders has been hampered by the complex, costly and sporadic procedures for land registration. As a result, less than 20% of land is registered under three formal land tenure systems (freehold, leasehold and a quasi-form of freehold, termed Mailo, that allows for lawful and bona fide occupants to co-exist with registered owners). The bulk of the land, which falls under the customary tenure (about 80%) is not registered, although the tenure system has been formalized since the promulgation of the 1995 constitution. There is evidence that the most occurring types of land disputes with potential to erupt into social strife are either boundary related (30%) and/or encroachment based (26%) [12]. Uganda's delayed response to critical land administration issues is evidenced by socio-economic problems, including land fragmentation, low agricultural productivity, land disputes, loss of forest cover and environmental degradation [13].

Uganda recognized that a feasible approach for achieving 100% coverage of land registration in a reasonable time and at affordable cost is adopting a FFP approach to land administration in the spatial, legal and institutional terms. Indeed, the country has had more than 10 years of piloting FFPLA through scattered pilot projects across the country, and is now ready to upscale it to the national level. However, a FFPLA approach requires an implementation strategy, if it is to be up-scaled from site-specific projects to a national level [14]. Such a strategy should define the interventions, time and cost required to transform the existing formal (western type) land administration system into an administration system that is based on FFPLA principles [15]. Without planning for such a country implementation strategy, there is limited guarantee that all the tenure security issues in the country will be addressed in a contextualized, consistent, cost-effective and timely manner.

The term "fit-For-purpose" means applying the spatial, legal, and institutional frameworks that are most fit for the purpose of providing secure tenure for all [16]. This approach will enable the building of national land administration systems within a reasonable time and at affordable cost. The systems can then be incrementally improved over time [15]. The concept includes guiding principles for building the three frameworks in a flexible and participatory way, that is responsive to the country context. Flexibility is a key component in terms of using aerial imagery and visible boundary rather than complying with costly and high accuracy regulations [17], and by including both legal and legitimate tenure forms rather than just ownership titles. By introducing participatory processes of land recordation at community level, the implementation can be carried out in parallel throughout the country and the system can be maintained by decentralized institutions.

This paper, reviews three case studies to investigate how lessons from pilot projects informed a FFPLA country implementation strategy in Uganda. The country context in terms of land tenure systems is presented in Section 3. This is followed by a review of the evolution of the land administration concept and the nature of customary tenure system in Uganda, where most of the FFLA pilots have taken place. A description of the three case studies is provided under Section 4, while the lessons learnt from the case studies are analyzed in Section 5 in relation to building the spatial, legal and institutional frameworks, representing the key component of the FFPLA concept. This provides the basis for developing a FFPLA national strategy as presented in Section 6, followed by discussions and conclusions in Sections 7 and 8, respectively.

2. Material and Methods

The National Government in Uganda recognizes the overall lack of secure land rights as a major problem in relation to economic growth and social and environment sustainability. This paper explores ways and means for addressing this problem. The paper starts by presenting the three types of customary tenure and the attempts by the Government to enable tenure security through certificates of customary ownership (CCO). The paper uses a qualitative case study approach and qualitative analyses for addressing the problem of tenure insecurity in Uganda. This methodology as described in [18] enables investigation of the boundaries between the phenomenon (the FFP approach) and the environment (the land tenure situation in Uganda).

The review is based on data collected during the development of the FFPLA strategy in which the authors directly participated. The data collection methods included document review, and interviews with purposively selected respondents from the institutions that had piloted FFPLA in Uganda. The visited institutions with FFPLA experience included Mityana District Land Office, which was being prepared to become a Ministry Zonal Office (MZO), Mbarara Ministry Zonal Office and Kabale District Land Office. In Mbarara MZO, the Government was piloting FFPLA for conversion of customary tenure to Freehold under a World Bank funded project, while in Kabale district, the Government was implementing a UN Habitat funded project to register CCOs. The authors also visited the project for documenting occupancy rights on Mailo land being funded and implemented by GIZ (German Development Agency) in Mityana and Mubende districts. The project had by then, documented more than 20,000 parcels using FFPLA techniques. Key data captured during field visits included:

(i) How each project contextualized the FFPLA generic principles,
(ii) How the existing legal framework supported or constrained the process,
(iii) How the existing land institutions (formal and informal) were involved and roles of each institution,
(iv) Mechanisms for dispute resolution,
(v) The recordation tools used, including their inputs and output data and formats,
(vi) The field team composition,
(vii) Issues that were encountered during the process of mapping and documentation of land rights,
(viii) Stakeholders that were engaged,
(ix) Average cost of mapping and recordation per parcel,
(x) Number of parcels registered,
(xi) Experiences and lessons learnt, and
(xii) How to ensure that the solutions are scalable.

The review was further informed by regional experiences borrowed from Rwanda and Ethiopia. Rwanda had implemented a country-wide project and issued titles for about 10.4 million parcels within a period of less than four years at a cost of six to eight dollars per parcel [17] while Ethiopia had used Fit for Purpose Land Administration approach to implement land tenure security projects for some parts of the country [19]. The international experience was derived from review of published material about success stories on FFPLA implementation in many countries. Further literature was derived from academic and professional documents published by GLTN, FIG and the World Bank.

3. Land Tenure Concepts and FFPLA Pilot Projects in Uganda

3.1. Evolution of the Land Administration Concept

Land administration is not a new discipline but has evolved out of the cadastre and land registration areas providing information systems with specific focus on security of land rights ([19,20]). A couple of decades ago, land administration was referred to as "the processes of determining, recording, and disseminating information about ownership, value, and use of land when implementing land management policies" [21]. This focus on information is still present, but within recent years, the type and quality of information needed has changed and pushes the design of land administration systems (LAS) towards an enabling infrastructure for implementing land policies in support of sustainable development.

LAS designed this way, enables the management of four key functions including land tenure (securing and transferring rights in land and natural resources); land value (valuation and taxation of land and properties); land use (planning and control of the use of land and natural resources); and land development (implementing housing schemes, infrastructures, and construction works). These four functions ensure proper management of rights, restrictions, and responsibilities in relation to property, land and natural resources. LAS designed this way will enable the implementation of land policies to fulfil political and social objectives and to achieve sustainable development [20]. However, the basis or the backbone of such systems is the land tenure component establishing the relation between people and land [3].

From this global perspective, LAS act within adopted land policies that define the legal regulatory pattern for dealing with land issues. LAS also act within a country's specific institutional framework that imposes mandates and responsibilities on the various agencies and organizations. LAS should service the needs of individuals, businesses and the community at large, as they contribute to delivering detailed information and reliable administration of land from the basic level of individual land parcels to the national level of policy implementation [20].

In most developed countries, security of tenure is taken for granted. Over centuries, these countries developed mature land institutions and laws that protect the people to land relationship and provide the services needed for supporting an efficient land market and effective land use management. However, an educated estimate indicates that for 70 per cent of the world's population, this is not the case [22]. In most developing countries, people cannot register and safeguard their land rights, or it may be too costly. The majority of these people are the poor and the most vulnerable in society.

Over recent years, LAS has developed to also capture and include more informal and social types of tenure. This is enforced through development of concepts such as the continuum of land rights [20], the social tenure domain model [21], and the aspects of responsible governance of tenure ([22,23]). Eventually, these efforts were conceptualised into the FFPLA approach designed to meet the challenges of providing secure land rights at scale [15]. The concept includes three interrelated frameworks that work together to deliver the FFP approach: the spatial, legal and institutional framework. The spatial framework supports the way land is occupied and used. The scale and accuracy of this representation are not determined by rigid regulations, but by the demand for meeting the purpose of securing the various kinds of legal rights and tenure forms recognised through the legal framework. The institutional framework is designed to manage these rights and the use of land and natural resources and to deliver inclusive and accessible services. The approach is flexible, affordable, and participatory, and the outcome is upgradeable over time [16].

3.2. Overview of Tenure Types in Uganda

The constitution of the Republic of Uganda allows for four land tenure systems, namely: freehold, leasehold, Mailo and customary tenure [23]. However, it leaves out informal occupants on registered land and gazetted forests and wetlands, who are regarded as squatters and at a risk of eviction. Figure 1 (left) shows the spatial location of the tenure systems in Uganda.

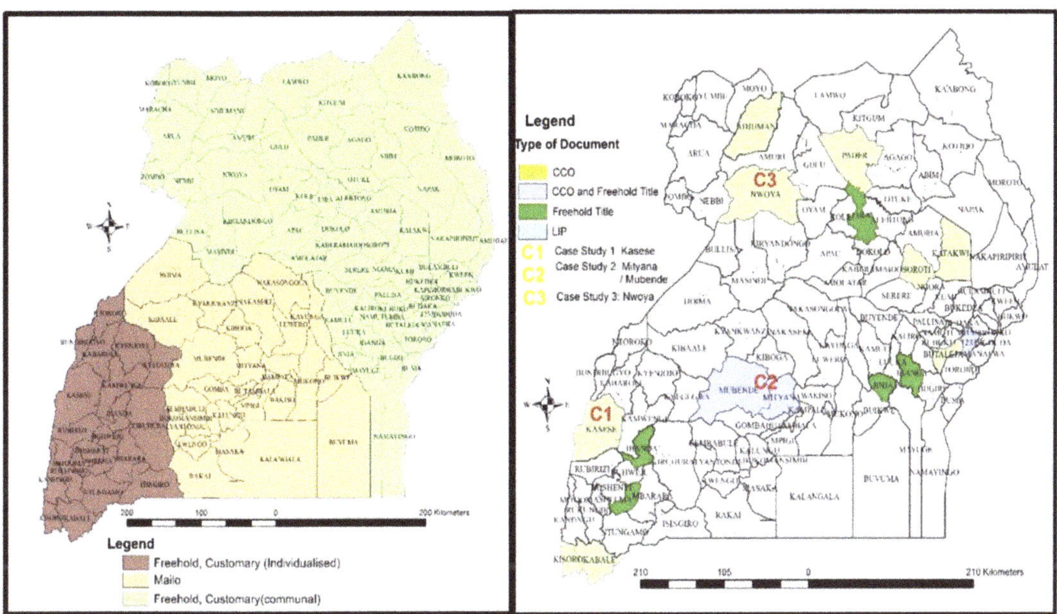

Figure 1. Spatial distribution of the major land tenure systems (**left**) and FFPLA pilot projects in Uganda (**right**).

(1) Freehold is the common term for perpetual ownership of real property, or land, and all immovable structures attached to such land.

(2) Leasehold is the right to use property granted by the owner (lessor) to the user (lessee) for a specified period, under agreed terms and conditions. The property and immovable structures return to the lessor at the expiry of the lease. In Uganda, leases offered by the Government range from 49–99 years. Leases can be created on any of the other tenure types (freehold, mailo and customary).

(3) Mailo tenure is a form of freehold specific to Uganda, but introduced by the British colonialists in 1900, and predominant in Central Uganda. Under Mailo tenure, ownership is in perpetuity, but is subject to the rights of lawful and bona fide occupants. Ownership rights are possessed by a registered owner who holds a Mailo land title. The occupant can transfer user rights to a descendant (heir) but requires permission from the registered owner in order to transfer user rights to a non-family member. Mailo tenure presents one of the major land issues in Uganda [24].

(4) Customary tenure, was first recognized as a formal system in the 1995 constitution. Customary tenure systems are inherently unique to the localities in which they operate and are thus difficult to characterize by generalities [25]. Mindful of such limitations, an attempt to characterize customary tenure regimes in Uganda may yield two generic forms as explained below.

Customary tenure, predominantly individualized: Land is held customarily through inheritance (mainly) but individuals or small family units have full autonomy to decide on its use and may mortgage it or transfer their rights to community or non-community members, without consulting community leaders. The role of community leaders is limited to dispute resolution in case there are disagreements on land rights between members of the family, or between families or individuals. This type is an example of the new customary tenure systems that are emerging across sub-Saharan Africa [26]. It has many features of freehold tenure except for the fact that land rights are not documented, boundaries are not clearly demarcated and most of the dealings on the land remain informal. This type is predominant in western Uganda.

Customary tenure, predominantly communal: Under this regime, land is owned by the community comprised of people with a common identity such as a clan. Community leaders such as clan heads, elders and family heads are responsible for allocating land use rights to the members of the community. Each family is responsible for a specific portion of the community land but is not permitted to transfer use rights to non- family or non-community members without approval of the clan leaders. This type of customary tenure is predominant in the Eastern and Northern parts of Uganda.

3.3. *Evolution of FFPLA in Uganda*

Uganda's Land Act, 1998 recognized the need to document customary land rights and issue legal documents, which would serve as conclusive evidence of customary rights to land. By documenting customary land and bringing it to a formal register, this would not only secure the tenure of customary land rights holders but would enable the Government to understand the dynamics on customary land, so as to plan better for its contribution to national development. Given that most customary land owners and rights holders are peasants or small-holder farmers who could not afford the lengthy and costly procedure for obtaining certificates of title, the solution was to simplify procedures and requirements for obtaining certificates of customary ownership (CCO). Regardless of the simplified requirements and procedures, the intention was to give the CCO a legal value comparable to a freehold title. Under S.8(2) of the Land Act 1998, a holder of a CCO is permitted to: lease land or part of it; permit a person to hold usufructuary rights; mortgage or pledge all or part of the land; subdivide the land; create an easement on the land; sell the land or part of it; or dispose of the land by will.

Whereas, the law gives full rights to a holder of a CCO to transact in land as indicated above, it may be unlikely that the customary rights holders would be able to enjoy these rights in real terms [27]. This seems to make sense given that transactions on customary land are subject to undocumented customs and traditions, more especially on customary land that is communally owned. Indeed, some studies undertaken in Uganda, for example [28], have identified that beneficiaries of CCOS that were issued around the year 2010 were worried about the acceptability of the documents by financial institutions as collateral for loans. On the other hand, experience from other pilot areas for CCO registration in Uganda (e.g., Kasese District) undertaken around the years 2015–2016 indicate that some financial institutions had gone ahead to accept CCOs as collateral for loans. It is expected that with time, most of the limitations imposed by customs and traditions will give way to full transaction on customary land by individuals or small family units.

In Uganda, the simplification in the CCO registration procedures and requirements introduced by the Land Act 1998 included:

(i) Adjudication to be undertaken by a land committee located at the parish level as opposed to district level. No academic qualifications are required for one to be appointed to the committee, as long as one has extensive knowledge of land issues in the area.
(ii) Measurement and mapping to be undertaken by the land committees, using eye judgement, pacing or measuring tape. A sketch map drawn by hand (see Figure 2) was sufficient to issue a CCO [29].
(iii) The previous legal requirement for measurement by a qualified land surveyor was removed for the purpose of registering customary land.
(iv) Issuance of a certificate was charged with a recorder located at the subcounty level, as opposed to a registrar at the ministry headquarters.
(v) The Land Registry (for both first registration and subsequent transactions) was placed at the subcounty level, down from the ministry headquarters.

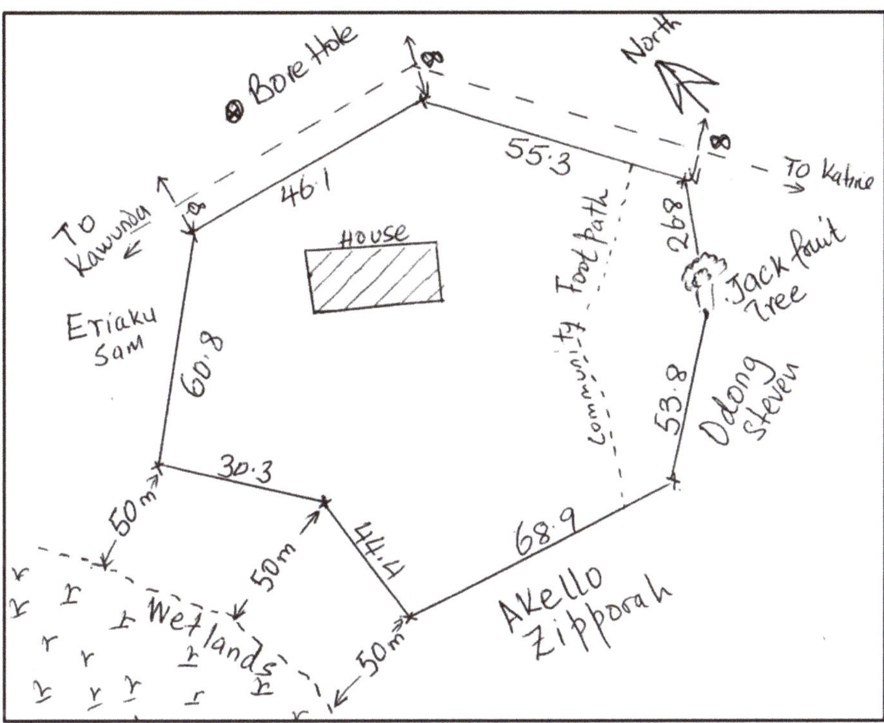

Figure 2. Example of Sketch map sufficient for issuance of CCOs in Uganda [29].

The simplified procedures and requirements were not immediately implemented because of a number of technical and operational challenges. It appears that the Government had under-estimated the cost of establishing and facilitating the new lower-level land institutions across the entire country. Eventually, when the new office bearers such as members of district land boards, members of the area land committees and recorders were recruited in a few districts, they were not trained to handle the new functions assigned to them. They also lacked logistics and tools to use in their new offices. In response to this capacity gap, an amendment to the Land Act in 2004 elevated the land committees, one administrative step higher, from the parish to the subcounty level, and they were renamed, area land committees. Another constraint in the implementation was that professionals such as land surveyors were skeptical about the usability of a sketch as a replacement for a cadastral plan. Whereas as a cadastral plan drawn to scale using surveyed/measured boundaries would facilitate re-tracing of boundaries in case of disputes, the land surveyors wondered how this would be achieved with the use of a hand drawn, not to scale sketch map.

Overall, implementation of the simplified procedures therefore hit a setback essentially because of lack of capacity to implement at the local government level and lack of capacity to supervise at the central government level. Because of these and many more challenges, CCOs were not registered anywhere in the country, until 2010 when Kasese district in South Western Uganda made some initial attempts. The area land committees in the district used a combination of measuring tapes, pacing and eye judgement to produce sketch parcel maps, which would be attached to applications for the district land board to approve issuance of CCOs. However, because of default on many legal procedures and standards, the process was halted by the Ministry of Lands, Housing and Urban Development.

Another effort for CCO registration is traced back to the same period and attributed to the District Livelihood Support Program in Uganda supported by the International Fund for Agricultural Development (IFAD). The project's land management component aimed at processing CCOs and freehold titles for customary land rights holders in 13 districts scattered across the country [28]. Although some few manual CCOs were processed in APAC district, the process was halted in Masindi district because of the need to standardize procedures for CCO registration across the country. No CCOS were processed in the remaining districts.

The first version of modern FFPLA was introduced by FAO in 2015 under a project to operationalize the Voluntary Guidelines on the Responsible Governance of Tenure (VGGTs). Under the project, more than 4000 CCOs were registered on customary land using Sola Open Tenure, a fit for purpose tool for documenting land rights. The experience from the FAO-supported project led to a multiplicity of other small-scale projects for registering customary land rights and formalization of tenure for occupants on registered mailo land using recordation tools. As of today, more than 150,000 parcels have been documented through such pilot project using FFPLA tools in various parts of Uganda (see Figure 1, right).

4. Description of Case Studies for FFPLA in Uganda

FFPLA pilot projects in Uganda number more than ten. However, in this paper, only three case studies have been reviewed to provide lessons for developing a country implementation strategy for FFPLA. To provide a balanced review, the first case study was selected from a region of predominantly communal customary tenure (Nwoya in Northern Uganda), the second case study was selected from a region of predominantly individualized customary tenure (Kasese in Western Uganda) and the last case study was selected from a region of customary tenants on registered mailo tenure (Mityana/Mubende in Central Uganda.

4.1. Case Study C1—Registration of CCOs in Nwoya District, Northern Uganda

Nwoya district is part of Acholi sub-region in Northern Uganda. The district is a center of attraction for the numerous natural resources that it has, such as the Murchison falls on the River Nile, Murchison Falls National Game Park, Lake Albert and the oil rich Western Rift Valley—the Albertine Graben. All in all, 91.1% of the total population in Nwoya district is engaged in subsistence crop farming, making land a very vital asset, crucial for livelihood [30].

In March 2015, ZOA, a civil society organization based in Uganda received funding from a private foundation in the Netherlands to support a community-led land dispute mediation and customary land tenure registration. The project had two specific objectives, namely: (1) more farmer-households to feel secure about their land rights for investing in agriculture and intensify production; (2) government institutions, civil society and community leaders take steps to ensure that customary tenure registration contributes to productive land use and does so in an inclusive and equitable manner. Subsequently, ZOA signed a memorandum of understanding with the Makerere University School of the Built Environment to provide technical assistance in the implementation of the project. Makerere University used its experience from implementation of a similar project in Kasese District to guide the processes for adjudication, demarcation, mapping and issuance of CCOs.

Parcel demarcation was based on a participatory approach [16,31], while taking into consideration the legal requirements for adjudication of rural land in Uganda. It was undertaken by area land committee members (formal local land institutions), Rwot Kweri's (local chiefs with knowledge of family land boundaries), elders (community representation), local councils (government village committees), ZOA field staff (back up trainers and a group of young educated persons from the district (volunteers) who supported the above institutions in the fieldwork and use of technology. (Note that Makerere University trained locally recruited volunteers and ZOA staff, who later become trainers for area land committees during fieldwork). Demarcation work would start after payment of a nominal fee of 10,000 Uganda shillings (2.5 US cents) and filling an application form. Data capture was planned to follow simple field procedures, borrowing from regional and international experiences [17,31,32]. The project used SOLA Open Tenure (see Figure 3), previously customized for a similar project in Kasese district. More than 1000 CCOs were generated and issued to the beneficiaries. Dispute resolution was based on tested alternative dispute resolution mechanisms [33,34] to avoid the costly and lengthy option of litigation. It was accomplished, through mediation by the Acholi Religious Leaders Peace Initiative, previously formed to deal with war conflicts between the Government and the Lord's Resistance Army of Joseph Kony [35].

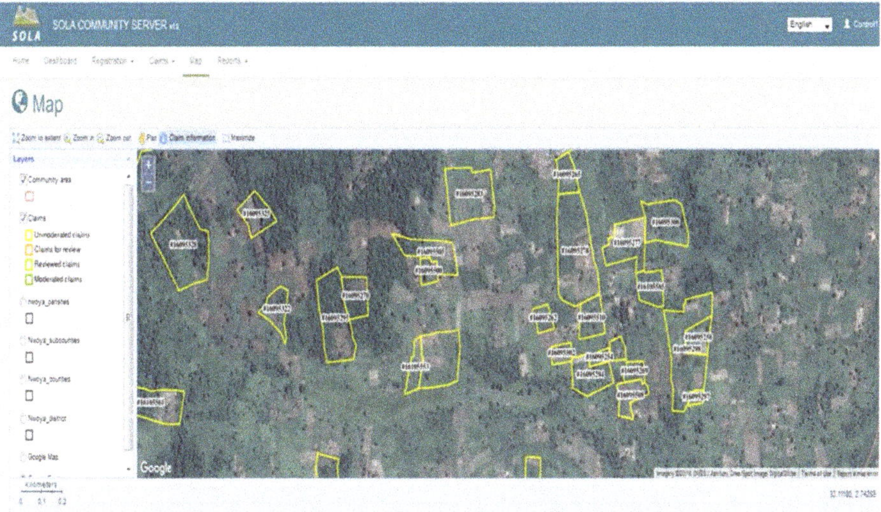

Figure 3. SOLA Community server showing parcels mapped in one of the villages in Nwoya district (Makerere University File Photo).

The success of FFPLA tools in securing tenure rights of customary land rights holders in Nwoya was instrumental in opening up land tenure regularization projects in Northern Uganda. Apart from scattered freehold titles which had been issued to privileged individuals in the region, there had not been any effort to issue any other legal documents on customary land rights in Northern Uganda. Given that northern Uganda was a post-conflict zone, having been affected by a civil way for more than two decades, land disputes involving returning citizens were widespread. The war had disorganized family structures, some members had fled the country, while others had relocated. There was a thinking among some local civil society organizations that registration of land tenure rights and issuance of legal documents should delay for another 50 years to allow for the families to re-organize. These sentiments were shared in the Northern Region Land Platform meeting that took place in Lira in 2016. There was also fear among potential actors in land regularization that Northern Uganda was a no-go zone for any land regularization program. The project therefore opened gates to many subsequent fit for purpose land administration projects, which have greatly improved land tenure security of customary land rights holders in Northern Uganda.

4.2. Case Study C2—Registration of CCOs in Kasese District, Western Uganda

Kasese district is situated in western Uganda next to DRC between the Kazinga Channel, Lake George and Lake Edward. Agriculture is the primary economic activity in Kasese District and employs over 69% of the total population with the majority of farmers (65%) in the district practicing small scale subsistence agriculture [30]. Land in Kasese was traditionally held in trust for the community by ridge leaders referred to in the local vernacular as "Mukulu wa Bulhambo". The most common form of land ownership today is by individual families (usually the nuclear family) with a few cases of land being jointly owned by groups siblings (the extended family) and referred to as "Clan Land".

Kasese District Local Government took advantage of the provisions in the Land Act 1998 and Local Government Act 1997 to embark on CCO registration independent of the Central Government, as provided for under the law. Given that the Ministry of Lands, Housing and Urban Development had not extended capacity building support to the district, the district lacked skills to independently register CCOs. Therefore, the Ministry of Lands, Housing and Urban Development, using its supervision and monitoring role under the Local Governments Act [36] S 97, halted the procedure in order to review and ascertain that the generated CCOs conformed to the required standards. The FAO supported project, therefore, came in at the request of the Ministry of Lands, Housing and Urban Development to improve the existing process of CCO registration, by substituting crude methods of generating a sketch map with IT-based tools built in SOLA open tenure land rights documentation tool. Furthermore, the FOA supported project aimed at raising awareness on VGGTS [37] and capacity development of local land administration institutions at subcounty and district level for CCO registration.

In partnership with Makerere University School of the Built Environment, the project trained members of the area land committees, members of dispute resolution committees, the recorders and locally recruited volunteers in CCO registration, using a customized version of SOLA open tenure. Essentially, the project replicated the methodology explained under Section 4.1 above. SOLA open tenure [38] incorporated many fit for purpose land administration tools such as ability to use a tablet and satellite image to map parcels and record land rights data by low skilled persons (see Figure 4).

Figure 4. The field team uses a tablet to map land rights for a woman-headed household in Kasese District in 2015: Makerere University file photo.

Each field team responsible for data collection comprised of a member of the area land committee (government representative), a student surveyor from Makerere University (technology transfer trainer) and a locally recruited volunteer (back up trainer). The volunteer was necessary, given that members of the area land committee who are mandated by the law, to undertake adjudication were in many cases, illiterate or too old to withstand the harsh field conditions. The data generated was subjected to quality checks by Makerere University, Ministry of Lands, Housing and Urban Development and the District Land Office staff, before submission to the District Land Board and subcounty for issuance of CCOs.

Through this project, more than 4000 CCOs were processed and issued to the beneficiaries. The project did not cover the entire district because of financial scope although the demand for land tenure security was very high among the land owners. The project was instrumental in demonstrating that fit for purpose tools could be used to improve the sketch map hence producing parcel maps to support CCO registration. The project also demonstrated that generic FFPLA tools could be customized to align with a country's legal requirements for land administration.

4.3. Case study C3—Land Inventory Protocol (LIP) in Mityana and Mubende Districts, Central Uganda

Since colonial times, and after independence, the most intractable policy issue facing the Government of Uganda was undoubtedly the future of Mailo land [14]. Whereas the Land Reform Decree of 1975 took a radical approach to address the mailo tenure issue, the Land Act (1998) acknowledged the legal and practical difficulties of pursuing a similar trend. The Land Act (1998) recognized the lawful and bona fide tenants while the Uganda National Land Policy (2013) further provided for rights and responsibilities for both the tenants and landlords on the Mailo land Tenure system to ensure an amicable relationship on the dual ownership of land use rights. The land policy further provided a basis for various options to unravel the complexity of dual rights over the same land. These options include: buying out, sharing, leasing and registration of occupancy (bona fide or lawful). Unfortunately, very few Ugandans living on mailo land are knowledgeable about the available provisions in the National Land Policy of 2013, the most un-informed being women and vulnerable groups in the rural areas. Since the Land Act was formulated in 1998, many interventions had been made by government in partnership with development partners to secure land rights across the different tenure types. However, no interventions had been conducted on private mailo land tenure and yet it hosts most of the complicated overlapping land rights.

GIZ (German International Development Agency) with support from EU and the German Federal Ministry of Economic Cooperation and Development (BMZ) under the Special Initiative "One World, No Hunger" [39] implemented a project to improve land governance in Uganda. Project activities were implemented in partnership with Makerere University School of the Built Environment, Ministry of Lands, Housing and Urban Development and the respective local governments. The project was implemented in Mityana and Mubende districts in central Uganda, with plans to extend it to more districts.

The land inventory was undertaken by field teams comprised of University students (as trainers), area land committee members (local land institution), members of local councils (government village councils), locally trained land administration assistants (technical assistants), and locally trained paralegals (for dispute resolution) [24]. Mobilization and sensitization of communities was undertaken by a local civil society organization while overall monitoring and policy support was provided by the Ministry of Lands, Housing and Urban Development. GIZ provided overall technical and management support. Land rights documentation was accomplished using a land rights documentation and mapping software tool named CRISP, developed by GIZ but customized to the land administration system in Uganda. CRISP, which is an acronym for Cadaster and Rights Inventory Saving Paper, has both social and spatial data collection components. The social components of the tool conform to the concept of continuum of land rights [40]. They were used to collect information on tenants and parcels which included: the tenants' biodata, land use, and number of people living on the parcel. The tool also captured the name of the landlord, the nature of tenant's occupancy and encumbrances on the parcel, (if any). CRISP spatial components in conjunction with the Global Navigation Satellite System (GNSS) receivers (EMLID type) were used to draw parcel boundaries and to input additional information such as the village, parish, subcounty, county and district to which the mapped parcel belonged. CRISP also offered provisions for generating the final land inventory documents—the LIP and a geo report. Figure 5 demonstrates the CRISP graphical screen for generating parcels in the pilot project undertaken in the Mityana District, in Central Uganda.

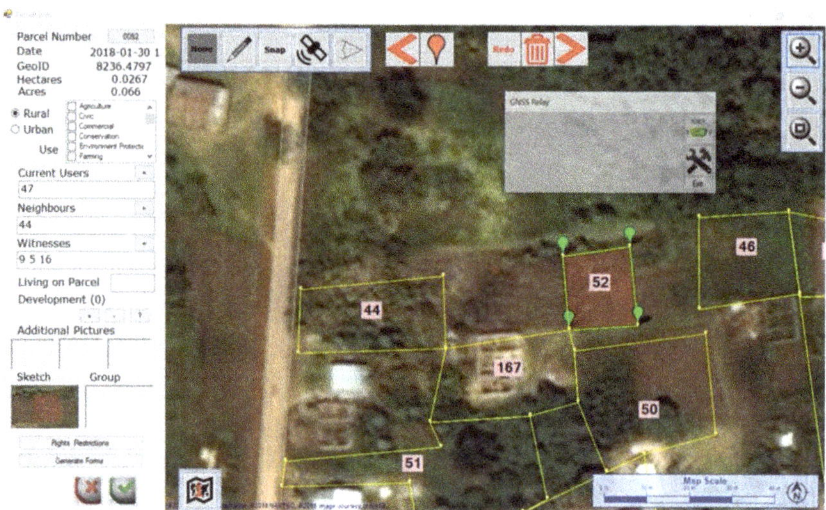

Figure 5. Parcels generated using CRISP software in Mityana district, Central Uganda.

Once parcels for all consenting tenants in a village were mapped, and disputes resolved through mediation by the locally recruited and trained paralegal, a village map would be generated, displayed on village noticeboards and verified by all the occupants and interested parties. This would be followed by issuance of social documents named land inventory protocols (LIPs) to occupants upon paying Uganda shillings 10,000 (2.5 US Cents) per LIP/parcel. Figure 6 shows one of the village maps being verified by occupants in the Mityana District, central Uganda. It should be noted that the use of paralegals for mediation replaced the need for professional lawyers and fitted well within the Principles of FFPLA [15,16].

Figure 6. Verification by occupants during a village map display in Mityana (Makerere University file photo).

The project documented more than 30,000 parcels, which resulted into improved relationships between landlords and tenants; because of this exercise, landlords got to know who their tenants were and vice versa. The benefits and detailed evaluation of the inventory approach for land rights recordation are well document, for example see [24].

5. Lessons Learnt from Pilot Projects in Uganda

5.1. Lessons Learnt for Building the Spatial Framework

The approach is easy to implement: One of the international lessons from implementation of FFPLA as documented in [14] is that the approach should be easy to implement. The experience from the case studies in Uganda, has also revealed that the approach for generating the spatial framework is not only very easy to implement, but also easily understood by local communities. In all the three case studies, it has been established that locally recruited land administration assistants (within subcounties), with high school education, can be trained in a period of five days to generate parcel maps and land rights descriptive information using FFPLA data collection equipment and software. The easy-to-use FFPLA graphical tools such as those built in, Sola Open Tenure and CRISP recordation software provide an opportunity for low-skilled persons to draw and view parcels as vector overlays on high resolution satellite images or ortho-rectified aerial images while in the field. Additionally, some of the recordation tools allow for drawing of parcels on hardcopy images while in the field and later digitizing them in the office. The possibility of displaying parcel maps and enabling community members to view the shapes and location of their parcels while in the field, makes it easy to obtain community buy-in.

Opportunity to cover the entire country within a few years at a low cost: The FFPLA approach has enabled mapping and capturing land rights information in a rapid manner. In Kasese and Mityana/Mubende case studies, where parcel sizes are smaller (2 acres on average), each field team was able to map and capture land rights information for 15–30 parcels per day. However, in Nwoya case study where parcel sizes are larger (50–200 acres) due to communal ownership, each field team was able to capture land rights information for 5–15 parcels per day. The variation in daily outputs can be explained by variations in the terrain, size of parcels and weather conditions.

There is potential to increase the daily output, if a systematic procedure of documenting all the parcels in an administrative unit, is adopted. Under a systematic approach, a field team builds a spatial framework by starting from one location and follows an orderly schedule, documenting all the adjoining parcels in a universal manner, before moving to another administration unit [41], p. 690. However, in the reviewed case studies, the pace was affected by the legal requirement for consent by individuals/communities before documenting their parcels [42]. Such endorsements require extensive public information and a communication campaign which slowed down the pace of field teams.

It was also established that despite the extensive sensitization through radios, community meetings, mobile loud speakers and house to house mobilization, a few land owners were still reluctant to participate. Such land owners would be skipped and left to decide as the teams moved ahead to document parcels for consenting land owners in the same village. This eventually made the process semi-systematic, hence falling short of some of the key benefits of a systematic approach to land titling [43]. In each of the three pilot areas, five teams were deployed to work concurrently and a supervisor (either a land surveying student or graduate land surveyor) was recruited to monitor the quality of the generated spatial framework. Despite the limitations with a semi-systematic procedure, the daily output for the five teams in each pilot area was 50–150 parcels, and 150–300 across all the three pilot areas. This compares with daily outputs in other countries where FFPLA has been piloted [16]. These results demonstrate that using a national approach, it will be possible to work in parallel throughout the country and hence complete the entire a country in a short time.

Multiplicity of various recordation tools: Many innovative tools have been developed for recordation of land rights worldwide [44]. This presents both opportunities and challenges. For the three case studies reviewed in this paper, two recordation tools, namely Sola Open Tenure and CRISP were identified. Both tools can run on tablets, hence presenting advantages of directly entering land rights data into the database [45], p. 16. These two tools have variations in the data formats and outputs, which makes it cumbersome to eventually collate and generate a consistent spatial framework. Variations were observed in data input requirements, data types and data output formats and products. For a country like Uganda, which has already invested in establishing a national land information system, generating parcels without following a uniform and consistent standard defeats the purpose for the heavy investment in system development. The ongoing efforts to establish a CCO working group in the Ministry of Lands, Housing and Urban Development is considered to be a good step towards building a consistent spatial framework that is hinged on FFPLA principles

New spatial framework comparable to the National Cadastral Database: The spatial framework generated under the three case studies was based on simplified FFPLA techniques. In all the three cases, either geo-referenced satellite images or ortho-rectified aerial images were used to digitize the parcels. Such images are highly recommended for generating a spatial framework because of the low lost [45]. When compared with the existing cadastral maps with their constraints [46], the new spatial framework presents better accuracy in adjacency although the absolute accuracy is still lower, but could be upgraded over time. Incremental upgrade is indeed a key principle of the FFPLA approach [15], implying that, at a later date, this framework may be improved to support many other functions, should need arise. The framework in its current form may support functions such as rapid physical planning, land use planning, environmental planning/management and preliminary infrastructure planning. Furthermore, given that land owners or rights holders visually verified the shapes and location of their parcels during village display, the acceptability of the framework is much higher than the current cadastral database.

5.2. Lessons Learnt for Building the Legal Framework

Improved possibilities under the existing legal framework: The implementation of pilot projects on FFPLA in Uganda benefited from relatively recent laws that were enacted after the 1995 constitution. Such laws include the Land Act (1998), the Local Government Act (1998) and the Local Council Courts Act of 2006. These laws incorporated some aspects that favored FFPLA implementation, though not comprehensively. The laws provided for: registration of CCOs using FFPLA-like simplified procedures; establishment of customary land registries at subcounties (2004 amendment) which are closer to the people; dispute resolution on customary land to be handled by a village local council as the first court of instance; and land management to be a responsibility of decentralized and semi-autonomous local governments. On the other hand, some laws such as the Survey Act 1939, though old, included flexible provisions that gave powers to the Commissioner of Surveys and Mapping to decide on procedures and standards for measurement and documentation of parcels. Such provisions made it possible to use FFPLA tools to generate spatial frameworks for other land tenure systems such as freehold, which are not well covered under the land Act 1998. However, the laws will require revision, if they are to support full realization of the benefits of a fit for purpose approach to land administration.

A systematic approach is a must: As mentioned in the paragraph above, pilot projects on FFPLA are largely implemented within the framework of existing laws. Both the Land Act 1998 and the Registration of Titles Act 1965 prescribe a sporadic approach to land registration. The major assumption under the two laws, and their respective regulations, is that one applicant is handled at a time from land inspection up to the production of the final certificate. This approach is not only inconsistent with FFPLA principles [15] but is slow, expensive, discriminatory and does not help to cover the country in a short time. In all the three case studies under review, the implementers suffered delays resulting from absence of laws that prescribed a systematic approach to land adjudication. The field teams were able to cover many parcels in a day, but the approval procedure required preparation of individual files for each parcel, ensuring that all the necessary legal attachments were included, and separately processing each application up to final approval by the District Land Board. Such duplications can be avoided if a new law that provides for mass data collection, mass processing and mass approval is put in place.

Handling of existing and new land disputes: In Uganda, the legal framework for land dispute resolution includes formal courts, local councils and informal institutions such as traditional leaders [33]. In all the three case studies under review, disputes were resolved through mediation, which is the preferred alternative dispute resolution (ADR) mechanism under the Ugandan legal framework. The more complicated long-standing disputes and those where parties could not reach an agreement were referred to the court for settlement. Furthermore, in each of the pilot projects, dispute resolution committees were formed at the commencement of the project. In the Kasese case study, the dispute resolution committees comprised of members of the local councils and respected persons selected from the communities. In Nwoya case study, the committees comprised of elders, traditional chiefs (Rwot Kweris) and members of the Acholi Religious Leaders Peace Initiative (ARLPI). ARLPI is an umbrella organization for the major religious denominations in Northern Uganda. Its goal was to pursue peaceful conflict resolution with the Lord's Resistance Army. While in Mityana/Mubende, the committee comprised of local councils, locally recruited and trained paralegals, and staff from a local civil society organization. The effectiveness of ADR was manifested in the escalation of disputes at the beginning of the project and substantial reduction towards the end. For example, in Kasese district, 33 disputes were reported in 2015 at the commencement of the project, 39 disputes were reported in 2016 mid-way the project and only six disputes were reported in 2017, towards the end of the pilot phase [47].

5.3. *Lessons Learnt for Building the Institutional Framework*

Decentralisation of land services: The institutional framework for land adminstration in Uganda includes national institutions, local institutions and informal but legitimate institutions. Having a mix of formal and informal (but legitimate) institutions in a land administration system promotes flexibility and good land governance rather than bureaucratic barriers [48], hence conforming to the principles of a FFPLA approach. In each of the three case studies, the success of the pilot project was hinged on the availability of land institutions at the subcounty level and the district level. The subcounty level institutions included the area land committee and the recorder. The area land committees were responsible for receiving applications, land inspection and compiling information that would help the district land board to make a decision on whether to grant a CCO or not. The information compiled by the committees included a duely filled and signed application form (The form is filled by the applicant and endorsed by area land committee members. The committee may help applicants who are illiterate); recommendation and minutes by the area land committee; an inspection report; and a sketch plan of the land which was the subject of the application. The law requires that at least three of the five members must inspect the land before compiling a report. This would imply that only one adjudication team should be deployed in one subcounty to carry out land inspection at any one given time. Meeting this legal requirement became cumbersome given that under the pilot projects, outputs were required in a very short time. Eventually, in total disgard of the legal provisions, and in consultation with the Ministry of Lands, Housing and Urban Develoment, five teams were deployed to work concurrently, in each subcounty, with a representation of only one member of the area land committee on each team. For purposes of future systematic registration projects, it will be necessary to review the provision that requires all committee members to be present during inspection. In any case, the level of transparency under the systematic approach is adequate to prevent any likely corrupt or fraudurent tendecies that the law intended to mitigate.

The recorder is resposnsible for registration and issuance of the final CCO. The subcounty chief (Senior Assistant Administrator) is the Head of Civil Service in a subcounty but was assigned the recorder role under the Land Act of 1998. The recorder is the equivalent of a Registrar of Titles under the Registration of Tiles Act. The recorder manages land records at the subcounty including generating the first registration records and managing subsequent transactions such as subdivisions, transfers, mortgages, and caveats, etc. The benefit of assigning recordation roles to a government officer at the subcounty is the easy accessibility to land services by the local people. However, the biggest limitation is the enormous investment required in training recorders and establishing/equipping functional land registries at the subcounty level. As of 2006, there were 943 subcounties in Uganda and this translates to 943 recorders and 4715 (5 × 943) members of area land committees to be trained.

Another critical issue encountered in the case studies was lack of legal provision for the age limit and qualifications of the members of area land committees. Whereas one needed to be an adult with a wide knowledge of land issues in the subcounty to serve on the committee, there was no requirement for age limit. Indeed, most members of the area land committees were elderly and either illiterate or semi-illiterate. It was not possible for the elderly to walk for long distances under harsh terrain and weather conditions, to adjudicate land rights and generate the daily targets of 10–30 parcels. Furthermore, the illiterate committee members required support to review and validate what the applicants had filled in the application forms. It is because of such limitations that the pilot projects solicited additional support from locally trained young persons capable of taking on the load to meet the daily targets. For the future systematic registration projects, there will be a great need to legalise the involvement of such young localy recruited persons in the institutional framework for CCO registration.

Requirement for strong political will as well as the support of key senior civil servants: The FFP approach is a national, top-down approach and requires strong political will and the support of key senior civil servants [14]. Indeed in all the three case studies, the initial permission to work in the project area was granted by the Ministry of Lands, Housing and Urban Development Political Head (Cabinet Minister) and the Technical Head (Permanent Secretary). The role of the top political and senior civil servants was to ensure that the project goals and deriverables fitted into the national development goals and government policy frameworks. At the distrcit local government level, support was sought from the District Political Head (LCV Chairperson), the Head of Civil Service in the District (Chief Administrative Officer—CAO) and the Central Government Representative in the District (Resident District Commissioner—RDC). In addition to the above, consent was sought from the respective members of parliament who would in turn build the trust of the local communities, by explaining the benefits of the CCO registration project. At the subcounty local government level, consultations were made with the Political Head (LC III Chairperson) and the Head of Civil Service (Subcounty Chief).

The project structure in each of the three case studies included a monitoring committee comprised of senior officials from the ministry headquarters, officers from the land office and the district and academicians from Makerere University, who provided technical back-stopping/support. The monitoring committee would visit the project site once a month. Another lower level committee included techncail officers from the district land offices and resident project coordinators (land surveyors) who provided direct support to the field teams, once a week. The role of the implementing a partner such as a civil society organisation or development partner organisation, was to provide financial support, build capacity and provide technical support to enable government institutions run the process in accordance with the legal provisions.

Ensuring sustainability of the fit for purpose approach to land administration: The longevity of projects will best be achieved through planning for sustainability from the start, including planning for long-term financial health, e.g., assessing total cost of ownership [49], p. 76. The need for ensuring that maintenance/updating takes place from day one is a lesson that informed the country implementation strategy. In the three case studies, sustainability was a factor put into consideration, the usual limitations of a project-based approach notwithstanding. All the case studies included components of training as a means of empowering land institutions to handle all the processes at the expiry of the pilot projects. In addition, the projects levied nominal charges for land services, which funds would be paid directly to the local governments. Indeed, applicants were required to pay mandatory application fees and issuance fees [42] totalling to Uganda shillings 20,000 (US 5 cents) per application/parcel. The pilot projects had however, not developed any guidelines for handling post registration transactions, which are considered necessary for keeping the registry updated. This omission became a lesson to consider while developing the FFPLA country implementation strategy.

6. Developing a FFPLA Strategy for Country Implementation

The development process for the country strategy for implementing FFPLA in Uganda was informed by the lessons learnt from the pilot phases and followed the phases below.

6.1. Identification of the Stakeholders

It became clear from the case studies that without involvement of key stakeholders, any efforts to implement FFPLA would be futile. Therefore, the first step in developing the Uganda implementation strategy for FFPLA was to identify stakeholder institutions and their anticipated roles in implementing the strategy. The roles were determined based on the provisions in the existing land laws and the experiences from the accomplished or ongoing FFPLA pilot projects. Stakeholders are required for financial mobilization, granting political approvals; collecting baseline information; mobilizing district and subcounty local support; training of stakeholders; community sensitization and mobilization; adjudication; dispute resolution; undertaking field measurements; quality control of data, processes and products; approving applications; and final processing and issuance of CCOs or other legal land tenure security documents to beneficiaries.

The identified stakeholders included politicians, ministries (Lands, Housing and Urban Development, Finance and Economic Planning, and local government), development partners and donors, professional bodies (e.g., the Uganda Institution of Surveyors), universities and training institutions, civil society organizations, district level government, the district land board and land office, area land committees, and traditional/religious institutions. The above-mentioned stakeholders played an important role during the pilot phase and were, therefore, considered essential for the implementation at national level.

6.2. Designing the Guiding Principles

The guiding principles serve as the basic foundation for making decisions on the provisions for the spatial legal and institutional components of the strategy. The international guidelines [15] provide a good basis for defining the national guiding principles. However, when combined with experience from specific pilots undertaken in the country, this provides more practical tested guidelines for the strategy. In the case of Uganda, the pilot projects had already tested the most acceptable methods for generating the spatial framework, the acceptable dispute resolution mechanism and composition of ADR committees, the gaps in the legal framework, and the necessary adjustments to institutional framework in order to support FFPLA. Most of the included guiding principles had been previously discussed with the major stakeholders during workshops organized by the implementers of the pilot projects. The role of the framers of the strategy was to confirm acceptance of the principles during the final phases of presenting the draft strategy.

6.3. Deciding on key Actions in the Strategy

Key actions of the strategy are those interventions necessary to transform the existing spatial, legal and institutional framework to comply with FFPLA principles. The actions, therefore, depend on the extent to which the current frameworks deviate from the FFPLA principles. In Uganda, the pilot projects had, to a large extent defined the desired standards for the spatial framework, including which technologies were feasible, what data would be collected on each parcel, how parcels would be represented graphically and so on. It was, therefore, easy to identify which actions were necessary to achieve those standards.

Furthermore, the pilot projects had revealed gaps in the legal framework that would affect the implementation of a FFPLA approach. During the pilot phase, the existing laws made processes such as systematic adjudication, mass processing of CCOs, and dispute resolution very slow and more cumbersome. The proposed actions under the legal framework component were therefore about reviewing the current laws to support the functioning of the spatial and institutional framework under FFPLA principles. On the other hand, the institutional framework was largely compliant with the FFPLA principles. No major proposals were made to change the setup of the existing land institutions, except formalizing the role of locally trained land administration assistants and providing for more inter-institutional collaboration. Most of the actions were aimed at building the capacity of the institutions so as to make them more efficient, effective and sustainable.

6.4. Deciding on Phasing and Costing of the Strategy

Decisions for phasing and costing of the country implementation strategy were primarily based on the country experience, while taking into consideration the best practices from international experience. The full implementation was organized into four phases, in order to facilitate learning, reflection, monitoring, and evaluation. An initial phase of one year focused on getting started, providing the infrastructure and technologies, revision of laws and regulations, testing various processes and recordation of about three million parcels. Phases Two and Three of three years each, were planned as the production phases aiming at covering about 18 million parcels in total. The fourth phase was planned for completion and sustaining the systems for future operations.

Regarding the costing of each parcel, the international experience has shown that amounts are usually in the range from US $1 as in the case of Ethiopia [50,51], US $6–8 as in the case of Rwanda [51], US $7 in general [45] but should not exceed US $20–30 [52]. In Uganda, experience from the pilot projects point to a range of costs from US $10–US $25 depending on the size of the parcel, its location and the technology used for measurement. However, by streamlining the processes of mapping and recordation, the costs are estimated as $10 USD per parcel, equivalent to around $230 million USD for covering 23 million parcels.

Experience from the pilot projects has also identified the need to establish basic infrastructure at the sub-counties in order to support and manage the records generated from the registration exercise. The infrastructure includes strong rooms and filing cabinets for storing manual records, electricity and computers for digital processing of data. Further costing relates to capacity building activities, awareness campaigns, and various managerial issues related to drafting of manuals, guidelines, supervision, monitoring and evaluation. These additional support costs are estimated at US $270 million. The full implementation is then designed in four phases over a 10-year period for a total cost of $500 million USD.

6.5. Soliciting Stakeholder Input and Endorsement

The final process in the development of the strategy was stakeholder input and endorsement. The identified stakeholders were invited for a workshop in which the draft strategy was presented. Some of the stakeholders had participated in previous workshops separately organized by implementers of the pilot projects. In addition, some of the stakeholders had participated in the pilot projects under various roles as data collectors, supervisors, quality controllers or respondents during baseline data collection. The strategy drafting team therefore took less effort to obtain stakeholder endorsement. For special stakeholders such as land surveyors, two separate workshops were organized targeting the institution of surveyors of Uganda and government surveyors, respectively. In both workshops, the major issues advanced by the surveyors were on quality of data generated and the role of surveyors under the new strategy. Given that land surveyors were the major drivers of the pilot projects, it was easy to build their confidence about their continued involvement and leadership under the new strategy, although the actual field work would be carried out by local teams as under the pilot projects. The cover page and summary outline of the strategy is demonstrated in Figure 7.

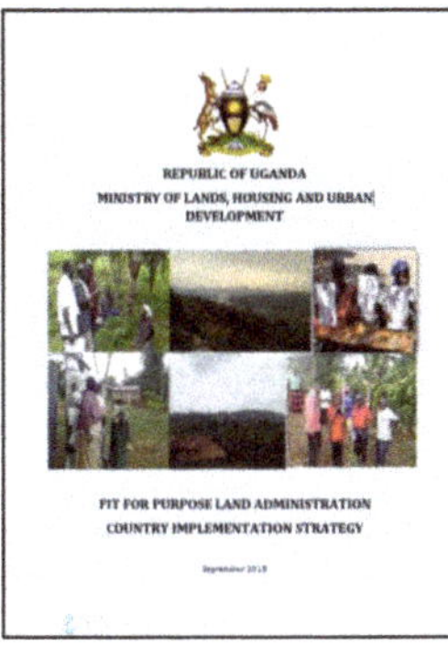

1. Introduction (1)
2. The FFPLA Concept (6)
3. Status of Land Administration in Uganda (18)
4. FFPLA Implementation Strategy (31)
4.1 Purpose and objectives (31)
4.2 Contextual principles for FFPLA in Uganda (32)
4.3 Spatial Framework Requirements (33)
4.4 Legal Framework Requirements (39)
4.5 Institutional Framework Requirement (41)
5. Capacity Development Strategy (46)
6. FFPLA Implementation Risks and Mitigation (49)
7. Phasing, Costing and Financing (52)

Figure 7. Uganda FFPLA Country Implementation Strategy [53,54].

6.6. Strategy Approval and Implementation

The draft strategy has to go through various stages before it is gazetted as a national strategy for FFPLA implementation. Initially, the strategy has to be reviewed by a technical committee comprised of government surveyors at the ministry and in the ministry zonal offices, and this has already been accomplished. Thereafter, the draft should be presented to the top management team of the ministry, chaired by the Permanent Secretary, Ministry of Lands, Housing and Urban Development. Finally, other stakeholders including Ministry of Justice and Constitutional Affairs, Ministry of Local Government, and professional bodies will provide input before the Minister of Lands, Housing and Urban Development signs it off as a national strategy.

7. Discussions on Developing a Country Strategy for Implementing a Fit for Purpose Approach to Land Administration in Uganda

Discussions on the Provisions for a Spatial Framework: The spatial framework is the basic, large-scale map showing the way land is divided into spatial units [15]. It accounts for the largest portion of the initial costs for building a land administration system. In Uganda, the FFPLA country strategy comes at a time when there is no updated law that comprehensively guides the compilation of a spatial framework. The Survey Act of 1939 and its subsidiary law, the survey regulations are out of touch with modern techniques have already been put aside [55], implying that survey observation, checking and plotting are not fully guided by the law. The current spatial framework is, therefore, full of errors such as overlaps in parcel boundaries. On the other hand, the FFP approach for generating a spatial framework in which visual, as opposed to measured boundaries have been advocated [56] and proven to be pro poor [14,31]. In Uganda, the FFP approach for generating a spatial framework is, to a small extent, embedded in the Uganda Land Act of 1998, but is restricted to regularization of customary tenure and lawful or bona fide occupancy rights on registered land. This approach has already been improved through the various FFPLA pilot projects in Uganda.

The three case studies reviewed in this paper have demonstrated that it is easy to generate a spatial framework for the entire country within a period of 10 years, targeting 23 million parcels. This is possible at low cost given that Uganda has already invested in a country-wide base map of high resolution ortho-rectified images, with spatial resolution of 30 cm in the rural areas, and 20 cm in the Urban areas. In addition, a geodetic reference frame consisting of more than 400 passive benchmarks and 12 CORS has been constructed. Such a network is important for the continued upgrade of the spatial framework depending on need. The generation of a FFPLA consistent spatial framework in Uganda is guided by the standards set by the land regulations of 2004 [42] which provide standardized application forms and formats of CCOs and other legal documents. Indeed, all the generic land rights recordation tools have had to be customized to comply with these standards. Furthermore, the National Land Information System, based on the land administration data model [57] has set additional requirements for a standard data model and exchange format which all the recordation tools must comply with in order to compile a consistent national spatial framework.

Discussions on the Provisions for a Legal Framework: The legal framework aims to provide security of tenure through recognition of legitimate rights and recording the corresponding evidence of rights on a national register that is publicly accessible [15]. In Uganda, customary and occupancy rights were not recognized and hence could not find their way to the national register. However, the Land Act of 1998, which is considered to the most important piece of legislation since the Land Reform Decree of 1975 [58], recognized the customary tenure, occupancy rights, gender rights and hence paved way for their inclusion in the national register. In line with FFPLA principles, the Land Act 1998 located customary land registration registers at the subcounty level to enable accessibility by the people. By locating the registers at the subcounty, this did not only reduce the bureaucratic procedures associated with registration of land rights in a national register, but also reduced the cost of registration. In principle, the Land Act closely conforms to design elements of the pro-poor land recordation system [59].

From the three cases studies reviewed, it became evident that registers based at sub-counties were easily accessible by the local people, although this put additional demands on capacity development and financing. Furthermore, the preference for addressing land disputes through alternative dispute resolution mechanisms (ADR), with involvement of local traditional institutions and individuals, as in the case of Nwoya (Case Study 1) and Mityana (Case Study 3) improved access to justice and led to settlement of the cases in a short time. It should be noted that though court annexed mediation is now a requirement before litigation in commercial courts in Uganda [33], the process has been frustrated by some stakeholders such as advocates who prefer to pursue litigation [60]. Other limitations to the implementation of the FFP systematic approach are the provision in the Land Act 1998 and the Land Sector Strategic Plan [13] respectively, that require individual submission and processing of application for registration of CCOs and those that make the process demand driven. Strict adherence to these provisions creates duplications in processes and paperwork. Likewise, lack of compulsion for enforcing systematic land adjudication (or at least for the purposes of data collection) as advocated for in the international best practice [21] increases the cost and time to obtain consent by all land owners in order to implement a systematic approach.

Finally, it was clear from the case studies that the approach has been to exploit any avenues in the current legal framework to implement FFPLA at pilot level. This approach created duplications in procedures and unnecessary documentation. In some of the complicated situations such as representation of at least three members of area land committees during land inspection, there was total default on the law to enable achievement of the daily targets of adjudicated and mapped parcels and households. The pilot projects have been implemented under a rigid legal framework but have helped to identify the gaps in the law; they have demonstrated the need to address the legal framework in order to make the FFP approach yield the desired outcomes of flexibility, reduced cost and time for securing tenure rights for all. These lessons are important for informing the contextualized principles and actions while developing Uganda's country implementation strategy.

Discussions on the Provisions for an Institutional Framework: The institutional framework in support of the FFP approach relates to good land governance, policy frameworks, institutional arrangements, organizational structures, deploying resources locally, partnerships, distribution of responsibilities, and establishing efficient, accountable government workflows for making the systems operational [15]. As previously mentioned, most of the above requirements have been included in the Uganda Land Act, Land Policy and the Land Sector Strategic Plan 2013–2023. Furthermore, two studies on capacity assessment undertaken in Uganda [61,62], have identified glaring gaps in the capacity of the current land institutions to implement the land policy and the land act. The challenges range from lack of funding, through inefficient structures, skills deficiency, to staff motivation. These challenges have been observed in the case studies where the institutions lack basic facilitation and skills to perform their basic roles. The donor funding associated with the pilot areas was instrumental in bridging some of the capacity gaps, but could not address some of the longstanding issues such as staff motivation and institutional development. These require a long term approach coordinated at national level. The FFPLA country implementation strategy therefore included provisions for addressing institutional capacity gaps in order to develop a sustainable environment for FFPLA implementation. Finally, the strategy included provisions for engaging all key stakeholders that are key in the implementation of the strategy. These include politicians, senior government civil servants, professionals (such as land surveyors and advocates), country leaders and the general public.

8. Conclusions

A review of the three case studies in Uganda has revealed the benefits of implementing a fit for purpose approach to land administration as a means to secure tenure rights in a fast, cheap, universal and non-discriminatory manner. The review has demonstrated that pilot projects are beneficial in identifying gaps in the legal and institutional frameworks and testing approaches and technologies, but are also avenues for explaining benefits to obtain the necessary political, community and stakeholder support.

On the other hand, the review has identified that uncoordinated pilot projects are potential sources of inconsistencies in data and products, which may be cumbersome to harmonize at a national level. In order to implement a fit for purpose approach at a national level, it is necessary to consolidate the lessons leant from pilots into a unified country implementation strategy for a fit for purpose approach to land administration.

A country implementation strategy for fit for purpose land administration builds both, on the benefits of the new approach and limitations of the existing approach to develop guiding principles, interventions, timelines and costs for transforming the spatial, legal and institutional components to align them with FFPLA principles. Such a process requires support of all stakeholders in government, the private sector, professional associations and the communities.

Finally, there is now considerable literature and country success stories that demonstrate the benefits of fit for purpose land administration as a new approach for securing land tenure rights in a rapid, cost effective and comprehensive manner. Incorporating FFPLA principles in national policies, laws and regulations is a guaranteed way of advancing the innovation across the developing world. Whereas a project-based approach helps to translate the generic FFPLA principles in a given county context, it does not guarantee adoption of FFPLA as means for national level implementation. A country implementation strategy, if developed as a result of a national dialogue and consensus between all stakeholders is a promising way of advancing the FFPLA concept.

Author Contributions: Conceptualization, M.M., S.E.; methodology, M.M., S.E.; software, M.M.; formal analysis, M.M.; resources, M.M., S.E., S.P.M.; data curation, M.M.; original draft preparation, M.M.; writing, M.M:; review and editing, M.M., S.E., S.P.M.; visualization, M.M.; project administration, M.M.; funding acquisition, M.M., S.E., S.P.M.; All authors have read and agreed to the published version of the manuscript.

Funding: The study is based on the findings of the UN-Habitat/GLTN funded project in Uganda as well as previous pilot project supported by Food Agriculture Organization of United Nations, and German International Cooperation (GIZ). Funding of the publication costs for this article has kindly been provided by the School of Land Administration Studies, University of Twente, in combination with Kadaster International, the Netherlands.

Institutional Review Board Statement: Not applicable.

Informed Consent Statement: Not applicable.

Data Availability Statement: The data presented in this study are available on request from the authors.

Acknowledgments: The authors would like to acknowledge the contribution of the Government of Uganda and the local land authorities in Districts of Nwoya, Kasese, Mityana and Mubende. We would like to thank our project partners UN-Habitat/GLTN for their support in data collection.

Conflicts of Interest: The authors declare no conflict of interest.

References

1. Uganda Bureau of Statistics (UBOS). *The Uganda National Household Survey (UNHS) 2012/13*; GLobal Land Tool Network (GLTN): Nairobi, Kenya, 2014.
2. GOU. *Uganda Vision 40*; GOU: Kampala, Uganda, 2013.
3. FAO. What is Land Tenure? In *Land Tenure and Rural Development*; FAO: Rome, Italy, 2002.
4. Bohn, H.; Deacon, R. Ownership Risk, Investment, and the Use of Natural Resources. *Am. Econ. Rev.* **2000**, *90*, 526–549. [CrossRef]
5. Arnot, D.C.; Luckert, K.M.; Boxal, P.C. What Is Tenure Security? Conceptual Implications for Empirical Analysis. *Land Econ.* **2011**, *87*, 297–311. [CrossRef]
6. Alban Singirankabo, U.; Willem Ertsen, M. Relations between Land Tenure Security and Agricultural Productivity: Exploring the Effect of Land Registration. *Land* **2020**, *9*, 138. [CrossRef]
7. Benjaminsen, T.A.; Holden, S.; Lund, C.; Sjaastad, E. Formalisation of land rights: Some empirical evidence from Mali, Niger and South Africa. *Land Use Policy* **2009**, *26*, 28–35. [CrossRef]
8. Ali, D.A.; Deininger, K.; Goldstein, M. Environmental and gender impacts of land tenure regularization in Africa: Pilot evidence from Rwanda. *J. Dev. Econ.* **2014**, *110*, 262–275. [CrossRef]
9. Deininger, K.; Jin, S.; Nagarajan, H.K. Land Reforms, Poverty Reduction, and Economic Growth: Evidence from India. *J. Dev. Stud.* **2009**, *45*, 496–521. [CrossRef]
10. Mabike, S.; Musinguzi, M.; Antonio, D.; Sylla, O. Land Tenure and its Impacts on Food Security in Uganda: Emperical Evidence from Ten Districts. In Proceedings of the 2017 World Bank Conference on Land and Poverty, The World Bank, Washington, DC, USA, 20–24 March 2017.
11. MLHUD. *Uganda National Land Policy*; MLHUD: Kampala, Uganda, 2013.
12. Oput, R. *Piloting of Systematic Adjudication, Demarcation and Registration for Delivery of Land Admnistration Services in Uganda*; FIG: Copenhagen, Denmark, 2004.
13. MLHUD. *Uganda Land Sector Strategic Plan 2013–2023*; MLHUD: Kampala, Uganda, 2013.
14. Enemark, S.; McLaren, R. Secure Tenure for All Starts to Emerge: New Experiences of Countries Implementing a Fit for Purpose Approcah to Land Admnistration. In Proceedings of the World Bank Conference on Land and Poverty the World Bank, Washington, DC, USA, 25–29 March 2019.
15. Enemark, S.; McLaren, R.; Lemmen, C. *Fit-For-Purpose Land Administration Guiding Principles*; United Nations Habitati Global Land Tools Network (GLTN): Nairobi, Kenya, 2015.
16. Enemark, S.; Bell, K.C.; Lemmen, C.; Mclaren, R. *Fit-For-Purpose Land Administration*; FIG Publication No. 60; FIG & World Bank: Copenhagen, Denmark, 2014; ISBN 9788792853103.
17. Byamugisha, F.F.K. Land Experiences and Development Impacts of Securing Land Rights at Scale in Developing Countries: Case Studies of China and Vietnam. *Land* **2021**, *10*, 176. [CrossRef]
18. Yin, R.K. *Case Study Research and Applications: Design and Methods*, 6th ed.; Sage Publishing: Newbury Park, CA, USA, 2017.
19. FIG. *The FIG Statement on the Cadastre*; FIG Publication No. 11; FIG: Copenhagen, Denmark, 1995.
20. Williamson, I.; Enemark, S.; Wallace, J.; Rajabifard, A. *Land Administration for Sustainable Development*; Esri Press: Redlands, CA, USA, 2010; ISBN 9781589480414.
21. United Nations Economic Commission for Europe (UNECE). *Land Administration Guidelines with Special Reference to Countries in Transition*; UNECE: New York, NY, USA; Geneva, Switzerland, 1996.
22. McLaren, R. How Big Is Global Insecurity of Tenure? GIM International: Vuurtorenweg, The Netherlands, 2015.
23. MLHUD. *Uganda Land Act*; MLHUD: Kampala, Uganda, 1998.
24. Musinguzi, M.; Huber, T.; Kirumira, D.; Drate, P. Assessment of the land inventory approach for securing tenure of lawful and bona fide occupants on private Mailo land in Uganda. *Land Use Policy* **2020**, 104562. [CrossRef]
25. Freudenberger, M. *The Future of Customary Tenure: Options for Policy Makers*; USAID: Washington, DC, USA, 2013.
26. Chimhowu, A. The 'new' African customary land tenure. Characteristic, features and policy implications of a new paradigm. *Land Use Policy* **2019**, *81*, 897–903. [CrossRef]

27. Mwebaza, R. How to integrate statutory and customary tenure? The Uganda case. In Proceedings of the DFID Workshop on Land Rights and Sustainable Development in Sub-Saharan Africa at Sunningdale Park Conference Centre, Berkshire, UK, 16–19 February 1999.
28. UNHABITAT. *Certificates of Customary Ownership: Experiences from the District Livelihood Support Programme in Uganda*; United Nations Human Settlements Programme: Nairobi, Kenya, 2015.
29. Becker, H.-G. A Fit-for-Purpose Approach to Register Customary Land Rights in Uganda. In Proceedings of the FIG Working Week, Geospatial Information for a Smarter Life and Environmental Resilience, Hanoi, Vietnam, 22–26 April 2019.
30. UBOS. *The National Population and Housing Census 2014—Area Specific Profile Series*; UBOS: Kampala, Uganda, 2017.
31. Chigbu, U.E.; Bendzko, T.; Mabakeng, M.R.; Kuusaana, E.D.; Tutu, D.O. Fit-for-Purpose Land Administration from Theory to Practice: Three Demonstrative Case Studies of Local Land Administration Initiatives in Africa. *Land* **2021**, *10*, 476. [CrossRef]
32. Nkrunziza, E. Implementing and Sustaining Land Tenure Regularization in Rwanda. In *How Innovations in Land Administration Reform Improve Doing Business*; World Bank: Washington, DC, USA, 2015; pp. 10–19.
33. Kakooza, A.C. Arbitration, Conciliation and Mediation in Uganda: A Focus on the Practical Aspects. *SSRN Electron. J.* **2012**, *7*, 268–294. [CrossRef]
34. Sullivan, E.; Solomou, A. Alternative Dispute Resolution in Land Use Disputes—Two Continents and Two Approaches. *Urban Lawyer* **2011**, *43*, 1035–1059.
35. Apuuli, K.P. Peace over Justice: The Acholi Religious Leaders Peace Initiative (ARLPI) vs. the International Criminal Court (ICC) in Northern Uganda. *Stud. Ethn. Natl.* **2011**, *11*, 116–129. [CrossRef]
36. MLG. *Uganda Local Governments Act, Laws of Uganda Ch 243*; Ministry of Local Government: Kampala, Uganda, 2006.
37. FAO. *Voluntary Guidelines on the Responsible Governance of Tenure of Land, Fisheries and Forests in the Context of Food Security*; FAO: Rome, Italy, 2012.
38. FAO. Sola Suite Website. Available online: http://www.fao.org/tenure/sola-suite/en/ (accessed on 29 May 2021).
39. BMZ. *Special Initiative ONE World—No Hunger*; BMZ: Berlin, Germany, 2018.
40. Mabikke, S.B. Historical Continuum of Land Rights in Uganda. *J. L. Rural Stud.* **2016**, *4*, 153–171. [CrossRef]
41. Hanstad, T. Designing Land Registration Systems for Developing Countries. *Am. Univ. Int. Law Rev.* **1998**, *13*, 647–703.
42. MLHUD. *The Land Regulations, 2004*; Ministry of Lands Housing and Urban Development: Kampala, Uganda, 2004.
43. Larsson, G. *Land Registration and Cadastral Systems: Tools for Land Information and Management*; Longman Scientific and Technical: Essex, UK, 1991.
44. Lengoiboni, M.; Lemen, C.; Zevenbergen, J. Innovative Tools for Tenure Recordation: Documentation of Overlapping and Secondary Land Rights. *GI Int.* **2017**, *31*, 41–43.
45. Zein, T. Fit for Purpose Land Admnistration: An Implementation Model for Cadastre and Land Admnistration Systems. In Proceedings of the World Bank Conference on Land and Poverty, The World Bank, Washington, DC, USA, 14–18 March 2016.
46. Wabineno, L.M.; Musinguzi, M.; Ekback, P. Land Information Management in Uganda: Current Status. In *The First Conference on Advances in Geomatics Research*; Gidudu, A., Otukei, J.R., Musinguzi, M., Eds.; Makerere University: Kampala, Uganda, 2011; pp.
47. MAK-SBE. *Support for the Implementation of the Voluntary Guidelines for responsible Governance of Land Tenure in Uganda—Project Reference: GCP/GLO/347/UK—Final Report*; Makerere University: Kampala, Uganda, 2018.
48. Enemark, S.; Mclaren, R.; Lemmen, C. Building Fit-For-Purpose Land Administration Systems: Guiding Principles. In *FIG Working Week 2016: Recovery from Disaster*; FIG: Christchurch, New Zealand, 2016.
49. UN-Habitat. *Fit for Purpose Land Administration Guiding Principles for Country Implementation*; UH-Habitat: Nairobi, Kenya, 2016.
50. Abza, T.G. Experience and Future Direction in Ethiopian Rural Land Admnistration. In *Annual World Bank Conference on Land and Poverty*; World Bank: Washington, DC, USA, 2011.
51. Byamugisha, F. *Securing Land Tenure and Easing Access to Land*; African Center for Economic Transformation (ACET) and Japan International Cooperation Agency Research institute (JICA-RI): Washington, DC, USA, 2016.
52. Wehrmann, B.; Lange, A. *Secure Land tenure for All—A Condition for Sustainable Development*; GIZ: Eschborn, Germany, 2019.
53. MLHUD. *Fit-For-Purpose Land Administration Country Implementation Strategy*; Ministry of Lands Housing and Urban Development: Kampala, Uganda, 2018.
54. Musinguzi, M.; Enemark, S.; Naome, K.; Mwesigye, S.P. A Country Implementation Strategy for Fit-For-Purpose Land Administration. The case of Uganda. *Afr. J. L. Policy Geospatial Sci.* **2020**, *3*, 1.
55. Musinguzi, M.; Kisakye, M. Discrepancy Between Survey Practice and Legislation in Uganda. In *The First Conference on Advances in Geomatics Research*; Makerere University: Kampala Uganda, 2011.
56. Enemark, S. Fit-for-Purpose: Building Spatial Frameworks for Sustainable and Transparent Land Governance. In *Land and Poverty Conference 2013*; The World Bank: Washington, DC, USA, 2013.
57. Lemmen, C.; van Oosterom, P.; Bennett, R. The Land Administration Domain Model. *Land Use Policy* **2015**, *49*, 535–545. [CrossRef]
58. Coldham, S. Land Reform and Customary Rights: The Case of Uganda. *J. Afr. Law* **2000**, *44*, 65–77. [CrossRef]
59. Zevenbergen, J.; Augustinus, C.; Antonio, D.; Bennett, R. Pro-poor land administration: Principles for recording the land rights of the underrepresented. *Land Use Policy* **2013**, *31*, 595–604. [CrossRef]
60. Kirabo, J. *Kyobe Court Annexed Mediation and Case Backlog Reduction*; Makerere University: Kampala, Uganda, 2015.
61. Musinguzi, M. *Piloting Capacity Assessment Tool for Land Policy Implementation in Uganda*; UN Habitat/GLTN: Kampala, Uganda, 2017.

62. DAI. *Study on Capacity Assessment of the Land Sector to Support Implementation of the National Land Policy;* DAI: Kampala, Uganda, 2016.

 land

Article

Fit-for-Purpose Land Administration from Theory to Practice: Three Demonstrative Case Studies of Local Land Administration Initiatives in Africa

Uchendu Eugene Chigbu [1], Tobias Bendzko [2,*], Menare Royal Mabakeng [1], Elias Danyi Kuusaana [3] and Derek Osei Tutu [4]

1 Department of Land and Property Sciences, Faculty of Natural Resources and Spatial Sciences, Namibia University of Science and Technology, Windhoek 9000, Namibia; echigbu@nust.na (U.E.C.); rmabakeng@nust.na (M.R.M.)
2 Chair of Land Management, Faculty of Aerospace and Geodesy, Technical University of Munich (TUM), 80333 Munich, Germany
3 Department of Real Estate and Land Management, SD Dombo University of Business and Integrated Development Studies, Wa, Ghana; ekuusaana@uds.edu.gh
4 Lands Commission, Cantonments, Accra, Ghana; derekostus27@gmail.com
* Correspondence: tobias.bendzko@tum.de

Citation: Chigbu, U.E.; Bendzko, T.; Mabakeng, M.R.; Kuusaana, E.D.; Tutu, D.O. Fit-for-Purpose Land Administration from Theory to Practice: Three Demonstrative Case Studies of Local Land Administration Initiatives in Africa. *Land* **2021**, *10*, 476. https://doi.org/10.3390/land10050476

Academic Editors: Stig Enemark, Robin McLaren and Christiaan Lemmen

Received: 25 March 2021
Accepted: 28 April 2021
Published: 2 May 2021

Abstract: Land is a critical factor of production for improving the living conditions of people everywhere. The search for tools (or approaches or strategies or methods) for ensuring that land challenges are resolved in ways that quickly respond to local realities is what led to the development of the fit-for-purpose land administration. This article provides evidence that the fit-for-purpose land administration—as a land-based instrument for development—represents an unprecedented opportunity to provide tenure security in Africa. The article presents case studies from three sub-Saharan African countries on local-level experiences in the applications of fit-for-purpose guidelines as an enabler for engaging in tenure security generating activities in communities. These case studies, drawn from Ghana, Kenya, and Namibia, are based on hands-on local land administration projects that demonstrate how the features of the fit-for-purpose guideline were adopted. Two of the case studies are based on demonstrative projects directly conducted by the researchers (Ghana and Kenya), while the other (Namibia) is based on their engagement in an institutional project in which the Global Land Tool Network (GLTN) and other local partners were involved. This work is relevant because it paves a path for land administration practitioners to identify the core features necessary for land-based projects.

Keywords: customary tenure; fit-for-purpose; land administration; land inventory; land management; land tenure; mobile-based applications; pro-poor; land surveying; tenure security

1. Introduction

Access to land means gaining physical availability to land and making decisions on its use or exercise of the rights embedded therein [1,2]. Access to land is fundamental for creating livelihood opportunities for many people living in developing countries [3,4]. It is also a determinant factor on how land is put into sustainable use countries [5,6]. Important to land access, is the issue of land tenure security. It is impossible for individuals without land tenure security—i.e., the rights individuals and groups have to effective protection by the state against forced eviction—to take advantage of opportunities to improve their living standards [7–9]. Land tenure security is linked to ending poverty in all its forms everywhere. The lack of secure tenure "creates significant instabilities and inequalities in society and severely limits citizens' ability to participate in social and economic development" and "also undermines better land use and environmental stewardship and deters responsible private investment, due to the associated land risk" [10] (p. 2). The role of land tenure security in the socioeconomic development of nations includes ending hunger, achieving food security, and improved nutrition [11]. Others include promoting sustainable use of terrestrial ecosystems and sustainably managing forests, combating desertification, halting, and reversing land degradation in addition to halting biodiversity loss [11,12]. Unfortunately, both access to land and tenure security remain serious challenges in many parts of the world.

Land is a critical factor of production in the economies of most developing nations in Africa, Asia, and North and South America. A crucial role of the governments from these nations is the development of their land administration system (LAS) with the goal of delivering sustainable development [7]. This study focuses on African nations because they are facing challenges of all kinds that are linked to land tenure insecurity (or the lack of land tenure security) in the use of their land and natural resources. Population growth and the absence of appropriate land policy initiatives (or poor implementation of such policies where available) in Africa renders access to land unguaranteed for women, youth, and men [13–16]. Contestations in land access makes land tenure security a big concern in the continent today. This is due to pressures posed by the diverse needs (and uses) for land and unstable land tenure systems in the continent. Hence, land tenure security tops the list of the most burning social and ethical issues in Africa [17].

Intense pressures in demand for land have rendered land rights of poor and marginal groups volatile—heightening the need for their security. The poor and marginal bracket is currently youth and women that are mostly affected by land tenure insecurity [9,16]. Besides being excluded from critical decisions over land that has a bearing on their future legacy, they are also denied or granted limited access to land to enable them to meet their livelihood needs. The available and most recognized mechanism for accessing and securing land for all persons remains the formal or conventional land registration system [18,19]. However, the objective of land registration to deliver tenure security to all has not been realized [20]. The outcomes have not been impactful in informal urban settlements and rural areas, due to the complexities of the customary tenure, which is the most dominant tenure type in rural areas, and hybrid tenures in informal settlements countries [6]. Owing to the limited impact of formal land registration in bringing changes that benefit the poor, scholars and practitioners engaged in land administration and management are re-orienting their thoughts towards pragmatic approaches that are responsive to the local realities. Concerted efforts have been put, and remain ongoing, towards innovating or producing pro-poor and participatory approaches to achieving land tenure security in Africa. These efforts are necessary for enhancing pro-poor growth, sustainable land use, and peaceful co-existence among communities through policy development aimed at increasing land tenure security for local resource users [10,20–22]. For success to be achieved, it is important to put in place effective land administration systems (LAS) in African countries.

LAS (i.e., the way in which land tenure is conceptualized, applied, and operated) provides countries with the enabling environment for implementing land-related policies and strategies. It provides a platform for socioeconomic and environmental development. However, in most developing sub-Saharan African countries, most land parcels and the people who use them are outside of the formal LAS [10,23–25]. Most of these people outside the formal systems comprise of poor and other disadvantaged groups of people [14,20]. Several efforts to engage in locally realistic approaches to land administration (LA) have been made to ensure more locally realistic impacts in the effort to improve land access and tenure security [24–27]. These efforts have not always yielded their required results. In proposing a renewed approach to engage in tenure security improvements, Enemark et al. [10] (p. 2) noted that:

"Attempts to introduce conventional (western style) land administration solutions to close the security of tenure gap have lacked success. New innovative solutions are required to build affordable, pro-poor, scalable, and sustainable systems to identify the way all land is occupied and used. The fit-for-purpose (FFP) approach to land administration has emerged as an opportunity for developing countries in this regard. It offers a viable, practical solution to quickly and affordably provide security of tenure for all and to enable control of the use of all land."

This article gives credence to Enemark et al.'s [10] call to adopt locally realistic approaches for affordable, pro-poor, and sustainable solutions to land challenges in developing countries. The case studies presented in this article provide evidence that the Fit-For-Purpose land administration (FFP-LA), as a guideline for development, represents an unprecedented opportunity to provide tenure security in Africa. The article presents case studies from three sub-Saharan African countries—Kenya, Ghana, and Namibia—on local-level experiences in the guideline. The following section of the article deconstructs the key element of the FFP-LA (i.e., *purpose*) from the perspective of the surveying profession. It highlights the power of *purpose* in the LAS of developing countries and describes how the FFP-LA is a *repurposing* tool for enhancing land tenure security within the LAS of developing countries. This is followed by a description of the procedures for gathering and analyzing the experiences (methodology), presented in the form of case studies. Then, the discursive demonstrative case study is presented, before our conclusion.

2. FFP-LA in Theory

2.1. The Power of "Purpose" in Surveying and Land-Related Professions

Surveying or any other land-related profession aside, purpose under any situation is the dimension or the core or essence of what people do and why activities are undertaken. In life, *purpose* is that which provides a "profound sense of who we are, where we came from, and where we are going, the quality we choose to shape our lives around, and the source of energy and direction" [28] (p. 1). Simply put, *purpose* is that one element of which shapes human actions and the direction of human actions. Hence, there is power in *purpose*. The evolution of purpose in land-related activities is a natural one. In everyday activities, the commitment to a *purpose* for living requires people to ask *'Why embark on an activity? Why embark on one activity instead of the other?'* In the same way, land professionals (from its earliest beginning) have always sought to answer the same question before embarking on any land-based intervention. This evolutionary trend leading towards our understanding of FFP today is best illustrated from the context of surveying. To understand what surveying is about, Bowie [29] (p. 545), over 100 years ago, explained that it is " ... one of the oldest of the geophysical sciences. Originally it embraced only a limited field, consisting of the determination of the shape of the earth and its size and how this was done." Today, the subject of surveying is very closely allied with other matters beyond Bowie's [29] description of the earliest practitioners. There are various arms of surveying that exist, each designed to tackle different aspects of the needs of humanity. However, for the sake of illustrating the emergence, power, and application of purpose as an intrinsic element in the surveying profession, we focus on three key aspects of surveying (i.e., land, estate/valuation, and quantity surveying).

Surveying—a profession for surveyors—is the notable land-focused profession in all developing countries. "Surveyor", as a term, is derived from the term *sur voir*, which is translated to mean "overseer" [30] (p. 2). In a broad sense, surveying historically relates to the profession of overseeing land. However, the scope of the term has stretched to include specific activities related to geographic positioning, valuation, and measurement of land (including the development of land) information and the administration/management of land. Of course, we recognize that "the surveying profession may be subject to different governance structures in various jurisdictions around the world, but the responsibilities of a surveyor are reasonably consistent in most countries" [29] (p. 1). In this article, we use the land, estate/valuation, and quantity surveying to represent the three broad traditional surveying professions found in many developing countries. All these surveying professions (and others not mentioned) carry out land-related actions that are based on the responsibilities and functions meant to improve the living conditions of people in societies [31]. The land surveyor is a professional engaged in a basket of services, including cadastral and mapping services. The estate/valuation surveyor is broadly engaged in land economy, estate management, property valuation (including plant and machinery valuation, among many other tasks), facilities management, conflict resolution, and land compensation appraisals [32]. The quantity surveyor tackles engineering estimations/costing, cost monitoring, engineering construction management, and related tasks.

Society is always the core beneficiary of the surveyors' professional actions, and therefore, *purpose* is central to surveying. *Purpose* is the working intension of surveyors. It is this intention that determines their actions. Even though their work is challenging, due to multiple tenure regimes and complex overlay of interests, surveyors strive to manage spatial needs. Their actions usually lean towards causing positive changes (or development) through activities and decisions based on functional land information infrastructure and physical land parcel management and administration systems. Land parcel management/administration roles in surveying relate to valuation, real estate, property management, and the preparation of plans or maps that represent land parcels, and to collect data [33]. These are the intensions for the *purpose* of land surveying. Based on a particular *purpose* in land surveying, different methods can apply, and various outputs. The estate/valuation surveyor engages in valuation (i.e., the assessment of property values), knowing that "the proper meaning of 'value' cannot be determined without reference to the purpose of the valuation" [34] (p. 159).

2.2. FFP-LA Is about the "Purpose" in LA to Become Fit for Developing Countries

As has been stressed above, surveying and all other land-related professions are purpose-driven and purpose-oriented. The need for improving land tenure security and its importance to global development is well highlighted in the Sustainable Development Goals (SDGs) [35]. The SDG-1 is focused on ending poverty in all its forms everywhere. The SDG-1 is based on a target (SDG-1.4) of ensuring that "all men and women, in particular the poor and the vulnerable, have equal rights to economic resources, as well as access to basic services, ownership and control over land and other forms of property, inheritance, natural resources, appropriate new technology, and financial services, including microfinance" by 2030. The indicator for measuring the SDG-1 is to ensure that the "proportion of total adult population with secure tenure rights to land, (a) with legally recognized documentation, and (b) who perceive their rights to land as secure, by sex and type of tenure" (SDG-1.4.2). Land tenure security is crucial for the achievement of the SDG-1 because how people and communities gain access to land and natural resources is defined and regulated by societies through land tenure systems. The role of LA is that it serves as an instrument for organizing tenure systems so that they can determine who can use what, which resources can be used, in which location the resource is located, for how long, and under which and what conditions it can be used. The procedures for achieving this vary and may largely be influenced by the sociopolitical and economic conditions of a particular country, and its colonial historical influences. That is why land tenure systems can be based on written and unwritten land policies and laws. In much of Africa, emerging characteristics of land tenure, whether secure or insecure, may be shaped by national-level legislation, customary regulations, and reforms.

Improving land tenure security in communities is a highly recognized means of reducing or alleviating poverty because it enables productive use and enjoyment, ownership, and securitization and enjoyment of land rights (including exercise of obligations, privileges, and restrictions) [36]. However, the way it has been done over the past decades is questionable. For instance, "less than a quarter of the countries in the world maintain complete LAS and about 4 billion of the world's 6 billion tenures operate outside formal governance arrangements" [37] (p. 3). To improve land tenure security in developing countries, realistic FFP-LA systems need to be established in these countries. Setting up FFP-LA systems in developing countries requires adopting a system that responds to the needs of the people and which are inclusive, equitable, efficient, sustainable, and pro-poor. Improving the land rights of the poor remains a major challenge to the industry, especially where processes and costs of LA are applied across the board for all land uses, for both urban and rural areas, and for all classes of persons. It also requires a change towards locally realistic and flexible approaches that can be handled within existing local-level hard and soft technologies and human resource proficiency. All of these explain why Enemark et al. [10] sought to find an approach to land interventions that is fit for purpose in tackling local land challenges in developing countries. Hence, the birth of the FFP guideline. To fully understand how the development of FFP-LA has contributed to altering the paradigm of thinking and practice in LA, it is important to revisit the meaning of land administration more vividly. This, in our view, will give a broader foundation to the proposition and application of the FFP-LA and streamline its targeting in developing countries.

LA entails "the processes of determining, recording and disseminating information about the ownership, value, and use of land when implementing land management policies" [38] (p. 108). This definition emphasized formally registered land ownership and excluded most people without land documents, such as those with customary rights, informal tenures, customary share tenants, and those with land use rights in the form of licenses. FFP-LA was developed to support access to, and the tenure security of these types of rights, but it did more than this, and has raised several questions of social exclusion and expropriation. It positively altered the how-to aspect of achieving these LA objectives, especially for many Africa countries with dominant customary land tenure and who are compelled to run parallel (hybrid) systems of LA.

The FFP-LA also has a spatial management dimension to its use, adoption, and application. The FFP-LA is based on three building blocks, including a spatial framework required to support "recording the way land is occupied and used" [10] (p. 108). The spatial frame allows for operationalizing FFP-LA to achieve pro-poor tenure security. However, this is only feasible when LA initiatives embrace the use of visible (physical) boundaries rather than fixed boundaries, aerial/satellite imagery rather than field surveys, a focus on accuracy that relates to the purpose rather than technical standards; and a demand for updating/upgrading ongoing improvements [10,22–24]. The spatial planning and management dimension of FFP-LA allows it to serve as a veritable tool in actualizing objectives of land use planning and the socioeconomic development of human settlements (urban, peri-urban, and rural). This role of FFP-LA is, however, not possible without the presence of a locally realistic spatial management system that deals with the land information and spatial data infrastructure.

FFP-LA also aligns with the idea that land rights evolve in a continuum, and an overlay of a plethora of entitlements. Continuum of land rights implies that rights to land lie along a continuum. This means that at one end of the continuum are formal land rights and at the other end of the continuum are informal rights, and between these two extremes are a wide range of other rights [39] (p. 12). The FFP-LA guideline applies in all cases (be it land titling, strengthening hybrid forms of land rights, or securing tenure under African customary laws).

It is pertinent to state that the development of FFP-LA is set in the context of the history of unrealistically exclusive western-oriented LAS in former colonies of western countries. These systems, with their associated high technology land information management systems, are not inclusive in nature. They generally exclude poor people, and regard customary tenure as retrogressive in nature. The FFP-LA is a response to finding locally realistic solutions to poor LAS in developing countries and ensure that the peculiar local needs are addressed inclusively. It is an outcome of the re-examination of the traditional land administration processes regarding their ability to support the interest of the poor in developing countries. In the words of Enemark et al. [10] (p. viii), FFP-LA is purpose-driven because:

> *"This new approach is focused mainly on the 'what' in terms of the outcome of security of tenure for all and, secondly, it looks at the design of 'how' this can be achieved. The 'how' should be designed to be the best 'fit' for achieving the purpose ('the what'). In this regard, the phrase 'As little as possible—as much as necessary' perfectly reflects the FFP approach."*

From the above statement, the key characteristics of FFP-LA can be described to include a focus on purpose, flexibility in land administration, responsiveness to local needs, and adoption of incremental improvements in the efforts towards improvements in land tenure security. The FFP-LA's development is one of the few innovations that has contributed to "altering the power relations and bringing about social change in parts of the global land industry; and to supporting tenure security for the poor in a number of local communities" [40] (p. 6). Its development contributes to the extension of LAS to improve the land tenure security of the poor for most people in developing countries. However, a concept cannot make a change unless it is transformed into a problem-solving tool. Hence, Enemark et al.'s transformation of the FFP-LA concept into FFP-LA guidelines [10,23,24]. This raises the question of how and where the FFP-LA guideline has been put to practice, and what impact it has made. It is essential to highlight some of these applications of the FFP-LA guideline, and to draw critical lessons for the future.

2.3. The Features of FFP-LA

As a concept, the FFP-LA embraces three interrelated aspects (i.e., the spatial, the legal, and the institutional frameworks) that form its practice [24]. As Enemark et al. [10]. (p. viii) clearly note:

> *"The spatial framework supports recording the way land is occupied and used. The scale and accuracy of this representation should be sufficient for securing the various kinds of legal rights and tenure forms recognized through the legal framework. The institutional framework is designed to manage these rights and the use of land and natural resources and to deliver inclusive and accessible services."*

To ensure that the FFP-LA building blocks are well-framed to achieve their purpose for various situations, four core principles have been devised for each of its three frameworks. These principles are crucial for putting the FFP-LA guideline into practice. However, they are also embedded within the general features (that is, the attributes that make it different from other LA approaches) of the FFP-LA. To make any LA approach FFP-LA compliant, such an LA approach must adapt to all or some of these FFP-LA features. Based on Enemark et al. [10] (pp. 22–80), the features of FFP-LA which can be partially or wholly adopted in local LA interventions in any developing country include the following:

- Building the spatial framework: Availability of flexible non-statutory framework reflecting the overall spatial situation on the area under investigation.
- Visible (physical) boundaries rather than fixed boundaries: Reliance on established community or neighborhood-wide accepted physical boundaries (such as fences, ditches, hedges, seasonal and non-seasonal water bodies, and walls, etc.), rather than geodetically fixed boundaries, can provide sufficient evidence of the occupation and the attribution to tenure and land rights.
- Aerial imagery rather than field surveys: The use of a range of scales of satellite/aerial imagery as the spatial framework to identify and record visible boundaries.
- Relating accuracy to purpose rather than to technical standards: Aspiring to achieve accuracy of the land information as a relative issue related to the use of this information, rather than being driven by technical standards that are often rigid and works against the purpose for pro-poor tenure.
- Demand for updating, upgrading, and ongoing improvement: Ensuring that LA activities are not a one-off process, but rather as a process in constant flux that requires updating, upgrading, and incremental improvement whenever necessary for fulfilling land policy aims and objectives.
- Flexible framework, designed along with an administrative framework, rather than judicial lines: LA activities (such as recording and registering rights, etc.) should be based on administrative rules/regulations where and when possible, instead of legal institutions.
- Continuum of tenure rather than individual ownership: Concerning its tenure security objectives or outcome, FFP-LA embraces the continuum of land rights concept and

practice. The continuum of land rights recognizes the existence of a diversity of tenure arrangements in practice, encompassing both legal and socially derived rights. Apart from the legal recognition of land rights, social recognition of land rights matters because it can protect the *de facto* land rights of local people and assign legitimacy to such rights.

- Flexible recordation rather than only one register: FFP-LA supports pro-poor land rights documentation to support the building of locally-based land recordation systems capable of serving local purposes and being able to run parallel with the national strategy or as separate activities in support of local needs.
- Gender equity for land and property rights: Irrespective of any form of progress that might have been made concerning gender issues, women's rights to land and secure tenure remain a challenge in many developing countries. The FFP-LA recognizes gender as an opportunity to create equality in the access, use, and exercise of tenure security among women, men, and youths.
- Building the institutional framework: Putting FFP-LA activities into practice requires a supportive institutional framework. This entails having supportive land policy, organizational structures, supportive resources, institutional networks, and institutional arrangements, organizational, among many other institutional supports.
- Good land governance rather than bureaucratic barriers: FFP-LA activities are, by way of both principle and practice, to be based on good governance. This implies embracing issues of accountability, control of corruption, political stability, and the rule of law, among many others.
- Integrated institutional framework rather than sectorial silos: In many developing countries, governments are still managing their land and natural resource assets in silos with limited interaction and coordination across sectors. FFP-LA supports coordination and collaboration across the land sectors.
- Flexible ICT approach rather than high-end technology solutions: FFP-LA recognizes the dependence on digital development as a key element in the efficient and sustainable management of land. It requires the best practices into technology-enabled activities in LA—so far, they are pro-poor and freely available for use by all.
- Transparent land information: FFP-LA encourages the provision and availability of open, transparent access to land information, subject to the protection of privacy.

In the following sections of the article, we present some case studies that highlight local-level adoptions of the FFP-LA guideline as an enabler for engaging land tenure security generating activities.

3. Mixed Cross-Case Studies as a Research Method

We use three demonstrative case studies to present experiential evidence that reflects on the application of FFP-LA from three regions of Africa (i.e., Eastern, Southern, and Western Africa). These case studies are drawn purposively from Ghana, Kenya, and Namibia. We use the demonstrative case study approach because it relies on presenting the situational analysis of multiple cases [40]. It is also suitable when aiming to understand the social or technical worlds of a specific development scheme using a discursive narrative to analysis. The demonstration reveals how the FFP-LA guideline was used in three case studies that involved different project arenas, land administration authorities, administrative levels, and how land administration events interact [41]. Two of the case studies are based on demonstrative projects directly conducted by the researchers (Ghana and Kenya), while the other (Namibia) is based on their engagement in an institutional project in which the Global Land Tool Network (GLTN) and other local partners were involved. We provide a precursor to the case studies below, followed by the FFP-LA focused presentations.

The Ghana case study was conducted during a two-month period, between September and November 2015. It was carried out in *Juaben-Atia*, a small rural community in the Juaben Municipality, one of the forty-three districts in the Ashanti Region, Ghana. The main economic activity of the populace is farming, with most of the inhabitants involved in oil palm plantations. *Juaben-Atia* lies within a customary tenure area, where the land is owned and used by local families as usufructs, while community chiefs are entrusted with allodial rights. The area has well-defined borders. This is because it benefitted from the multilateral Ghana land administration projects in 2008. During this time, its customary boundaries were demarcated as part of a piloted community mapping. However, most of the individual farm parcels were not surveyed, and individual land rights were undocumented (neither by way of informal recordation nor formal registration). The proximity of *Juaben-Atia* to the proposed inland port at Boankra resulted in increased land values which created a pressure of tenure insecurity within its customary land tenure systems. The need for immediate mapping and documentation of individual land rights to secure the rights of the people in the area was hindered by their inability to afford the services of professionally skilled surveyors with high-end tools. Their inability to afford such services excluded most of them, especially the poor and marginalized, from securing their land against speculators and landgrabbers. We approached the chiefs of the area to engage in a participatory land inventory, using the FFP-LA, to improve tenure security. With their permission, we engaged in piloting an inventory process in the area to explore the possibility of using standalone single frequency GPS (GNSS) receivers in smartphones and tablets to capacitate the local youths to do the mapping and documentation as a step towards securing their land. The capacity development and cost-effectiveness aspects of this case study have been reported by Osei Tutu et al. and Bendzko et al., respectively [42–44]. Our use of this case study, as is shown in the next section of this article, is to demonstrate the FFP-LA context of that exercise.

The Kenya case study was conducted in 2018 over the course of four weeks inside the *Ngerenyi* area in *Wundanyi/Mbale* in Taita, Taveta County. The entire research area covers approximately 4.7 km^2 and is mostly rural. The closest city is Wundanyi Town, located 4.5 km away from the mountainous research area. Its high altitude is one of the main reasons for its water accessibility, due to an annual average rainfall of between 600 and 2300 L per m^2. Most economic activities in this area are closely connected to its availability of water resources. The entire area of the Taita Hills is considered one of only five water sources in Kenya. Here we sought to understand the land tenure, land use, and tenure security issues by applying FFP-LA. Ngerenyi area was selected, due to its importance to Kenyan agriculture and dairy farming activities which yields are distributed all over the country. Tenure insecurity emerging from the use of outdated cadastral information was the main challenge in the community (worsened by increasing high land values). We engaged in a small-scale FFP-LA initiative that involved the mapping of parcels with the support of local youths in cooperation with the local Taita Taveta University. This also involved digitizing analog maps of the local mapping authorities to create a digital land information system, including up-to-date parcels and boundaries. In total, four individual teams of at least two people each were engaged in mapping the parcels and conducting the interviews in a local dialect. Although large parts of the area were mapped more than 50 years ago, only very few view updates have been added to the existing cadastral maps, due to the high cost connected. Most surveyed parcels were owned by families under customary tenure, relying on titles that were issued to deceased family members several decades ago.

The Namibian case is based on local land administration work conducted to facilitate tenure security in the Freedom Square Informal Settlements (FSIS). The researchers participated in participatory planning, data collection, and flexible surveying methods that were used over a period of eight years from 2013 to 2020 to deliver secure land rights. Gobabis is a small town in the eastern part of Namibia with an estimated population of over 20,000 in 2011 [45]. About 40% of the population reside in informal settlements [45]. Like many other informal settlements in sub-Saharan Africa, the FSIS (being one of the six informal settlements in Gobabis, Namibia) suffers from insecure tenure. The tenure insecurity in FSIS manifested in the forms of lack of access through statutory channels, lack of documented evidence of tenure, and dependence on varieties of informal tenures. This means that "their occupation of land and/or housing is usually either illegal, quasi-legal, tolerated or legitimized by customary or traditional laws, which can either be recognized or simply ignored by the authorities" [46] (p. 6). The local authority initially planned to relocate the residents, as they occupied a prime piece of land planned for a formal housing project. To halt the process, community members spearheaded an enumeration and land recordation exercise. Namibia Housing Action Group (a local NGO—supporting a Shack Dwellers Federation of Namibia, a community-based organization) initiated an informal settlement profiling and enumeration process. Data on households was used to negotiate for land rights and inform in situ upgrading. A professional land surveyor surveyed the external boundaries of the settlement, while the internal subdivisions were completed by the community with a Technical Land Surveyor, according to Flexible Land Tenure System [47].

To determine how the case studies aligned with the FFP-LA guideline, we analyzed them by deconstructing the LA processes in each case study and then matched them against a matrix consisting of the FFP-LA features outlined by Enemark et al. [10]. By using these FFP-LA features to create an assessment matrix, it was possible to demonstrate an understanding of the case studies (comprising of local LA interventions) related to or adopted (or aligned to) FFP-LA.

4. The FFP-LA in Practice: Three Demonstrative Cases Studies

4.1. The Use of Mobile-Based Application for Youth-Led Land Inventory in Juaben-Atia, in Ghana

As part of the demonstrative FFP-LA exercise, three (3) young males from the case study area were trained to be able to assist in mapping and documenting the land rights of community members within a selected area. These trained assistants administered open and close-ended questionnaires to the 53 farmers. Collected land information data that included parcel sizes, types of interests held in land, and nature of tenure. In mapping the land parcels, the youths were led by the farmers to inspect the land parcels in the presence of a witness who, in most cases, are neighboring farmers. The farmers, in the context of Ghana's customary land tenure system, are users of land because the land is held for the community by the Chief.

Land information acquisition and storage was made using a mobile-based application which allowed for the acquisition of graphical and attributed data about all land parcels, with assurance from the local authorities that the database created can be used for land registration under proper legal provisions. The team and local technicians (youth) conducted a land survey/inventory comprised of the following equipment: A smartphone, open-source software (*GIS360* and *ODK Collect*), laptop, portable printer, paper, ink cartridge, power-bank, and wireless GPS-device. These materials/equipment were used by the youths who participated in the exercise as para-surveyors. Table 1 illustrates the functions of these materials in the process.

This exercise demonstrates that individuals not educated in surveying can affect FFP-LA if capacitated with relevant knowledge and given necessary equipment and materials. The process of conducting the land survey/inventory involved four main stages/phase of activities, including participatory enumeration, data processing, output of data, and data management. These stages represent a cyclic procedure (see Figure 1).

Table 1. Equipment and materials for the youth-led land survey/inventory Juaben-Atia (Source: Authors'—based on materials for land use inventory exercise).

Material/Equipment	Description and Use
Smartphone/tablet	Single handheld mobile phone and tablets with integrated GPS sensors for mapping land parcels. Smartphone/tablet is also used for taking photos attached to a documented evidence of tenure referred to as a *land passport*.
Software (*GIS360* and *ODK Collect*)	Open-source software Q-GIS to enable processing and outputting.
Laptop	For data storage and processing.
Wireless GPS-Device	For geo-tracking and georeferencing (mapping) land in real-time
Portable Printer, desk, paper, and ink cartridge	For printing—e.g., maps on A4 sheets of paper and any printable needs that may arise.
Power-Bank/electricity	For power storage and usage for charging smartphones and laptops.

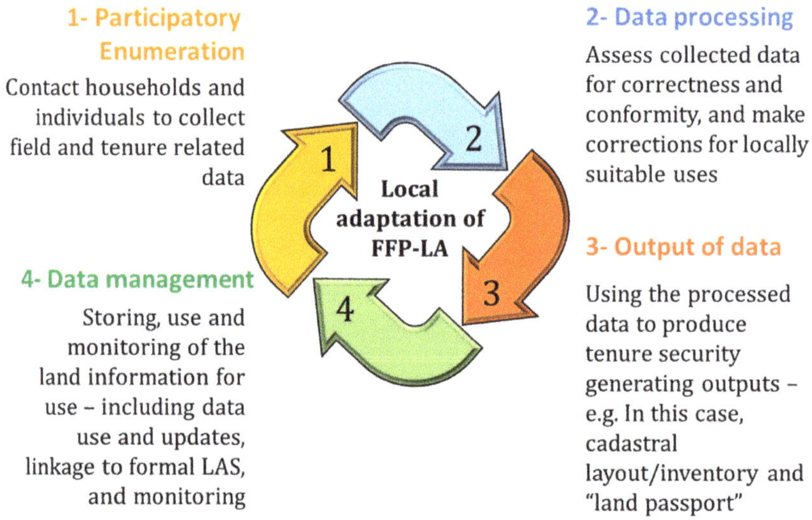

Figure 1. The process of mobile-based application for land survey/inventory in Juaben-Atia (Source: Authors').

The four main stages of implementing FFP-LLA in a local land inventory in Ghana (Figure 1) are described below.

Participatory enumeration: This is the data capture stage. Enumeration means "to count" or "to list down" or "to ascertain the number of" [48] (p. 7). Participatory enumeration is a data-gathering process jointly conducted by the people who are being surveyed. In this case, the local youths were engaged starting from the initial consultation through other activities, such as boundary identification, and household interviews, tenure recordation, and public data verification. This stage also involved general land tenure studies about the country and the local tenure situation.

Data processing: At this stage of the exercise, we plotted the data from the field in QGIS. Where it was impossible to rectify parcel identification problems, such parcels were referred to the field technicians for re-mapping. These para-surveyors helped to map, as well as carried out data processing under our training and supervision.

Output of data: Adopting the FFP-LA guideline led to two major outputs: Cadastral layout/inventory of parcels and the *land passport* (a terminology we devised to represent a parcel user or land use identification card meant as proof of use right).

Cadastral layout/inventory: The cadastral layout generated in the study comprises all the mapped parcels in the case study area. It contains spatial information on the land parcels, as well as the tenure attributes of the parcel, such as the user, the parcel size, the land use, and the land right held by the users. Each parcel was assigned a unique number that indexed it to the name and attributes of the land user. Fifty (50) mapped land parcels were generated after the data processing stage. These 50 parcels represent a total area of approximately 1.05 sq. km, and a boundary length of about 32 km.

Land Passport and Customary Land Register: Family heads and all individuals identified as users of the land were enumerated (documented) as holders of interests in the parcel. Where all identified persons were present, they were photographed, and their photos used as evidence on the land passport. The *land passport*—a document received by the family head or farmer after enumeration—is a text/photo documentation of the land parcel, the land rights, and rightsholders of the parcel and signed by the relevant parties. In all cases of enumeration, the farmers using the land were members of his/her household. The *land passport* is a document that gives spatial information, and a profile of an individuals' and households' land rights over a parcel in writing and signed by users of neighboring parcels with a paramount chief acting as a witness. The purpose of issuing a *land passport* is to provide documented evidence of the households' and farmers' interests in a land parcel. Data on the *land passport* are extracted from the cadastral layout/inventory developed after the output of data. Information put on a *land passport* includes a unique parcel identifier, land-use type, location, name of the farmer, sex, age, a snapshot of photo-ID card, the right thumbprint of the farmer (family head or individual farmer where applicable), and their witness. These details form the front matter of the *land passport*. Although the head of households and landowners were provided with photo-ID cards, all enumerated data were stored in a customary database reflecting all identified rights in a parcel and the rightsholders and put in the custody of the Chief. To ensure that all information on the land passport and customary land register were correct, efforts were put to ensure that family members with interest on land were not excluded by ensuring that their interests are identified in the. Where only one person was met at home during the enumeration (e.g., a son), the information supplied was validated by neighbors at first, and finally by the Chief at the point of signing. This approach requires honesty on the part of participants in identifying legitimate household members. However, the Chief had the authority to request verification of data from village heads and neighboring households, before signing as a witness. Appended to the *land passport* is a plan of the land parcel. While not a legal document, the land passport is a paralegal certificate that can serve as evidence of ownership in the court of law in situations of dispute where none of the disputants have a full legal document to the land under contestation. Legal issues apart, it can serve as a fully recognized evidence of land use and the user within the customary governance system in which it was issued (such as in *Juaben-Atia*) because of the signature of the Chief as a witness on the land passport. It is enough to provide a socially recognized level of tenure security for farmers and their households at the local-level. Because it has the seal or signature of the local chief, it can deliver tenure security against unlawful eviction and entitles the occupant usufruct right to restitution/compensation under the power of an eminent domain.

Data management: In the context of this inventory exercise, this stage involved the storage and management of data. Data review and updating are crucial activities at this stage. During this stage of the land inventory, potential boundary disputes were identified and discussed to ensure they do not (re)occur or even escalate in form. Considering that we relied on general (and not fixed) boundaries in our mapping, activities in the data management stage, including updates or reviews of land information, can lead to new enumeration. Potentially, this stage is crucial as it may involve the (re)validation of data for linkage into the formal land registration system (in the case of a transfer or formalization.

4.2. Land Use and Land Tenure Inventory in the Taita Hills, in Taveta Town, Kenya

The Kenyan case study area, the Ngerenyi area, is located in Wundanyi/Mbale in Taita, Taveta County (Figure 2). The entire research area covers approximately 4.7 km^2 and is mostly rural. To carry out the fieldwork, four groups of at least two trained young para-surveyors (six male and three female) were sent into the research area. The teams were selected based on their willingness to participate in the inventory exercise. The team then approached the residents who were willing to respond to the questionnaire and have their land mapped.

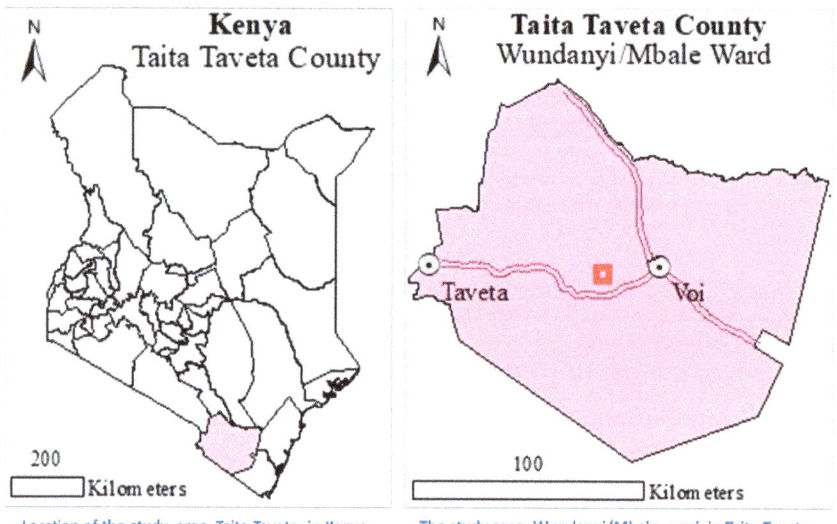

Figure 2. The location of the study area from national and local contexts (Source: Authors').

During the four weeks of data collection, the teams managed to map 347 parcels, covering large parts of the project area (Figure 3). They conducted interviews comprising to identify the users, types of tenure, incidences of tenure security, and land-uses. Following the interviews, the young para-surveyors instructed the interview partners to show the boundaries of their individual parcels.

Different from the Ghana case study, the mobile devices were equipped with the application *Mapit GIS* for the mapping purpose along with *ODK Collect* for the interviews. Daily, all collected data were gathered on one Laptop operating *QGIS* and creating an overarching database of all mapped parcels. After mapping each parcel, the teams were issued print-outs of the mapped parcels showing location, size, shape, name, and picture of the user of each parcel.

The same approach used in the Ghana case study was used in the Kenya case study. The equipment and materials used in Kenya were the same as those used in Ghana, except that *Mapit GIS* and *QGIS* software were combined with *ODK Collect*. It also followed a youth-led land survey/inventory strategy as applied in Ghana. The details of the process are as described below.

Figure 3. Mapped parcels in Ngerenyi in Wundanyi/Mbale in Taita hills, in Taveta town, Kenya (Source: Authors').

Participatory enumeration: Like the Ghana project, all participants were enumerated with each team using a specific number range of 200. Due to this unique enumeration, each mapped parcel and interview was possible to track back to the recording team and to avoid duplications of records. The enumeration process enabled a clear identification of each plot, which would not be possible with family names or other identifiers.

Data processing: After each field day, all para-surveyor teams meet for the data processing and exchange of collected data. First, the data were collected from all tablets, integrated, and merged inside *QGIS*. As the second step, the *ODK* data from the interviews were connected to each mapped parcel using the identifiers from the enumeration. In the final step, all teams received the current stage of mapping as an interactive map on their devices using the map import function of *Mapit*. During the six weeks of data collection, the para-surveyors were trained in the usage of QGIS to conduct the data processing independently. Additional to the daily data processing, conventional cadastral maps have been acquired, scanned, and digitized to receive an overview of the existing boundaries and registered plots.

Output of data: The processed data lead to three different results: The land passport for all individual users, the newly mapped cadastral inventory, and the digitized old paper-based maps from the cadastral office.

Land Passport and Customary Land Register: The land passport was documented following the same procedure as described in the case of Ghana. The procedure was replicated as stated earlier.

Cadastral Inventory: The generated land inventory comprises all the mapped parcels in the case study area, including all received data from the conducted interviews and satellite imagery helping with the general orientation. It contains spatial information on the land parcels, as well as the tenure attributes of the parcel, such as described in the land passport. Each parcel has its unique identifier can be assigned to all information collected by the questionnaires. In total, 347 individual land parcels were mapped, covering an area of about 180 ha owned by 331 different individuals.

Digitized paper-based map: In total, sixteen (16) paper-based maps of the entire area were received from the cadastral office, scanned, and brought into a digital form using rectifying tool in *QGIS*. For the rectifying process, at least thirty-two (32) points were georeferenced using remotely sensed images for each map to receive a highly accurate digital map for an area with very uneven terrain. The paper-based documents from the cadastral office only had very few updates and showed in all other cases the original map and boundaries from 1967.

Data management: Within the final stage of this case study and application of the FFP-LA, the results from the data processing were shared with local officials and the mapping authorities to help improve the accuracy of the current paper-based maps to the latest up-to-date technologies using GIS. To create a comprehensive digital cadastre, further efforts must be made to update the existing database and move towards a more efficient and cost-saving approach to land registration and administration. The fieldwork and the received data from the interviews showed a surprisingly high perceived tenure security which is either based on outdated documents or family graves located on the individual plots.

4.3. Adoption of FFP-LA in Implementing the Flexible Land Tenure System in Freedom Square, Gobabis, Namibia

The Flexible Land Tenure System (FLTS), introduced in 2012 in Namibia, offers an alternative form of tenure that addresses the core issues of imbalance and discrimination against low-income communities in Namibia. Informal settlements were envisaged to be the potential core beneficiaries of the FLTS application. The premise of establishing the FLTS is based on fit-for-purpose principles of ensuring flexibility and encouraging alternative approaches to tenure. Namibia's Flexible Land Tenure Act [49], the law that established the FLTS, was introduced with the objectives of:

- Creating alternative forms of land title that are simpler and cheaper to administer than existing forms of land title.
- Providing security of title for persons who live in informal settlements or who are provided with low-income housing.
- Empowering the persons concerned economically by means of these rights.

The tenure insecurity issues in FSIS (in Gobabis) provided a good test for the application of FLTS. About 40% of the population of Gobabis resides in informal settlements with no access to individual water connections, no toilets, while houses are mostly made from corrugated iron [49]. The residents have no documentation for the land they occupy. For the ten-hectare settlement, residents only had two communal water points for over 15 years. Like many informal settlements in Namibia, residents cannot invest in improving their houses, due to a lack of development rights [50] (p. 204). Residents feared eviction and were concerned about the lack of access to services. In 2012 the Gobabis Municipality shared information on a plan to relocate households from Freedom Square, as the land that was identified for formal housing. The eviction and relocation of residents from municipal land have been a normal occurrence in Gobabis, as residents have no formal rights to the land, many could not resist. However, in 2012 the community refused and proposed a better approach to avoid relocation, which will involve participatory enumerations and participatory planning [50]. They embraced these participatory approaches because of the "perceived and actual lack of land tenure security due to unclear planning regulations to citizens in the informal settlements" [51] (p. 30). The process of planning which was used in adopting FFP-LA in implementing the flexible land tenure system procedure involved the following stages.

Participatory enumerations and mapping: A team consisting of members from the community, Shack Dwellers Federation of Namibia (SDFN) representatives, with support from Namibia Housing Action Group (NHAG), started the enumeration with a two-day training exercise on enumeration and mapping. The profiling and enumerations are part of the Shack Dwellers of Namibia rituals to securing tenure and improved services for residents in informal settlements [52]. Each structure in Freedom Square was numbered and mapped using aerial images acquired from the Namibia National Planning Commission. To speed up the collection of the data, the ten-hectare settlement was divided into ten small manageable blocks [53] (p. 12).

Informal settlement data and analysis: Individual household data were collected by residents using questionnaire print-outs and aerial images. To guide the finalization of land rights registration, the project implemented in various phases used the Flexible Land Tenure Act and Regulations [54], for guidance. The legislation, which was gazetted in 2012 and regulations finalized in 2018, provided a legal framework for communities in informal settlements to access formal land rights as part of an association. Data on individual households were collected, and each structure was mapped. Data were later manually analyzed by the community using big spreadsheets. The summary data were presented to the community at various community meetings, at which the local authority representatives were present. The data were disaggregated by gender, focusing on total population, development priorities, and affordability levels of households to pay for development (land, water, and or electricity). The community analyzed the data and shared it with the local authority during a community meeting (Figure 4).

A proposal was made to re-block the settlement through participatory planning and upgrading. The participatory data collection, planning, and flexible surveying methods were used over a period of eight years from 2013 to 2020 to deliver secure land rights. Residents, mostly women and youth, were trained on mapping, enumeration, and data analysis.

Social Tenure Domain Mapping: The settlement was mapped using high resolution aerial imagery from the National Planning commission. Each structure was provided with a structure number that was later linked to the attribute data on the household. New structures not appearing on the map were drawn in with a handheld GPS. All field maps were later digitized using open-source GIS software (*gvSIG software* and *QGIS*) by students and officials of the NHAG. In 2014 there were only 700 households who resided in Freedom Square; this number increased in 2015 to 1100 households. This required an update on the initial enumerations and mapped data. STDM application was introduced to the local authority and the community representatives with support from GLTN. The purpose was to improve the data management and provision of certificates of recognition to residents, while awaiting the implementation of the FLTS [54,55]. Before the project was complete, land transactions were taking place in the settlement. Residents were transferring their shacks. A local youth was stationed at the local authority and used the data from the community to update the land information. Residents had to use their national identity documents to initiate a transaction. No documents were issued at this stage. The aerial images were captured using an Unmanned Aerial Vehicle (UAV), which ensured the acquisition was fast and reflected the state of the settlement [56,57]. The new settlement layout was overlaid on the images to identify households that need to be relocated.

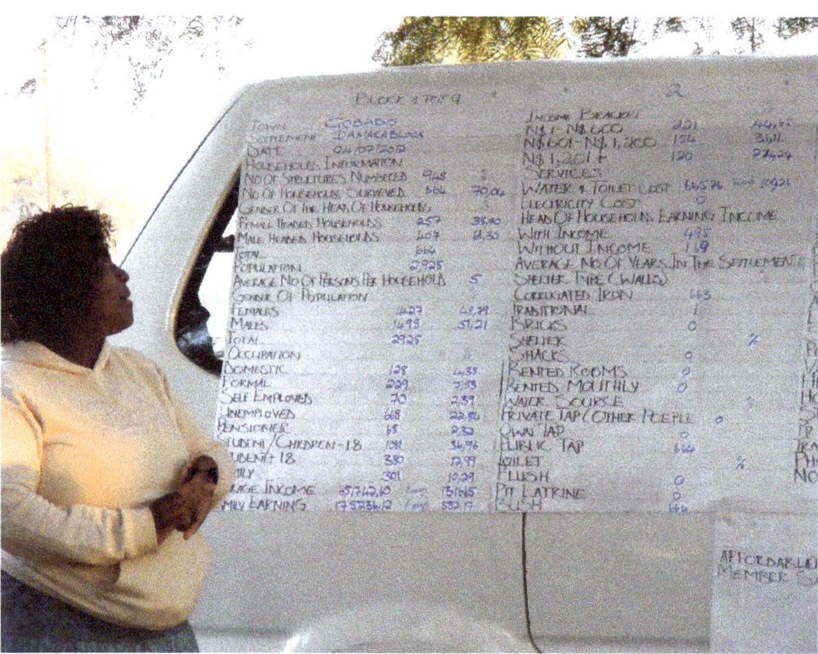

Figure 4. Community representative presenting data from enumerations (Source: Authors').

The internal boundaries in the settlement were measured by a technical land surveyor and maintained the accuracy requirements according to the Land Survey Act of the country, and the external boundaries of the settlement were surveyed by a professional land surveyor (in accordance with the high accuracy requirements under Land Survey Act 33 of 1993 of Namibia). Initial boundary recordation for Freedom Square was implemented using aerial images, and accuracy was improved upon relocation and installation of services. Implementing upgrading approaches using a continuum of accuracy would provide communities with an opportunity to complete the land inventory, and improvement can be made over time. The Flexible Land Tenure Act in Namibia provided the legal framework for using flexible approaches. The success of the project was a result of intense community participation and support from local and national government. The STDM application is ideal for communities to manage their information, however, this cannot be done in isolation. Local Authority support is ideal in this process, particularly because of their roles as a local planning authority.

Formalization: The process for secure land rights used innovative and flexible approaches, mostly with residents aiming to later construct houses [58]. The participation efforts, as part of securing tenure, are effective and can bring about results that speak to the context [59]. Residents in Gobabis started off using participatory enumerations that involved mapping, participatory planning, upgradable data collection approaches in inform the land rights registration. To proceed towards formalization, community, local authority, with the support of the Namibia University of Science and Technology, embarked on planning studios that focused on situational analysis, updating of household enumeration, and environmental impact assessment. The data collected was used to inform the town planning aspects of the upgrading [60]. A layout was designed by the community (see Figure 5) with guidance from spatial planners.

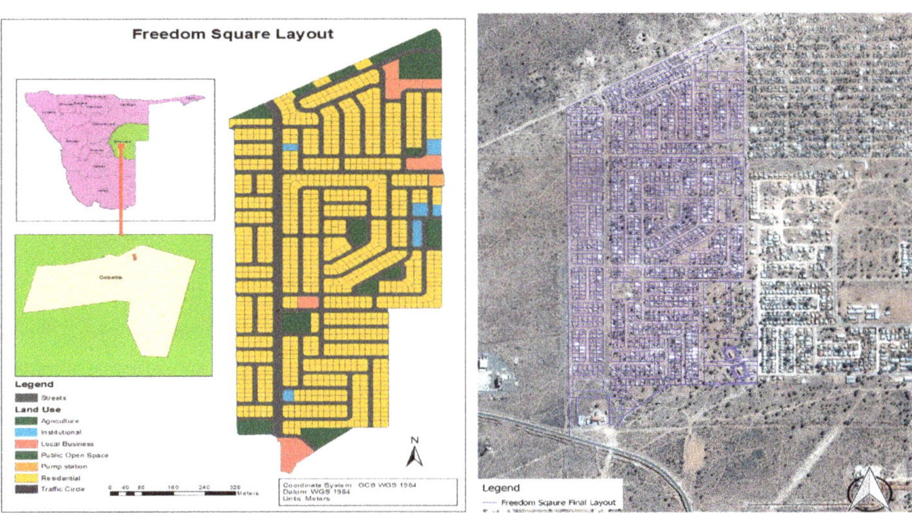

Freedom Square community layout (before) Freedom Square community layout (after)

Figure 5. Freedom Square community layout (Source: Authors').

The local authority approved the layout, and residents had to relocate within the settlement according to the layout. Public open spaces and business areas were catered for in the layout. The approved layout was overlaid on the latest imagery of the settlement and later informed the land surveying process. The survey of the settlement happened over two phases. External boundary surveys followed accuracy requirements according to the Land Survey Act 33 of 1993 [61,62]. The survey was completed by a professional land surveyor accordingly, while the internal subdivisions of the parcels followed the Flexible Land Tenure Act 4 of 2012 [47] (p. 10). The technical land surveyor applied Differential GPS (DGPS) that used the accuracy requirements according to the Land Survey Act 33 of 1993, the difference in the survey work done was the professional responsibility to complete the land survey. In the month of March 2021, some 1080 residents were presented with *land hold titles* through the FLT project. These land hold titles were certified by the local authorities after two years of verification of the data from the enumeration. Unlike in Ghana and Kenya, the land hold titles in Namibia are statutory documents. Owners of the land hold titles in informal settlements can exercise the right to develop their land and subdivide. Owners can also build permanent houses. The participatory enumerations led to improved tenure security for residents by firstly avoiding evictions and later accessing formal land documentation (i.e., land hold titles). Being a statutory title for informal settlers, the data from the enumeration is stored and managed by the Land Rights Office of the Ministry of Agriculture Water and Land Reform of Namibia.

5. Discussion—Adoption of FFP-LA Features in Practice in Ghana, Kenya, Namibia, and Beyond

A common concern in the three case studies described in this article is that they sought to secure land and property rights for people as a means for social inclusion and local economic development. A lack of access to land and insecurity of tenure epitomized a pervasive exclusion of informal settlers in FSIS (Namibia) from mainstream socioeconomic opportunities. The documentation of customary tenure to create documents in support of land tenure security in the social and paralegal environments was the key issue in the rural area of Ghana and Kenya. To address these problems, the FFP-LA guideline was adopted in both similar and differing ways to secure people's access to secure land. In Figure 6, we present a matrix indicating the critical features of the FFP-LA guideline that were adopted in the three case studies.

Adoptable FFP-LA features	Ghana (Rural)	Kenya (Peri-urban)	Namibia (Urban)
A spatial development framework (SDF)	✓ Ghana has SDF upon which all LA are based	X No SDF upon which all LA are based in Kenya	X FLTS served as spatial framework
Visible (physical) rather than fixed boundaries	✓ Natural features (trees) used as boundary marks	✓ Natural features (trees) used as boundary marks	X Fixed boundaries was used
Aerial imagery rather than field surveys	✓ Both aerial images & field surveys were used	✓ Both aerial images & field surveys were used	✓ Aerial imagery and field survey used
Accuracy of purpose rather than technical standards	✓ Tenure security as purpose. Neighbors served as witnesses	✓ Tenure security as purpose. Neighbors served as witnesses	✓ Tenure security as purpose. Community members as witnesses
Updating, upgrading and ongoing improvement	✓ Update & upgrading via land rights continuum	✓ Update & upgrading via land rights continuum	✓ Update & upgrading via land rights continuum
Legal and regulatory framework (L&R)	✓ L&R allow for statutory & customary land tenure	✓ L&R allow for statutory & customary land tenure	✓ L&R allow for social & statutory land tenure
Flexibility in administrative rather than judicial lines	~ Land administration via chiefs is flexible & rigid	~ Administrative flexibility improvised, but on-the-ground structure inflexible	✓ FLTS is designed to provide flexibility in LAS
Continuum of tenure rather than ownership	✓ Focused on ownership & subsidiary land rights	✓ Focused on ownership & subsidiary land rights	✓ Focused on ownership & subsidiary land rights
Flexible recordation rather than only one register	✓ Created a customary land register	✓ Created a land register	✓ Created social tenure register
Gender equity for land and property rights	~ Youth engagement irrespective of gender	~ Youth engagement irrespective of gender	✓ Women engagement based on inclusiveness
Building the institutional framework	✓ Emphasized both social and statutory tenure	✓ Emphasized both social and statutory tenure	✓ Emphasized both social and statutory tenure
Good land governance rather than bureaucracy	~ Bureaucracy reduced, but not avoided. chiefs follow customary rites	~ Bureaucracy reduced, but not avoided. chiefs follow customary rites	✓ Highly bureaucracy due to statutory demands
Institutional integrations rather than sectorial silos	~ Created database not linked to formal register	~ Created database not linked to formal register	✓ Created database linked to statutory register
Flexible ICT rather than high-end technology	✓ Use of standalone single frequency GPS (GNSS) receiver in smartphones	✓ Use of standalone single frequency GPS (GNSS) receiver in smartphones	✓ Use of STDM

Legend	Fully adopted ✓	Moderately adopted ~	Not adopted X

Figure 6. Matrix of case studies' adoption of the FFP-LA features (Source: Authors').

Figure 6 is a matrix showing the adoption of FFP-LA features. In the matrix, the green, blue, and yellow shades represent the qualitative extent of FFP-LA adoption in the case studies. Green represents *fully adopted*, blue entails *moderately adopted*, and yellow indicates *not adopted*. Focusing on the case study matrix, we discuss the *what* and *how* in relation to the adoption of the FFP-LA features identified by Enemark et al. [10]. These features were adopted at the local-level in the case studies in Ghana, Kenya, and Namibia, as described below.

- Building the spatial framework: Availability of a spatial development framework is crucial for ensuring a flexible non-statutory framework. The approaches used in Ghana depended on the existing SDF. Ghana has a national SDF [63], while Namibia and Kenya do not have any SDF.
- Visible (physical) rather than fixed boundaries: The Ghanaian and Kenyan case studies relied on natural landmarks as physical boundaries (e.g., fences, trees, ditches, hedges, seasonal and non-seasonal water bodies, and walls), rather than geodetically fixed boundaries. In Namibia, due to the statutory interest in the FLTS, it was essential to use fixed boundaries. External boundaries of blocks of parcels are fixed in the

Namibian case. The internal boundaries are also fixed; however, they are usually less accurate as they were done by technicians or para-surveyors.

- Aerial imagery rather than field surveys: In all cases, both aerial images were complemented with field surveys to collect data and map land parcels. Aerial images from the local authority helped in identifying and recording visible boundaries.
- Accuracy in relation to the purpose rather than technical standards: In all three cases, land tenure security was a core purpose. Pursuing technical accuracy was considered less important to achieving the purpose (i.e., tenure security).
- Updating, upgrading, and ongoing improvement: With a focus fixed on purpose rather than standards, all cases ensured that the LA activities were viewed as a continuum only efficient through regular updates and improvements, rather than a one-off process. For instance, although the FSIS in Gobabis ended in 2018, it evolved into the title to the land for some residents in early 2021. The continuum of land rights was the core principle for the enumerations and inventories. This allowed for updating and upgrading.
- Legal and regulatory framework: In all cases, there are legal and regulatory frameworks for FFP-LA. In Namibia, the FLTS is well designed for an FFP approach. In Ghana and Kenya, the land policies (including the statutory recognition of customary land tenure) provide a legal/regulatory framework for tenure security improvement as applied in these projects.
- Flexibility in administrative rather than judicial lines: The methods (e.g., land tenure enumeration and use of STDM) and the not-so-technical materials used in the land use inventories allowed for flexibility. The customary governance (in Ghana) and the commitment of the county administration (in Kenya) allowed for flexibility on the projects. The informal governance in FSIS (in Namibia) allowed for greater flexibility. The need to follow administrative rules/regulations (rather than legal institutions), as in the case of Namibia, is crucial because it helped to lay a foundation for social legitimacy in implementing FFP-LA. And this is because the participatory enumerations took place before the FLTS regulations were passed. Apart from the ease or flexibility of implementing the land-based activities, it also allowed the communities to own the process and be accountable for its success or failures. Administrative practice allows for social legitimacy or formalities, while legal frameworks help to ensure adherence to legislation.
- Continuum of tenure rather than ownership: In Namibia, the FLTS is a manifestation of Namibia's adherence to the continuum principle in land rights. In all cases, the approaches recognized the existence of a possibility of various tenure arrangements within the project areas. Property rights were envisaged beyond ownership.
- Flexible recordation rather than only one register: The three case studies involved flexible and pro-poor land rights documentation based on partnerships with government and local people (the youths in Ghana and Kenya, and women in Namibia). It also involved the use of easily accessible and usable land recordation systems (or equipment) capable of serving local purposes and being operated by local people.
- Gender equity for land and property rights: Focus on participation was put on the equal participation of men and women in Ghana and Kenya. This allowed for learning and equal capacity building for both men and women. In Namibia, the FSIS used enumerators who identified as female. This allowed for the inclusion of female voices, which may otherwise have been suppressed in the Namibian social context. The STDM was also used to capture and store gender (male and female only) disaggregated data for gender-responsive decisions in the informal settlements.
- Building the institutional framework: The three cases led to the development and the testing of new rules of tenure application (i.e., establishment of new land information systems for local LAS applications). They emphasized both social and statutory tenure and procedures for future tenure improvement processes by setting a platform for upgrading and updating the new LAS.

- Good land governance rather than bureaucracy: All three cases are based on appropriate governance, which is not perfectly *good* in the sense of *good land governance*. In the rural and peri-urban areas of Ghana and Kenya, respectively, reliance on chiefs and local authorities is still bureaucratic. In the informal settlements (Namibia), dependence on the FLTS was highly bureaucratic, due to the statutory requirements for issuing land hold titles.
- Institutional integrations rather than sectorial silos: The Kenyan and Ghanaian case studies involved the potential sector integration between customary and statutory tenure. The Namibian case is a complete integration of the formal and informal urban land sectors.
- Flexible ICT rather than high-end technology: In all cases, the governance approach and equipment used to allow for open and transparent access to land information.

Considering how these FFP-LA procedures took place in the three case studies, this study has broader implications elsewhere in other sub-Saharan countries and the global south. Lessons from these case studies indicate that the FFP-LA is not limitable by geography. Its purpose, flexibility of its engagement, and pro-poor expectations from its outcomes are the core determinants of the feasibility or viability of its adoption anywhere. This means that the different (yet highly similar) procedures followed in the three cases (Ghana, Kenya, and Namibia) can also be in other countries, such as India, Indonesia, or Fiji, if their context is similar. Such similarity may relate to situations where the purpose, procedure, and expected outcome are relatable.

6. Conclusions

These case studies demonstrate that adopting FFP approaches in local LAS is possible in different spatial contexts. They show that FFP-LA supports pro-poor land interventions. The cases support the understanding that securing tenure does not require accurate surveys [25] (p. 19). They also indicate that FFP-LA improves tenure security (important measures for achieving SDGs 1, 2, 5, 11, and 15) which are of crucial importance to Africa's development [35]. Cognizance of the technical nature of surveying, this article demonstrates that using a mobile application (that operates on a single frequency GPS) by trained local people (especially youths and women) enables mapping exercises that minimize cost and still enhance the land tenure security of people. All three case studies demonstrate the paradigm shift in LA which emerged because of the FFP-LA guideline and practice. Alignment towards pro-poor approaches as a means of securing land rights within a continuum concept, especially in rural (Ghana), peri-urban (Kenya), and informal urban settlements (Namibia) in Africa, is demonstrated. The study represented three simple, locally-adopted, and innovative approaches that focused on local experiences and allowed the local people to play critical roles in the process.

The application of the guideline of FFP-LA in Namibia provides an opportunity about how to secure tenure in informal urban settlements at scale and ensure improved living conditions for occupants in areas formerly under tenure insecurity. The case studies of Ghana and Kenya, though not at scale, reflect an example of how the FFP-LA can be adopted in rural and peri-urban areas, respectively. Securing land rights for informal settlement residents in Juaben-Atia (Ghana) and Taita Hills (Kenya) and FSIS in Gobabis (Namibia) relied on features of FFP-LA. Each aspect of the FFP implementation involved flexible, inclusive, participatory, affordable, reliable, attainable, and upgradable measures that fit in the local legal, spatial, and institutional frameworks [22] (p. 17). Moreover, the respective communities led the process to secure the tenure through land rights documentation and titling for over 1000 households in Gobabis Namibia. The Gobabis case study is noteworthy because its tenure documentation has evolved beyond the local (alternative documentation) into statutory land titles. In March 2021, households in FSIS became beneficiaries of land holding titles, under the Flexible Land Tenure Act 4 of 2012) [46]. This final recognition of land rights took more than five years to materialize. However, unlike the cases of Juaben-Atia (Ghana) and Taita Hills (Kenya), it demonstrates that FFP-LA techniques can do more than provide paralegal protection. It can lead to full tenure security if the processes are well managed, and participatory approaches are center stage.

It is essential to note (or rather reiterate) that this study is a demonstrative one. It has been conceived to show that FFP-LA as a concept can be applied in different forms, depending on where and how it is implemented. It confirms that FFP-LA outcomes should be done within an enabling environment that respects "administrative structures, policy frameworks, institutional and social settings, collective actions, and social learning" [64] (p. 12). To what degree it is adopted based on established principles. While the demonstrative case studies (i.e., evidence presented with a focus on procedures) are crucial for grasping how to adopt and implement FFP-LA in African countries, there is still a need to engage in research to provide substantive evidence to help land administrators resolve LA challenges relate to the performance of specific steps in the FFP-LA. This will enable land administrators to understand the impact and consequences of each of the steps (and principles of the FFP-LA).

Author Contributions: Conceptualization, U.E.C.; methodology, U.E.C., M.R.M., T.B., E.D.K., D.O.T.; validation U.E.C., M.R.M., T.B., E.D.K., D.O.T.; formal analysis, U.E.C., M.R.M., T.B., E.D.K., D.O.T.; investigation, U.E.C., M.R.M., T.B., E.D.K., D.O.T.; resources, U.E.C., M.R.M., T.B., E.D.K., D.O.T.; data curation, U.E.C., M.R.M., T.B., D.O.T.; writing—original draft preparation, U.E.C., M.R.M., T.B., E.D.K., D.O.T.; writing—review and editing, U.E.C., M.R.M., T.B., E.D.K., D.O.T.; visualization, U.E.C., M.R.M., T.B.; supervision, U.E.C. All authors have read and agreed to the published version of the manuscript.

Funding: This research received no external funding. The APC was funded by the German Research Foundation (DFG) and the Technical University of Munich (TUM) in the framework of the Open Access Publishing Program.

Institutional Review Board Statement: Not applicable.

Informed Consent Statement: Not applicable.

Data Availability Statement: Not applicable.

Acknowledgments: Acknowledgment is given to the Global Land Tool Network (GLTN), especially its Research and training cluster, for facilitating the study. The GLTN supported the Ghana and Namibia case studies. The DAAD (German Academic Exchange Service) supported the Kenyan case study. The article was prepared under the auspices of a research collaboration between the Namibia University of Science and Technology (Namibia), SD Dombo University of Business and Integrated Development Studies, Wa (Ghana), Technical University of Munich (Germany), and the Ghana Lands Commission.

Conflicts of Interest: The authors declare no conflict of interest.

References

1. Bendzko, T.; Chigbu, U.E.; Schopf, A.; de Vries, W.T. Consequences of land tenure on biodiversity in arabuko sokoke forest reserve in Kenya: Towards responsible land management outcomes. In *Agriculture and Ecosystem Resilience in Sub Saharan Africa;* Bamutaze, Y., Kyamanywa, S., Singh, B., Nabanoga, G., Lal, R., Eds.; Springer: Cham, Switzerland, 2019.
2. Chigbu, U.E. Masculinity, men and patriarchal issues aside: How do women's actions impede women's access to land? Matters arising from a peri-rural community in Nigeria. *Land Use Policy* **2019**, *81*, 39–48. [CrossRef]
3. Toulmin, C. Securing land and property rights in sub-Saharan Africa: The role of local institutions. *Land Use Policy* **2009**, *26*, 10–19. [CrossRef]
4. Gwaleba, J.M.; Chigbu, U.E. Participation in property formation: Insights from land-use planning in an informal urban set-tlement in Tanzania. *Land Use Policy* **2020**, *92*, 104482. [CrossRef]
5. Mas, J.-F. Monitoring land-cover changes: A comparison of change detection techniques. *Int. J. Remote Sens.* **1999**, *20*, 139–152. [CrossRef]
6. Owusu Ansah, B.; Chigbu, U.E. The nexus between peri-urban transformation and customary land rightsdisputes: Effects on peri-urban development in Trede, Ghana. *Land* **2020**, *9*, 187. [CrossRef]
7. United Nations Human Settlement Programme. *Secure Land Rights for All;* GLTN/UN-Habitat: Nairobi, Kenya, 2008.
8. Handayani, W.; Chigbu, U.E.; Rudiarto, I.; Putri, I.H.S. Urbanization and Increasing Flood Risk in the Northern Coast of Central Java—Indonesia: An Assessment towards Better Land Use Policy and Flood Management. *Land* **2020**, *9*, 343. [CrossRef]

9. Chigbu, U.E. Land, Women, Youths, and Land Tools or Methods: Emerging Lessons for Governance and Policy. *Land* **2020**, *9*, 507. [CrossRef]
10. Enemark, S.; McLaren, R.; Lemmen, C. *Fit-for-Purpose Land Administration Guiding Principles for Country Implementation*; UN Habitat: Nairobi, Kenya, 2016.
11. Gaafar, R. *Women's Land and Property Rights in Kenya. Center for Women's Land Rights*; Centre for Women's Land Rights-Landesa: Seattle, WA, USA, 2014.
12. Dachaga, W.; Chigbu, U.E. Understanding tenure security dynamics in resettlement towns: Evidence from the Bui Resettlement Project in Ghana. *J. Plan. Land Manag.* **2020**, *1*, 38–49. [CrossRef]
13. Catioiti, A. *Land Issue, a Challenge to the Social Mission of the Church: A Theological Socio-Cultural Study on Kenya Land Tenure*; Tangaza University College: Nairobi, Kenya, 1999.
14. Ntihinyurwa, P.D.; de Vries, W.T.; Chigbu, U.E.; Dukwiyimpuhwe, P.A. The positive impacts of farm land fragmentation in Rwanda. *Land Use Policy* **2019**, *81*, 565–581. [CrossRef]
15. de Vries, W.; Chigbu, U.E. Responsible Land Management–Concept and application in a territorial rural context. *Fub–Flächenmanag. Bodenordn.* **2017**, *79*, 65–73.
16. Kidido, J.K.; Bugri, J.T.; Kasanga, R.K. Dynamics of Youth Access to Agricultural Land under the Customary Tenure Regime in the Techiman Traditional Area of Ghana. *Land Use Policy* **2017**, *60*, 254–266. [CrossRef]
17. Mkuzi, T.H. Assessment of Land Tenure, Land Use and Land Cover Changes in Taita Hills Forest Fragments: A Case Study of Ngerenyi Forest Fragments in Taita Taveta County, Kenya. Master's Thesis, Pwani University, Kilifi, Kenya, 2020.
18. Ubink, J.M.; Hoekema, A.J.; Assies, W.J. *Legalising Land Rights: Local Practices, State Responses and Tenure Security in Africa, Asia and Latin America*; Leiden University Press: Leiden, The Netherlands, 2009; p. 618.
19. Chigbu, U.E.; Schopf, A.; De Vries, W.T.; Masum, F.; Mabikke, S.; Antonio, D.; Espinoza, J. Combining land-use planning and tenure security: A tenure responsive land-use planning approach for developing countries. *J. Environ. Plan. Manag.* **2016**, *60*, 1622–1639. [CrossRef]
20. Quan, J. Lessons learnt in land policy and land administration. Presented at the World Bank Electronic Conference, Online.
21. Zevenbergen, J.; Augustinus, C.; Antonio, D.; Bennett, R. Pro-poor land administration: Principles for recording the land rights of the underrepresented. *Land Use Policy* **2013**, *31*, 595–604. [CrossRef]
22. Enemark, S.; Hvingel, L.; Galland, D. Land administration, planning and human rights. *Plan. Theory* **2014**, *13*, 331–348. [CrossRef]
23. Enemark, S. Fit-for purpose Land Administration. *GIM Int.* **2013**, *27*, 26–29.
24. Enemark, S.; Bell, K.C.; Lemmen, C.H.; McLaren, R. *Fit-for-Purpose Land Administration*; International Federation of Surveyors (FIG): Copenhagen, Denmark, 2014.
25. Ameyaw, P.D.; Dachaga, W.; Chigbu, U.E.; De Vries, W.T.; Abedi, L. Responsible Land Management: The Basis for Evaluating Customary Land Management in Dormaa Ahenkro, in Ghana. In Proceedings of the World Bank Conference on Land and Poverty, Washington, DC, USA, 19–23 March 2018; pp. 1–18.
26. Antonio, D.; Mabikke, S.; Selebalo, C.; Chigbu, U.E.; Espinoza, J. Securing Tenure through Responsive Land Use Planning: An Innovative Tool for Country Level Interventions. In Proceedings of the FIG Working Week Christchurch, Christchurch, New Zealand, 2–6 May 2016.
27. Masum, F.; Chigbu, U.E.; Espinoza, J.; Graefen, C. The limitations of formal land delivery system: Need for a pro-poor urban land development policy in dhaka, bangladesh. In Proceedings of the World Bank Conference on Land and Poverty, Washington, DC, USA, 14–18 March 2016.
28. Leider, R.J. *The Power of Purpose: Creating Meaning in Your Life and Work*; Berrett-Koehler: San Francisco, CA, USA, 2004.
29. Bowie, W. Present status of geodesy and some of the problems of this branch of geophysics. In Proceedings of the First Annual Meeting of the American Geophysical Union, Washington, DC, USA, 23 April 1920.
30. Allred, G.K. The surveying profession. In Proceedings of the 6th FIG Regional Conference, Coastal Areas and Land Administration—Building the Capacity, San José, Costa Rica, 12–15 November 2007.
31. Allred, G.K. The Land Surveyor as a Public Officer. In Proceedings of the XX Congress of the Fédération Internationale des Géomètres, Surveying global changes, Melbourne, Australia, 5–12 March 1994.
32. Chigbu, U.E. Land tenure security in Nigeria. In *Land Tenure Security in Selected Countries: Global Report*; Kirk, M., Mabikke, S., Antonio, D., Chigbu, U.E., Espinoza, J., Eds.; UN-Habitat: Nairobi, Kenya, 2015; pp. 62–86.
33. Minchin, M. *Introduction to Surveying. Government of Western Australia*; Department of Training and Workforce Development: West Perth, Australia, 2016.
34. Nerlove, S.H. "Valuation of Property": A Review. *Univ. Chic. Law Rev.* **1939**, *6*, 157. [CrossRef]
35. United Nations. *The Sustainable Development Goals Report*; United Nations Organisation: New York, NY, USA, 2016.
36. Chigbu, U.E.; Ntihinyurwa, P.D.; de Vries, W.T.; Ngenzi, E.I. Why tenure responsive land-use planning matters: Insights for land use consolidation for food security in Rwanda. *Int. J. Environ. Res. Public Health* **2019**, *16*, 1354. [CrossRef]
37. Zevenbergen, J.; De Vries, W.T.; Bennett, R. (Eds.) *Advances in Responsible Land Administration*; CRC Press: New York, NY, USA, 2016; ISBN 978-4987-1959-9.
38. United Nations Economic Commission for Europe–UNECE. *Land Administration Guidelines with Reference to Countries in Transition*; United Nations: New York, NY, USA, 1996.

39. UN-Habitat; Global Land Tool Network (GLTN); International Institute for Rural Reconstruction (IIRR). *Handling Land, Innovative Tools for Land Governance and Secure Tenure*; UNON: Nairobi, Kenya, 2012.
40. Augustinus, C. Catalysing global and local social change in the land sector through technical innovation by the United Nations and the Global Land Tool Network. *Land Use Policy* **2020**, *99*, 105073. [CrossRef]
41. Vinge, H. Farmland conversion to fight climate change? Resource hierarchies, discursive power and ulterior motives in land use politics. *J. Rural. Stud.* **2018**, *64*, 20–27. [CrossRef]
42. De Vries, W.T. The influence and relevance of social values and believe systems in land readjustment processes. In Proceedings of the International Conference on Real Estate Development and Management (ICREDM 2020), Ankara, Turkey, 30 January–2 February 2020.
43. Osei Tutu, D.; Asante, L.A.; Appiah, M.N.; Bendzko, T.; Chigbu, U.E. Towards a pro-poor customary land rights security in rural Ghana: Land tenure inventory using mobile application by local youth. In Proceedings of the World Bank Conference on Land and Poverty, Washington, DC, USA, 14–18 March 2016.
44. Bendzko, T.; Ameyaw, P.D.; Chigbu, U.E.; de Vries, W.T.; Osei Tutu, D. Embracing the rubber-boot approach to securing customary land rights with focus on low-cost land-use inventory. In Proceedings of the World Bank Conference on Land and Poverty, Washington, DC, USA, 20–24 March 2017.
45. Namibia Statistics Agency (NSA). *Namibia 2011 Population & Housing Census-Main Report*; NSA: Windhoek, Namibia, 2013.
46. Payne, G.; Durand-Lasserve, A.; Geoffrey Payne & Associates. Holding on: Security of Tenure-Types, Policies, Practices and Challenges. Research Paper Prepared for the Expert Group Meeting on Security of Tenure Convened by the Special Rapporteur. 2012, pp. 22–23. Available online: http://www.ohchr.org/Documents/Issues/Housing/SecurityTenure/Payne-Durand-Lasserve-BackgroundPaper-JAN2013.pdf (accessed on 19 March 2021).
47. Republic of Namibia. Flexible Land Tenure Act of 2012. *Government Gazette*, 13 June 2012.
48. Global Land Tool Network. *Count Me in: Surveying for Tenure Security and Urban Land Management*; GLTN/UN-Habitat: Nairobi, Kenya, 2010.
49. Weber, B.; Mendelsohn, J. *Informal Settlements in Namibia: Their Nature And Growth*; Development Workshop Namibia: Namibia, Africa, 2017. Available online: https://www.researchgate.net/publication/322200418_Informal_settlements_in_Namibia_their_nature_and_growth (accessed on 14 March 2021).
50. Lankhorst, M. Land Tenure Reform and tenure Secuirty in Namibia. In *Legalising Land Rights Local Practices, State Responses and Tenure Security in Africa, Asia and Latin America*; Ubink, J.M., Hoekema, A.J., Assies, W.J., Eds.; Leiden University Press: Leiden, The Netherlands, 2000; pp. 1–21.
51. Chigbu, U.E. The Quest for "Good Governance" of Urban Land in Sub-Saharan Africa: Insight into Windhoek, Namibia. In *Land Issues for Urban Governance in Sub-Saharan Africa*; Home, R., Ed.; Springer: Cham, Switzerland, 2021.
52. Muller, A.; Mbanga, E. Participatory enumerations at the national level in Namibia: The Community Land Information Programme (CLIP). *Environ. Urban.* **2012**, *24*, 67–75. [CrossRef]
53. Shack Dwellers Federation of Namibia (SDFN); Namibia Housing Action Group (NHAG). *Participatory Planning for Informal Settlement Upgrading in Freedom Square, Gobabis*; NHAG: Windhoek, Namibia, 2014. Available online: http://sdfn.weebly.com/uploads/2/0/9/0/20903024/freedom_square_report_clip2.pdf (accessed on 12 March 2021).
54. Republic of Namibia. *Flexible Land Tenure Regulation*; Government of Namibia: Windhoek, Namibia, 2018. Available online: http://cms.my.na/assets/documents/p19dmn58guram30ttun89rdrp1.pdf (accessed on 14 March 2021).
55. Muller, A.; Mabakeng, R.; Gitau, J.; Selebalo, C. Experiences in Developing Business Process for Flexible Land Tenure Act Implementation in Gobabis Namibia. Available online: https://landportal.org/node/35829 (accessed on 17 March 2021).
56. Barnes, G.; Volkmann, W.; Muller, A. Drones in Support of Upgrading Informal Settlements: Potential Application in Namibia. Coordinates, November 2015. Available online: http://mycoordinates.org/drones?in?support?of?upgrading?informal?settlements?potential?application?in?namibia/ (accessed on 12 March 2021).
57. Intergovernmental Panel on Climate Change (IPCC). Summary for Policymakers. In *Climate Change 2013: The Physical Science Basis. Contribution of Working Group I to the Fifth Assessment Report of the Intergovernmental Panel on Climate Change*; Stocker, T.F., Qin, D., Plattner, G.-K., Tignor, M., Allen, S.K., Boschung, J., Nauels, A., Xia, Y., Bex, V., Midgley, P.M., Eds.; Cambridge University Press: Cambridge, UK; New York, NY, USA, 2013; pp. 1–30.
58. Delgado, G. *Land and Housing Practices in Namibia: Cases of Access to Land Rights and Production of Housing in Windhoek, Oshakati and Gobabis*; University of Cape Town: Cape Town, South Africa, 2019. Available online: http://hdl.handle.net/11427/31363 (accessed on 17 March 2021).
59. Delgado, G.; Muller, A.; Mabakeng, R.; Namupala, M. Co-producing land for housing through informal settlement upgrading: Lessons from a Namibian municipality. *Environ. Urban.* **2020**, *32*, 175–194. [CrossRef]
60. Namibia Housing Action Group (NHAG); Shack Dwellers Federation of Namibia (SDFN). *July 2019–June 2020 Annual Report*; Namibia Housing Action Group: Windhoek, Namibia, 2020.
61. Republic of Namibia. *Land Survey Act 33 of 1993*; Government of Namibia: Windhoek, Namibia, 1993.
62. Republic of Namibia. *Regulations of the Land Survey Act 33 of 1993*; Pububulic Legislation: Windhoek, Namibia, 2002; Volume 58, p. 1.

63. Government of Ghana. *Ghana National Spatial Development Framework 2015-2035-Space, Efficiency and Growth. Final Report;* Government of Ghana: Accra, Ghana, 2015; Volume I.
64. Auziņš, A.; Chigbu, U.E. Values-led Planning Approach in Spatial Development: A Methodology. *Land* **2021**, *10*, 461. [CrossRef]

Article

Transforming Land Administration Practices through the Application of Fit-For-Purpose Technologies: Country Case Studies in Africa

Danilo Antonio, Solomon Njogu *, Hellen Nyamweru and John Gitau

UN-Habitat/Global Land Tool Network, P.O. Box 30030, Nairobi 00100, Kenya; danilo.antonio@un.org (D.A.); hellen-nyamweru.ndungu@un.org (H.N.); john.gitau@un.org (J.G.)
* Correspondence: solomon.njogu1@un.org

Abstract: Access to land for many people in Africa is insecure and continues to pose risks to poverty, hunger, forced evictions, and social conflicts. The delivery of land tenure in many cases has not been adequately addressed. Fit-for-purpose spatial frameworks need to be adapted to the context of a country based on simple, affordable, and incremental solutions toward addressing these challenges. This paper looked at three case studies on the use of the Social Tenure Domain Model (STDM) tool in promoting the development of a fit-for-purpose land administration spatial framework. Data gathering from primary and secondary sources was used to investigate the case studies. The empirical findings indicated that the use and application of the STDM in support of the fit-for-purpose land administration framework is quite effective and can facilitate the improvement in land tenure security. The findings also revealed that the tool, together with participatory and inclusive processes, has the potential to contribute to other frameworks of Fit-For-Purpose Land Administration (FFP LA) toward influencing changes in policy and institutional practices. Evidently, there was a remarkable improvement in the institutional arrangements and collaboration among different institutions, as well as a notable reduction in land conflicts or disputes in all three case studies.

Keywords: fit-for-purpose land administration; spatial framework; tenure security; STDM; technology

Citation: Antonio, D.; Njogu, S.; Nyamweru, H.; Gitau, J. Transforming Land Administration Practices through the Application of Fit-For-Purpose Technologies: Country Case Studies in Africa. *Land* **2021**, *10*, 538. https://doi.org/10.3390/land10050538

Academic Editors: Stig Enemark, Robin McLaren and Christiaan Lemmen

Received: 29 March 2021
Accepted: 15 May 2021
Published: 19 May 2021

1. Introduction

Access to land for the majority of people, particularly in developing countries, is insecure and continues to pose risks to poverty, hunger, forced evictions, and social conflicts. The delivery of land tenure in many cases has not been adequately addressed, due to weak legal systems, complex procedures, and poor institutional arrangements [1,2]. Notably, in Sub-Saharan Africa, customary and informal land tenure systems dominate statutory forms of tenure and are generally not recorded in the official registry [3]. The growing competition for land is threatening customary land due to the population increase and pressure from large-scale land-based investment projects [4]. The risk of evictions and land grabbing is even higher with the increasing speculation of land that threatens the access and use of land among the poor and vulnerable groups, particularly women. Similarly, due to the high rate of urbanization, much of the growth is taking place within informal settlements linked to the challenge of affordability [5]. Most land administration systems have set aside the registration of informal, customary, and complex land rights due to ambiguities in the legal, socio-cultural, and technical standards, among others. Hence, access to land administration services for the poor is thus limited by the cost, capacity, complexity of procedures, and rigid technical requirements, depriving them the opportunity to access land that would otherwise enable them to invest toward improving incomes and livelihoods [6].

The benefit of well-defined, documented, and enforceable land rights cannot be underestimated. With the global commitment enshrined in the Sustainable Development Goals (SDGs), 'tenure security for all' is emphasized, especially for the poor and marginalized groups, including women. The United Nations General Assembly through New Urban Agenda (NUA) commitment number 35 promotes secure tenure for all and the recognition of the continuum of land rights approach through fit-for-purpose solutions with a particular emphasis on women [7]. Similarly, Agenda 2063 of the African Union (AU Goal 17) aims to give 20% of rural women the access to and control of land by 2023 [8]. The African Union, while resolving the commitment to poverty eradication, agreed to ensure equitable access to land and related resources among all land users, including the youth and other landless and vulnerable groups such as displaced persons [9]. The Voluntary Guidelines on the Responsible Governance of Tenure of Land, Fisheries and Forests in the Context of National Food Security (VGGT) also recommend the application of pro-poor and gender-responsive land policies to achieve equitable development [10].

1.1. Innovative Approaches for Addressing Tenure Insecurity

Currently, it is estimated that less than 30% of land parcels are formally registered [11]. To address the land registration gap, locally suited methodologies based on the continuum of land rights approach are being promoted as alternative, effective, and scalable solutions in favor of conventional practices in land administration [12]. The principles of the fit-for-purpose land administration (FFP LA) approach complement these methodologies as building blocks in the way we record and perceive tenure security [13]. The FFP LA approach recognizes the need to be flexible in the spatial, institutional, and legal frameworks without being limited by stringent standards of technical accuracy and legal requirements [14]. Their application leverages sustainable practices in land administration that are the foundation for good land governance [15].

Evidently, it has been accepted that relying on a single form of tenure to meet the demands of different social groups will be cumbersome [16]. This follows the argument of Payne regarding the range of actions, steps, and decisions on land that can be interpreted as having meaning to land tenure security [17]. Essentially, the analogy of tenure security can be viewed in many forms even in social claims, and the assumption on this study alludes to those interpretations along the continuum of land rights approach. The Continuum of Land Rights advances this further by recognizing that land rights are complex or overlap and can be assumed to be lying on a continuum often able to transform at different times [18]. The Continuum of Land Rights advocates for the incremental approach in recording and recognizing diverse forms of tenure that may be statutory or social claims at the time of recording. The new Urban Agenda commitments 35 promotes this approach in delivering security of tenure for all, recognizing the plurality of tenure types, and developing fit-for-purpose solutions to suit the context.

The FFP LA principles reinforce the application of the Continuum of Land Rights approach through the use of appropriate technology such as the STDM [19]. The FFP LA spatial framework requires that land recording is aligned to the local requirements based on a simple, affordable, and incremental approach [20]. The International Federation of Surveyors (FIG) considers this approach as 'fit for purpose,' as it is based on locally available resources and capacities [13]. The use of technology coupled with freely available geospatial data enables the creation of a suitable infrastructure for the coordinated production, access, and use of geospatial data among producers and users in an electronic environment [21].

In this paper, the focus is on the use of the STDM tool in promoting the application of FFP LA and the Continuum of Land Rights approach, particularly in the development of the FFP LA spatial framework. The STDM tool has evolved from conceptualization to implementation in the last 15 years. To date, the STDM tool has been implemented (and is being implemented) in at least 15 countries in various contexts such as informal settlements [22,23], customary [24], post-conflict, and formal land registration [25]. Various studies in the literature have pointed out the need for the STDM tool, particularly toward supporting the continuum of land rights approach and in improving the land tenure security of poor people [26,27]. This study sought to demonstrate, through empirical findings, that the STDM tool, with the participatory and inclusive approaches, facilitates tenure security improvement and other development outcomes. The paper also showed how the STDM tool, which is supporting the development of a spatial framework, is able to influence the incremental but transformative change in the policy and institutional aspects of FFP LA.

1.2. The Social Tenure Domain Model

The STDM is a concept and model that has been developed into an information tool to support FFP LA approaches and to overcome the barriers of conventional land administration systems [28,29]. The STDM model conforms to the global standards of the Land Administration Domain Model (LADM) and is meant to provide a flexible customization framework to suit local applications in different contexts [30]. Thus, the conceptual design of the STDM is robust and flexible to support FFP LA in the recording of both formal and informal tenure rights. The motivation has been to break the norm from the core cadastral thinking of parcels toward a range of spatial units and a broad view of tenure types, as perceived in the continuum of land rights. The development of STDM as a tool has been tested and accepted internationally for being practical, fast, and affordable in facilitating land tenure security especially for the poor and marginalized groups [31]. The conceptual model of the STDM is shown in Figure 1 below.

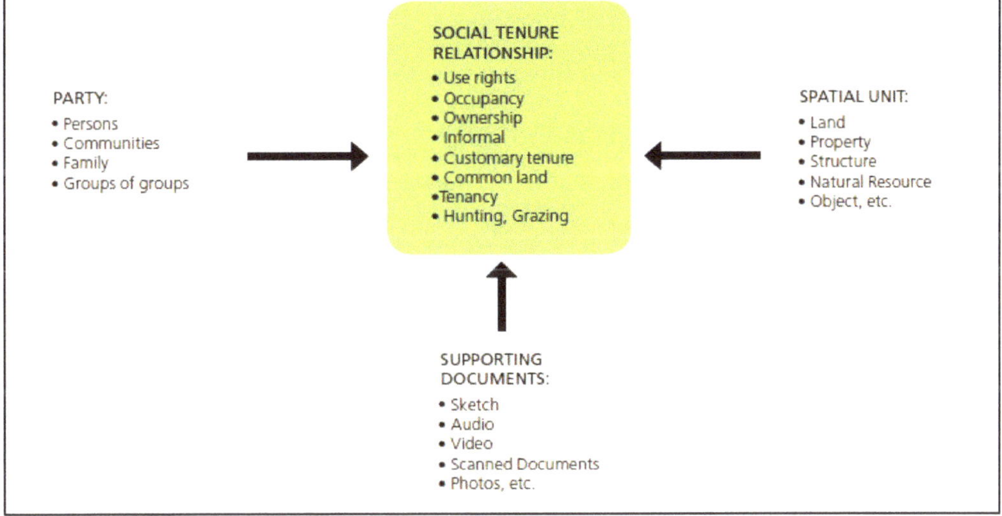

Figure 1. The conceptual model for the Social Tenure Domain Model (STDM).

The technical development of the STDM tool is being led by the Global Land Tool Network partners and UN-Habitat, and it is based on a popular and stable open-source QGIS and PostgreSQL/PostGIS database application. The authors in this paper have been involved in the design, testing, and implementation of the tool at the country level. Different formats of spatial and nonspatial data are supported using common exchange formats supported in many GIS and database applications. This also means that the STDM tool can be integrated with other applications and even formal land information systems through a common interface. The source code has been made available for wide access by developers to enhance the features and functionality. A stable version of the tool is maintained by UN-Habitat and the GLTN Secretariat, who also supports in the customization of the tool to suit different application contexts. The tool can also be used at various levels within an organization from a simple geo-information system to a complete land information system with a centralized data repository utilizing client–server architecture to serve multiple users in different departments in an organization.

Moreover, the design of the STDM tool is oriented toward a pro-poor approach to support applications even with limited resources and technical skills. The data collection methodologies rely on the use of handheld Global Positioning System (GPS) receivers, satellite imagers, and smartphone applications, among others. Thus, the tool is promoting the use of low-cost solutions particularly in the development of the FFP LA spatial framework. Essentially, the tool aims to complement the existing land administration system in addressing the land registration gap [29]. The notable impact has improved the perception of tenure security and the easy access to information for planning and decision making [28].

Hence, the STDM contributes to the development of the FFP LA spatial framework through its flexible recording of data, participatory approaches, and application of low-cost technologies. This research sought to investigate, through empirical evidence, the impact of the STDM tool on improving land tenure security along the Continuum of Land Rights approach and its contributions to changes in policies and institutional arrangements. To test these impacts, this study hypothesized that the STDM tool facilitates tenure security improvement, ultimately impacting policy implementation and changes in institutional practices. The arguments in the discussion reflect on the case studies' observations and primary data collection to test the hypothesis.

2. Materials and Methods

The primary objectives of this study were to determine whether the application of the STDM tool facilitates the improvement of land tenure security along the Continuum of Land Rights Approach and whether it contributes to influencing positive changes in policy and institutional arrangements. To evaluate these objectives, the research work developed these specific research questions: Whether the STDM tool facilitates land tenure security improvement, and if so, how? Whether the STDM application contributes to positive changes in policy and institutional arrangements, and how? Other research questions were open-ended and Likert-scale questions were meant to further understand the emerging outcomes and to reinforce the above questions.

To respond to the research questions, a combination of qualitative and quantitative research methodology was used. Three case studies were considered where the STDM tool was implemented and/or is being implemented. The authors are familiar with them and have played a substantive role in the development and implementation of the STDM tool, which contributed to the design and conduct of the research and the subsequent analysis of the findings. A questionnaire was developed and administered through primary data collection to gather quantitative data. A purposeful sampling technique was considered to target respondents that played a specific role in the case studies, and the selection was guided by project partners in each country. The questionnaire was administered using an online survey tool called Survey Monkey (https://www.surveymonkey.com, accessed on 30 March 2021). Detailed information about the respondents' role is presented in Section 2.2. The observations and results from the primary data collection were then presented in the spreadsheet format, they were analyzed, and reflections were offered to particularly respond to the research questions. The study also used secondary data and the literature review technique to validate the primary data. The analysis of secondary data assessed the case study reports, papers, articles testimonies, and video documentation evidence to present key achievements and results.

2.1. Case Studies

2.1.1. Introduction

The case studies section presents the projects in brief, the application context, the processes involved, and the key achievements in each country. The use of the STDM tool was central to the project and the research attempts to explore its contribution and impacts in relation to the research questions. The three case studies were from Sub-Saharan Africa, and the processes were driven by local communities and were supported by civil society organizations (CSOs) and local authorities in each country. The Figure 2 below shows maps of the three countries and the specific location where STDM was/ is being implemented.

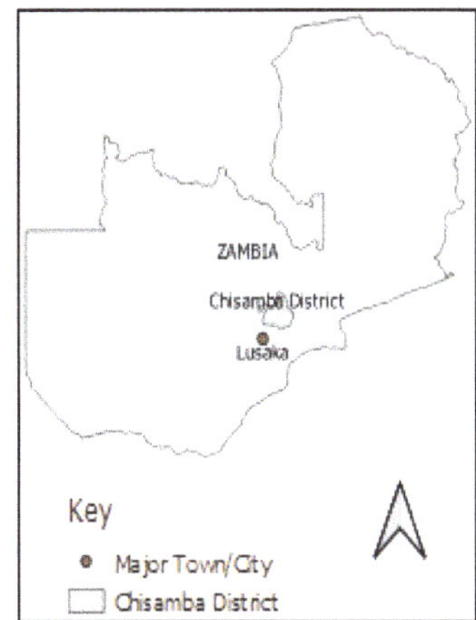

Figure 2. STDM case studies area in the three countries.

The initial activities involved planning and consultations with key implementing partners at the national and municipal levels to introduce the STDM tool and its objectives in the early stages of the project. A series of sensitization meetings with target communities and broader stakeholders, including joint need assessment missions, were conducted at the local level [23,31]. The concepts of the STDM tool were presented at the very start of projects, including the FFP LA principles. An enumeration questionnaire that captured access to land questions was developed and administered at the individual and household levels. The enumeration questionnaire also served as the tool for collecting specific information that was deemed relevant to the community, such as access to services, household information, and community priorities. The STDM tool was then customized to fit the local context. Communities were then trained to use satellite imagery and handheld GPS devices to demarcate settlement boundaries and land parcels (spatial units) in each context.

The data from the enumeration and participatory mapping processes were encoded and added into the STDM database. However, it should be noted that while this is a 'generic' procedure, the specific processes were always determined in consultations with partners and local communities. The data were handed over to local authorities for quality assurance and validation. Activities to support the basic maintenance of the database through the regular backup of data were performed by selected community members on a voluntary basis. UN-Habitat/GLTN also provided support through the provision of updated software whenever a new version of the STDM tool became available.

2.1.2. Uganda

Context and Interventions

Uganda is rapidly urbanizing. With an urban population growth rate of approximately 5.9% per annum as of 2019 and with close to 25% of the total population living in urban areas (approximately 11 million), urbanization in Uganda is set to further increase by over 30% by the year 2030 [32–34]. According to UN-Habitat, 48.3% of Uganda's population was living in informal settlements in 2018, a situation caused by the inadequate capacities of urban areas to absorb new populations who, in turn, found their own informal ways to establish themselves in the city [35]. The project 'Transforming Settlements of the Urban Poor in Uganda' (TSUPU) was launched by the Government of Uganda in 2011, through the Ministry of Lands, Housing and Urban Development (MLHUD) in partnership with the Cities Alliance, aiming to address urbanization challenges in cities and towns. Among the projects provided for under the TSUPU project was the piloting of the Social Tenure Domain Model (STDM) tool and, later, its implementation and scaling up. Bufumbo and Mission cells in Namakwekwe Ward, covering approximately 28.26 hectares in the Northern Division within Mbale City (then a municipality), were selected as pilot zones based on the accessibility and proximity to Mbale town [23]. Local project partners were ACTogether Uganda and the grassroots organization, National Slum Dwellers Federation of Uganda (NSDFU), affiliated with Slum Dwellers International (SDI), an international CSO. The capacities of local youth and selected village representatives were enhanced on the use of the STDM tool and technical know-how on data analysis through a series of training sessions.

Key Achievements

The Mbale Municipal Council provided office space for managing the data in the STDM tool equipped with a computer, a printer, cameras, and furniture. The office space/resource center created a platform for regular interaction between the community and the municipal authorities. The community through the Mbale Slum Dwellers Federation was able to interrogate the data to identify gaps related to community development projects, issues of security, and access to services. The STDM generated data were used to lobby for services and other pressing needs in the community. For instance, in the subsequent phases of the TSUPU project, the local community applied for a grant to address the gaps identified from the STDM generated data. This led to the construction of eight (8) water points, five (5) public toilets, one (1) school toilet, and two (2) stone pitched drainage channels, as well as the widening of 1 road and the improvement of street lighting by Mbale Municipal Council [36]. A community center was also constructed in Mission cell with sanitation and office/meeting facilities, managed by the local community.

Further, citywide profiling, enumeration, and mapping were carried out covering fourteen (14) secondary towns of Uganda (including Mbale City), in which more than 181,000 households residing in approximately 120 informal settlements were enumerated and mapped, and their data were captured using the STDM [37]. The lessons and experiences from the TSUPU project have informed more recent projects in Uganda on the use of the STDM tool toward the legal recognition of land tenure security. The STDM tool and the participatory processes have been used to demarcate land at the household level. The data generated from the STDM tool is being used to issue Certificate of Occupancy, Certificate of Customary Ownership (CCO), and the application of freehold land titles, as provided for in the Land Act of 1998 and in the National Land Policy of 2013 [32,37]. The MLHUD in 2019 issued a decree on the use of the STDM tool to issue digital CCOs [38].

2.1.3. Kenya

Kenya's urban population as a proportion of the total population increased from 8.8% between 1960 and 1970 to 20.9% between 2000 and 2010, and it is projected to exceed by 36% between now and 2040 [39]. The growing trend of urban informality defines much of Kenya's urbanization patterns often recorded within informal settlements as the towns and cities grow [40].

The case study in Kenya looked at an informal settlement area called the Kwa Bulo settlement located in Kadzandani ward in Mombasa County. The settlement measures approximately 36 hectares and is divided into four clusters/villages: EPCO; Kashani; Timboni; and Msufini [38]. The first registered proprietor of the plot was Emmanuel Bulo, after which the land was transacted on and passed to numerous entities. In 2005, approximately seven hundred houses were demolished in an eviction orchestrated by the Kenya National Assurance Company [39]. In May 2006, residents of Kwa Bulo went to court seeking legal redress on grounds of adverse possession but had to wait until 2011 when the Kenya High Court of Mombasa issued a decision allowing the community to access the land.

Kwa Bulo community was then advised by the Mombasa County Government (MCG) to conduct an enumeration exercise that emulated a similar exercise conducted in Mnazi Moja, an informal settlement at the edges of Mombasa City, to identify the actual occupants of the land. Enumerations began in December 2014 led by community members in Kwa Bulo, Pamoja Trust, and Muungano Wa Wanavijiji, a social movement of 'slum' residents and urban poor people in Kenya and a partner to Pamoja Trust. Officials from the government collaborated with the community in data collection and data verification.

The community members performed data entry for all the administered questionnaires into the customized STDM tool. The data were later analyzed to produce a list of all beneficiaries in the settlement, including their relationship to the land and structures. The Kwa Bulo settlement land committee worked with representatives of Pamoja Trust and Muungano Wa Wanavijiji to verify the accuracy of the beneficiary list [31]. The County government also performed a similar validation exercise that involved officials from MCG, accompanied by community leaders of Kwa Bulo and Pamoja Trust representatives. After a successful validation exercise, the team concluded and accepted about 800 parcels for further processing.

Key Achievements

The MCG agreed to issue certificates of occupancy to the occupants that had complete and uncontested data. A total of 944 certificates of occupancy (397 to women, 542 to men, and five jointly/both spouses) were issued to residents of the Kwa Bulo settlement in 2017, benefitting approximately 3722 family members [36,41]. The issuance of certificates of occupancy improved the tenure perception of the community, gave them a sense of identity, and protected them from forced evictions [20,41]. Improved confidence in tenure security is being witnessed through housing improvements and better construction materials (ibid). A community data center was also established with furniture and computers that hosted the STDM database. A team of youths trained on the STDM tool were responsible for data management in the center. Local advocacy efforts by the community using the STDM generated data led to the installation of a clean water supply in the settlement [41]. Additionally, the council authorities are planning to build an access road, extend water lines, and install public toilets in the settlement. To date, the MCG is finalizing the process of transferring the land from the private landowner to the council and is having some discussions with the community on the next steps toward further strengthening their tenure security.

2.1.4. Zambia

Zambia has a dual land tenure system namely, leasehold, or statutory tenure and customary tenure type. At independence in 1964, customary land was said to account for approximately 94% of all land in Zambia [42–44]. As of 2005, customary land was estimated at 84% [45]. About 10% of customary land has been converted to private leaseholds since the enactment of the Land Act in 1995 [46,47]. Customary tenure in Zambia is an indigenous form of land ownership that is governed by unwritten traditional rules and administered by traditional leaders/Chiefs and Chieftainess. The active occupation or use of a piece of land is the main evidence of ownership or an existing interest in the land. Access to customary land is also contingent upon ethnic, kin, or community membership, controlled by the Chief or Chieftainess. This form of tenure is, however, not secure, because it is largely undocumented, making inhabitants of customary land susceptible to forced displacements, abuse from traditional authorities, threats from investors due to the increased demand of land, and frequent land disputes among villagers, head persons, and even between chiefs pertaining to land boundaries.

The case study in Zambia looks at Chamuka chiefdom, which consists of 208 villages in Chisamba District, Central Province, about 100 km from Lusaka city, the capital city of Zambia. Each village is headed by a village head person. The chiefdom covers a spatial extent of approximately 300,000 hectares of land. The discussion on the STDM implementation started in 2016 between the People's Process on Housing and Poverty in Zambia (PPHPZ), the Zambia Homeless People Federation (ZHPF), and the traditional leaders of Chamuka Chiefdom led by His Royal Highness Chief Chamuka VI, in collaboration with government authorities and other stakeholders with support from the UN-Habitat Country Office in Zambia.

Chief Chamuka had picked the interest of the STDM tool and its participatory processes while implementation was ongoing in the neighboring Mungule chiefdom in 2015, and they were very eager to welcome this process in his Chiefdom that brought together the traditional authorities in Chamuka, the local community, as well as government officials from Kabwe municipality in the implementation phase [48].

Key Achievements

The implementation of the STDM in Chamuka begun in 2016 and, so far, twenty-seven (27) villages have been enumerated and mapped. A total of 1794 households have acquired certificates of customary land occupancy representing a population of approximately 10,812 people, out of which 5789 are women beneficiaries. There are 765 joint ownership certificates, which means that the certificates are in the name of both spouses [41]. A community resource center has been established in Chamuka chiefdom, to be operated and managed by youths trained on the STDM tool. The center will also serve as an anchor to all STDM activities in the chiefdom. Over 85 para-surveyors were trained to ensure the sustainability of the STDM and related processes in other villages in the Chamuka chiefdom.

The community is using these certificates as proof of actual occupancy, giving them a voice in negotiating with investors who have expressed interest in their land. For example, between 2018 and 2019, seven (7) families in Bulemu village occupying a 103-hectare piece of land successfully negotiated with a private firm for a solar project in a meeting presided over by Chief Chamuka and the headmen of the Chiefdom. The signing of agreements between the investor and the families was completed in April 2019.

The implementation process has also improved the women's participation by giving them a conspicuous role in spearheading data collection and facilitating discussions surrounding land rights with the traditional authorities. His Royal Highness Chief Chamuka has also introduced a policy declaring that fifty (50%) per cent of the land at any given time be reserved for women in all the 208 villages [49]. Additionally, every village committee is required to have women representation to ensure women voices are heard. Some youth members from Chamuka chiefdom have also benefitted by being allotted a piece of land for an aquaculture project by the Ministry of Livestock and Fisheries [50].

Chief Chamuka, headmen, and villagers alike observe that boundary disputes that were rampant before the STDM enumeration and mapping processes are now a thing of the past [49]. A total of 426 land disputes (intra-family succession disputes, village boundary disputes, etc.) were recorded and successfully resolved. Thus, the STDM tool and related processes have become instrumental in resolving long-standing disputes in the Chiefdom. Moreover, the experience from Chamuka chiefdom has greatly contributed to the further development of the draft National Land Policy. Specifically, the recommendations from Chamuka provided evidence on the practicality and versatility of fit-for-purpose land tools such as the STDM to enhance tenure security on customary land for all [25]. The final draft of the national land policy has prominently provided for the recognition of customary land rights through the Chiefs and the promotion of fit-for-purpose land administration approaches. The National Land Policy was approved and adopted by the Government of Zambia in May 2021.

2.2. Primary Data Collection

The primary data collection survey was conducted in October and November 2020 and targeted government authorities, CSOs, traditional authorities, members of enumeration, mapping teams, community leaders, and members within the case study areas. The intention of the survey was to gather views and opinions of the various beneficiaries and stakeholders around the use and application of the STDM tool, particularly in relation to its contribution to improving tenure security and any indications that demonstrate its influence on policy and institutional arrangements. Aside from online surveys, and due to limited internet access and the need for translation for some respondents, some questionnaires were filled manually by some respondents. The partner CSOs supported in reaching out to the target respondents and assisted in the administration of the questionnaires, particularly in translation into local languages.

From Table 1, a total of 82 participants responded to the survey in the three case studies and comprised 50% women and 50% men; and about 80% of the respondents were community stakeholders. All respondents participated in STDM interventions for a period ranging from two to eight years, and 80% of them are still involved in the ongoing work at the time of the survey.

Table 1. Respondents for the primary data collection.

Respondents	Country			Percentage
	Kenya	Uganda	Zambia	
Local government official/staff	0	1	2	4%
Traditional authority	0	0	4	5%
Community leader	8	4	5	21%
Community member	9	8	8	30%
Member of enumeration/mapping team	3	10	7	24%
Member of CSOs	4	6	3	16%
Total	24	29	29	100%

2.3. *Challenges and Limitations of the Primary Data Collection*

The observation from the responses was that they were quite consistent across the three countries. As the questionnaire was administered online, we could not interact with the respondents to follow-up on the research questions that would have improved the analysis of the quantitative data.

3. Results

The analysis of the survey entailed compiling the results using the survey monkey tool and then exporting the data into a spreadsheet format. The data were organized and categorized into a tabular format and were grouped according to the research questions. Conversion of the data into a spreadsheet allowed for more discrete analysis.

3.1. *The STDM Tool Facilitates Land Tenure Security Improvement*

We asked the respondents to state whether the STDM tool facilitates tenure security improvement. According to the results, 100% of the respondents from Zambia and Uganda agreed with the statement that the STDM tool facilitates tenure security improvement, while for Kenyan respondents, the endorsement of this statement was at 83%. From Table 2 below, 95% of all respondents from the three countries acknowledged the statement, with 42% of respondents strongly in agreement. Moreover, we analyzed the responses based on gender and found that over 97% of the men and 93% of women agreed that tenure security has improved in the target community, as a result of the STDM-related interventions.

Table 2. Responses on whether the STDM facilitates tenure security.

Description	Number of Responses			Percentage
	Kenya	Uganda	Zambia	
Strongly agree	10	12	12	42%
Agree	10	16	17	53%
Neither agree nor disagree	3	0	0	4%
Disagree	1	0	0	1%
Strongly disagree	0	0	0	0%
Total	24	28	29	100%

We also asked the respondents to indicate what could be the reasons for such perceptions of improved tenure security. There are at least three key common responses to this question across the countries: reduced land disputes and conflicts (81%), increase in the overall confidence and perception of tenure security (78%), and receipt of the certificate of occupancy (74%). For Zambian respondents, their top two responses were: increase in the overall confidence and perception of tenure security (96%), and receipt of tenure documents (82%). Meanwhile, in Kenya, the respondents chose reduced land disputes and conflicts (68%), and available/access to settlement and household data and maps (63%) as their main indications of the increased perception of tenure security. For Ugandan respondents, reduced land disputes and conflicts (96%) and a decreased risk of forced evictions in the community (88%) were the key elements of having an improved perception of tenure security.

3.2. *Use of the STDM Tool Contributes to Positive Changes in Policy and Institutional Arrangements*

We then sought to determine the respondents' view about the STDM tool being able to contribute to influencing changes in related land policies and institutional arrangements. The result indicated that 100% of all the respondents from Zambia and Uganda agreed with the statement that the STDM tool was able to influence related land policies and institutional arrangements. Interestingly, Kenyan responses only indicated an agreement of 74% with the question. Table 3 below represents the perception from the three countries. Overall, 93% of the total respondents agreed with the statement, of which 38% strongly agreed.

Table 3. Perceptions on the STDM ability to influence changes in policy and institutional arrangements.

Description	Number of Responses			Percentage
	Kenya	Uganda	Zambia	
Strongly agree	7	9	14	38%
Agree	10	19	15	55%
Neither agree nor disagree	5	0	0	6%
Disagree	1	0	0	1%
Total	23	28	29	100%

Tweaking the data to assess the responses by gender, the data showed that 57% of the women expressed confidence compared to 52% of men who also agreed that the STDM tool had an impact on influencing changes in policy and institutional arrangements.

Moreover, we tried to find out how the application of the STDM tool had influenced policies and institutional arrangements. The common responses (above 50%) included the following: marked improvement in the relationship between the community and authorities, increase in the demand for similar interventions in other areas, recommendations for the wider application of the STDM through potential changes and institutional arrangements and stakeholders, including traditional and local authorities being supportive of the STDM tool implementation and its replication/scaling-up. While conforming to the above trend of the responses, each country, however, had a different emphasis. For example, we found out that 66% of Kenyan respondents felt that having the different stakeholders being supportive of the STDM tool implementation and its replication is the most important manifestation that it is influencing policies and institutional arrangements. More than 80% of the respondents from Uganda agreed with the views from Kenya, but also included two other relevant factors: the improved relationship between the community and local authorities and the better engagement and support of local authorities in improving the settlement. In Zambia, more than 65% of the responses pointed out that there is a strong recommendation for wider application of the STDM tool application through changes in policies and institutional arrangements, as well as increased demands for similar interventions in other areas.

3.3. Other Results

We asked the respondents to give their opinion on whether the community is better off after the STDM-based interventions. The results in Table 4 below show that 95% of the respondents in the three countries agreed that the community is better off after the STDM-based interventions, with 41% of the respondents were strongly in agreement with the statement.

Table 4. Perceptions that the target community is better off after STDM interventions.

Description	Number of Responses			Percentage
	Kenya	Uganda	Zambia	
Strongly agree	10	12	12	41%
Agree	10	16	17	54%
Neither agree nor disagree	3	0	0	4%
Disagree	1	0	0	1%
Total	24	28	29	100%

Perception questions were asked and one of them referred to whether the respondents would recommend replicating or scaling up the STDM implementation in other areas. This question received a positive response of 95%, with 54% of respondents strongly in agreement. Another related question was asked on whether the respondents would consider the STDM tool as an innovative and useful tool in addressing land tenure and land governance. About 91% of the respondents agreed with the proposition, with 46% strongly in agreement. Finally, the respondents were asked why they thought the STDM was a useful and innovative tool. The respondents highlighted at least three justifications for their answers, namely: low-cost and affordability (73%); inclusive, gender-responsive, and fit-for-purpose (73%); and the process is empowering (70%).

4. Discussion

It can be clearly shown that the STDM tool contributed to facilitating improved perceptions of tenure security along the continuum, and to a significant extent, to influencing policy and institutional arrangements. It is important to note, however, that the implementation of the STDM tool was complemented by active community engagement and the use of other tools such as participatory enumeration, and the gender evaluation tool also developed by GLTN.

4.1. STDM Tool as Facilitating Tenure Security

To contextualize the discussion on tenure security in this research, we considered the continuum of land rights approach in interpreting the findings. The results indicated that through the STDM application, tenure security of the community has improved. This was mainly attributable to: the increased confidence through participation; the availability and accessibility of the data; the reduction in land disputes and conflicts; the issuance of tenure documents; the decreased risk of arbitrary forced evictions; the improved recognition and relationship with authorities. While these indicators are evidently present in all three countries, each case study has different hierarchical views and emphases, as indicated above. Potentially, the indicators of the improved perception of tenure security will evolve as new events, situations, or variables occur along the continuum of land rights approach.

Likewise, there was recognition that land tenure documents were issued at reduced costs compared to formal tenure documents (such as freehold titles), which are normally costly for the beneficiaries and for the government authorities. The issuance of the occupancy certificate provided a sense of legitimacy to the *de facto rights* of communities in that the rights are recognized by authorities and the certificates can be used to settle claims. It is important to note that women received the same tenure documents in a similar manner as men. We also noted the high perception of tenure security among women, and we attributed this to their active participation in the enumeration and mapping processes, including in the data validation. Hence, the process had large implications in their identity as bona fide occupants or claimants of the land.

There was also strong evidence that land conflicts or disputes were reduced. This could be attributed to the availability and accessibility of data at the settlement level and the strong community engagement and participation in the whole process. Furthermore, we observed that anyone in the community could check and verify, and that the preliminary data were validated by the community, which facilitated data acceptance, data accuracy, and the collective resolution of disputes. In Kenya, validation was done at three levels before the data could be approved for the issuance of occupancy certificates. In all three case studies, the disputes were resolved in the presence of traditional and local authorities, elders, and neighbors as witnesses, and all pertinent documentation was attached to the information and were recorded in the STDM database.

4.2. STDM Application as Contributing to Influencing Change in Land-Policies and Institutional Arrangements

4.2.1. Influence on Policy Development and Implementation

In Zambia, the case study experiences impacted the policy development process to a level that the FFP LA approach was included in the current draft national land policy to speed up customary land registration, among others. From the case study, we learnt that land policy consultations with traditional authorities had increased, and this was also shown in the survey finding through increased demand for similar interventions in other regions.

It is notable that the experiences of the STDM tool in the case studies resulted in its increased uptake in other contexts and the scaling up of the interventions in other areas. For instance, in Uganda, the roll-out of the tool from Mbale City to other regions, and later on, to customary land registration processes, has resulted in its adoption by the Government of Uganda in digitalizing the processes and the issuance of Certificates of Customary Ownerships (CCOs) and related processes.

In Kenya, after the issuance of Certificates of Occupancy to the residents, perceptions of tenure security increased, and even the housing conditions that resulted in the County Government increased investments in settlement upgrading plans and incremental developments, also building from the stronger relationship that was established with the local community. The slum upgrading process is guided by the national slum prevention and upgrading policy of Kenya. The policy supports community organization toward a common goal through the establishment of an information system that records community resources and rightful beneficiaries. This step has already been achieved. As a result, the MCG is now negotiating with the private landowner so that they can officially procure the Kwa Bulo land for future planning, the upgrading of the settlement, and the further strengthening of tenure security of the residents.

4.2.2. Changes in the Institutional Arrangements

The outcome of the country case studies clearly demonstrated positive changes in the institutional arrangements particularly between the communities and local and traditional authorities. The authorities' acceptance to offer office space for storing and managing community data demonstrates improved trust, confidence, and relationships among the local stakeholders. In addition, local and traditional authorities and community members have equal access to the information. Thus, access to data facilitated an open-door policy for interactions, dialogues, and consultations in the planning and decision-making processes.

Similarly, the use of the STDM tool improved accessibility to the data and gave the community the opportunity to negotiate on matters related to service delivery, local investment in infrastructure, settlement planning, and development projects with authorities. These opportunities further contributed to community empowerment. At the time of this research, the negotiated projects have already been completed or are ongoing in the three case studies. With the strong partnership with local and traditional authorities, the communities were able to advocate and articulate their issues and concerns through them, which translated to some concrete actions in influencing changes in policy and regulatory frameworks.

4.3. STDM Technology and Related Processes Supporting Community Empowerment

The use of the STDM tool has benefited the community in building collective knowledge of their settlement. It has also enabled them to organize themselves better in addressing community challenges such as identifying priorities, resolving land disputes, and strengthening their negotiation with authorities and with external stakeholders (e.g., investors).

It is also important to note that the participatory approach employed in the STDM tool implementation contributed to the 'ownership' of the projects through active involvement in the enumeration activities, community mobilization, and participatory mapping, thus contributing to the overall community empowerment. In addition, community members, particularly women, have a deeper sense of ownership of the processes, and it has raised their position to be part and parcel of decision making in the community. The process of localizing the related methodologies benefitted the community members, particularly the youth, from training and access to technology and information, which allowed them to manipulate and analyze the data in the STDM database.

In Zambia, the transformative change at the community level is clearly demonstrated, where several households were able to properly negotiate with investors on the planned development projects. Previously, such negotiations were concluded only by the traditional authorities without the consent and knowledge of community members. The use of the STDM tool has also scaled up in Zambia to urban areas, including increasing demands for capacity development interventions. Similarly, in the Kwa Bulo settlement, they were able to directly negotiate with the County Government to act on their behalf, issue occupancy certificates, and increase investments for settlement improvement.

5. Conclusions

The study investigated whether the STDM tool facilitated the improvement of land tenure security and whether it has contributed to influencing positive changes in policy and institutional arrangements in case study areas. This research has confirmed positive responses to these questions. There was remarkable improvement in the perception of tenure security in the three case studies linked to the STDM tool implementation along with the range of recommendations regarding the various actions, steps, and decisions on land that can be interpreted as having meaning to land tenure security. It is also clear that the use of the STDM tool is more effective and sustainable because it integrates participatory processes along with other tools in the implementation process. This approach, together with increased community empowerment and access to information for planning and decision-making, contributed positively to influencing positive changes in policy implementation and institutional arrangements. While the changes are not happening quickly, the indications are clearly demonstrating positive outcomes occurring in an incremental but steady manner that is sustainable at the local level.

The research has shown that technology such as the STDM tool can be effective toward the development of the FFP LA spatial framework and in facilitating the land tenure security of poor communities. The research also concluded that focusing on one pillar of FFP LA can influence positive changes in other frameworks of FFP LA through the use and application of technology such as the STDM tool. The study also contributes to the knowledge that the development of the FFP LA spatial framework can be achieved through the combined elements of adopting fit-for-purpose technology such as the STDM tool, awareness and capacity building, participation, engagement of local communities, and political support from local and traditional authorities. Therefore, there is a potential for more research to be carried out to demonstrate how one of the frameworks of FFP LA can catalyze changes in other frameworks. More related research on the topic is needed to further understand the dynamics of the various FFP LA frameworks and to strengthen the findings of this study.

Author Contributions: Conceptualization, D.A., S.N., H.N. and J.G.; methodology, D.A. and S.N.; software, S.N.; validation, D.A., S.N., H.N. and J.G; formal analysis, D.A. and S.N.; resources, D.A.; data curation, D.A. and S.N., writing—original draft preparation, S.N., D.A., H.N. and J.G.; writing—review and editing, S.N., D.A., H.N. and J.G.; visualization, S.N. and D.A.; project administration, S.N. and D.A.; funding acquisition, D.A. All authors have read and agreed to the published version of the manuscript.

Funding: Funding support for the research work including data collection is provided by UN-Habitat and Global Land Tool Network (GLTN) programme. The funding of the publication costs for this article has kindly been provided by the School of Land Administration Studies, Faculty ITC, from the University of Twente in combination with Kadaster International, the Netherlands.

Institutional Review Board Statement: The study was conducted according to the guidelines of the Declaration of Helsinki, and approved by the Institutional Review Board (or Ethics Committee) of UN-Habitat and the GLTN.

Informed Consent Statement: Informed consent was obtained from all subjects involved in the study.

Data Availability Statement: All the case study materials are referenced in the manuscript.

Acknowledgments: The study is based on the findings of UN-Habitat and the GLTN projects in the case studies in all three countries. The authors would like to acknowledge the contribution of the partner organizations in each country, Pamoja Trust in Kenya, ACTogether Uganda in Uganda, and People's Process on Housing and Property in Zambia, for their support in data collection.

Conflicts of Interest: The authors declare no conflict of interest.

References

1. Zevenbergen, J.; Augustinus, C.; Antonio, D.; Bennett, R. Pro-poor land administration: Principles for recording the land rights of the underrepresented. *Land Use Policy* **2013**, *31*, 595–604. [CrossRef]
2. Hull, S.; Whittal, J. Towards a framework for assessing the impact of cadastral development on land rights-holders. In Proceedings of the International Federation of Surveyors Working Week, Christchurch, New Zealand, 2–6 May 2016.
3. Toulmin, C. Securing land and property rights in sub-Saharan Africa: The role of local institutions. *Land Use Policy* **2009**, *26*, 10–19. [CrossRef]
4. German, L.; Schoneveld, G.; Mwangi, E. *Contemporary Processes of Largescale Land Acquisition by Investors: Case Studies from Sub-Saharan Africa*; Center for International Forestry Research, CIFOR: Bogor, Indonesia, 2011; Volume 68.
5. Augustinus, C. Social tenure domain model: What it can mean for the land industry and for the poor. In Proceedings of the XXIV FIG International Congress: Facing the Challenges, Building the Capacity, Sydney, Australia, 11–16 April 2010.
6. Deininger, K.W. *Land Policies for Growth and Poverty Reduction*; World Bank Publications: Washington, DC, USA, 2003.
7. United Nations General Assembly. *New Urban Agenda: Quito Declaration on Sustainable Cities and Human Settlements for All (71/256)*; United Nations General Assembly: New York, NY, USA, 2016.
8. African Union. *Agenda 2063: The Africa We Want*; AU: Addis Ababa, Ethiopia, 2013.
9. African Union. *Declaration on land issues and challenges in Africa*; AU: Addis Ababa, Ethiopia, 2009.

10. UN FAO. *Voluntary Guidelines on the Responsible Governance of Tenure of Land, Fisheries and Forests in the Context of National Food Security*; Food and Agriculture Organization of the United Nations (FAO): Rome, Italy, 2012.
11. McLaren, R. Crowdsourcing Support of Land Administration. In *Word Bank Conference on land and Poverty*; World Bank Publications: Washington, DC, USA, 2012; pp. 23–26.
12. Enemark, S.; Mclaren, R.; Lemmen, C.; Antonio, D.; Gitau, J. Scaling Up Responsible Land Governance: A Guide for Building Fit-for-Purpose Land Administration Systems in Less Developed Countries. In *Scaling up Responsible Land Governance: Annual World Bank Conference on land and Poverty*; World Bank Publications: Washington, DC, USA, 2016; p. 213.
13. Enemark, S.; Bell, K.; Lemmen, C.; McLaren, R. Building fit-for-purpose land administration systems. In Proceedings of the XXV FIG International Congress, 16–21 June 2014; pp. 16–21.
14. Enemark, S.; McLaren, R.; Lemmen, C. *Fit-for-Purpose Land Administration Guiding Principles*; Global Land Tool Network (GLTN): Copenhagen, Denmark, 2015.
15. Williamson, I.; Enemark, S.; Wallace, J.; Rajabifard, A. *Land Administration for Sustainable Development*; ESRI Press Academic: Redlands, CA, USA, 2010; p. 487.
16. UN-Habitat. *Secure Land Rights for All*; United Nations Human Settlements Programme/Global Land Tool Network: Nairobi, Kenya, 2008.
17. Payne, G. (Ed.) *Land Rights and Innovation*; ITDG Publishing: London, UK, 2002.
18. Whittal, J. A new conceptual model for the continuum of land rights. *S. Afr. J. Geomat.* **2014**, *3*, 13–32.
19. Antonio, D.; Marongwe, N.; Nyamweru, H.; Okello, J. Securing land rights within the continuum of land rights approach: Evidence from tenure security innovations in Kenya and Uganda. In Proceedings of the World Bank Land and Poverty Conference 2017, Washington DC, USA, 20–24 March 2017.
20. UN-Habitat. *Fit-for-Purpose Land Administration Guiding Principles for Country Implementation*; Report 2(2016); Nairobi, Kenya, 2016.
21. Huggins, C.; Frosina, N. ICT-driven projects for land governance in Kenya: Disruption and e-government frameworks. *GeoJournal* **2017**, *82*, 643–663. [CrossRef]
22. Antonio, D.; Gitau, J.; Njogu, S. Addressing the Information Requirements of the Urban Poor, STDM Pilot in Uganda. 2014. Available online: https://www.researchgate.net/publication/312125650_Addressing_the_Information_RequirementS_of_the_Urban_Poor_STDM_Pilot_in_Uganda (accessed on 17 May 2021).
23. Ouma, S.; Ochong, R.; Antonio, D.; Maina, D. Making the community land act effective: The case of Mashimoni settlement in Nairobi County and Kwa Bulo settlement in Mombasa County of Kenya. In Proceedings of the Annual World Bank Conference on Land and Poverty, Washington, DC, USA, 20–24 March 2017.
24. Kumwenda, M.; Shumba, F.; Ncube, N.; Antonio, D.; Nyamweru, H.; Gitau, J.; Chileshe, A. Innovations for securing tenure rights on customary lands through traditional authorities: Experiences from Chamuka Chiefdom, Zambia. In Proceedings of the World Bank Conference on Land and Poverty, Washington, DC, USA, 19–23 March 2018.
25. Njogu, S.; Ngumba, L.; Kakule, S.; Antonio, D. *Flexible Land Information System as Driver for Change, Peace and Development: The Case of Post Conflict DRC*; Conference Paper, FIG Working Week; 2019.
26. Antonio, D.; Ndungu, H.N.; Gitau, J.; Sylla, O. Innovative Approaches in Securing Land Rights and Enhancing Transparency in Sub-Saharan Africa: Good Practices and Lessons Learned from Four African Countries. In Proceedings of the Conference on Land Policy on Africa, Abidjan, Ivory Coast, 25–29 November 2019.
27. Augustinus, C. Catalysing global and local social change in the land sector through technical innovation by the United Nations and the Global Land Tool Network. *Land Use Policy* **2020**, *99*, 105073. [CrossRef]
28. Antonio, D. The Social Tenure Domain Model: A Specialisation of LADM towards Bridging the Information Divide. 2013. Available online: https://www.fig.net/resources/proceedings/2013/2013_ladm/15.pdf (accessed on 17 May 2021).
29. Lemmen, C.H.J. *The Social Tenure Domain Model: A Pro-Poor Land Tool*; FIG: Copenhagen, Denmark, 2010.
30. Lemmen, C.; Van Oosterom, P.; Bennett, R. The land administration domain model. *Land Use Policy* **2015**, *49*, 535–545. [CrossRef]
31. Stanfield, J.D.; Danilo, R.A.; Sylla, O.; Nyamweru, H.; Gitau, J.; Selebalo, C. Information is Power: Evidence on Community Based Land Information for Upgrading Community Tenure Security in Uganda and Kenya. In Proceedings of the Land and Poverty Conference, Washington, DC, USA, 20–24 March 2017.
32. Population Division, Department of Economic and Social Affairs, United Nations. *World Urbanization Prospects: The 2018 Revision (ST/ESA/SER.A/420)*; United Nations: New York, NY, USA, 2019.
33. Myers, C.; Suzuki, E.; Atamanov, A. The Demographic Boom: An Explainer on Uganda's Population Trends. World Bank Blogs. Available online: https://blogs.worldbank.org/africacan/demographic-boom-explainer-ugandas-population-trends (accessed on 18 February 2021).
34. Government of Uganda; Ministry of Lands, Housing and Urban Development. Statement by the State Minister of Urban Development Ministry of Lands, Housing and Urban Development, World Cities Day on 31 October 2020. Available online: https://www.google.com/url?sa=t&rct=j&q=&esrc=s&source=web&cd=&ved=2ahUKEwjB5obWpdHwAhXmz4UKHVmcDhkQFjAAegQIAxAD&url=https%3A%2F%2Fgggi.org%2Fsite%2Fassets%2Fuploads%2F2020%2F10%2FMinistry-of-Lands-Housing-and-Urban-Development-World-Cities-Day-Statement-2020.doc&usg=AOvVaw1mu7zLqlSS2p8eWt-uK8SW (accessed on 17 May 2021).
35. United Nations Human Settlement Programme. *Global Land Tool Network (GLTN)—Phase 2 End of Phase Evaluation*; UN-Habitat: Nairobi, Kenya, 2018.

36. STDM. Information is Power; STDM in Uganda [Video] You Tube. Available online: https://www.youtube.com/watch?v=NV7K21OtARY (accessed on 16 September 2014).
37. Cities Alliance. Programme Evaluation of Cities Alliance Country Programmes Ghana, Uganda and Vietnam. 2017. Available online: https://www.citiesalliance.org/resources/publications/cities-alliance-knowledge/programme-evaluation-cities-alliance-country (accessed on 17 May 2021).
38. Global Land Tool Network. Uganda Moves to Digital Certificates of Customary Ownership to Secure Land Rights and Improve Land Use. 2020. Available online: https://gltn.net/2020/10/08/uganda-moves-to-digital-certificates-of-customary-ownership-to-secure-land-rights-and-improve-land-use/ (accessed on 30 April 2021).
39. Hope Sr, K.R. Urbanization in Kenya. *Afr. J. Econ. Sustain. Dev.* **2012**, *1*, 4–26.
40. United Nations Human Settlement Programme. UN-Habitat Support to Sustainable Urban Development in Kenya. In *Addressing Urban Informality Volume 4: Report on Capacity Building for Community Leaders*; United Nations Human Settlement Programme: Nairobi, Kenya, 2016.
41. STDM. STDM Implementation; Informal Settlements in Kenya. [Video] You Tube. Available online: https://www.youtube.com/watch?v=FO46vGXkysg (accessed on 7 February 2019).
42. Brown, T. Contestation, confusion and corruption: Market-based land reform in Zambia. In *Competing Jurisdictions: Settling land claims in Africa, 79-102*; Brill Academic Publishers: Leiden, The Netherlands, 2005.
43. Mudenda, M.M. The Challenges of Customary Land Tenure in Zambia. 2006. Available online: https://www.fig.net/resources/proceedings/fig_proceedings/fig2006/papers/ts50/ts50_03_mudenda_0740.pdf (accessed on 17 May 2021).
44. Chileshe, R.; Shamaoma, H. Examining the challenges of cadastral surveying practice in Zambia. *S. Afr. J. Geomatics* **2014**, *3*, 53–63.
45. USAID. *USAID Country Profile. Property Rights and Resource Governance, Zambia*; USAID: Lilongwe, Malawi, 2010.
46. Veit, P. Custom, law and women's land rights in Zambia. *Gates Open Res.* **2019**, *3*.
47. Chitonge, H.; Mfune, O.; Umar, B.B.; Kajoba, G.M.; Banda, D.; Ntsebeza, L. Silent privatisation of customary land in Zambia: opportunities for a few, challenges for many. *Soc. Dyn.* **2017**, *43*, 82–102. [CrossRef]
48. STDM. Social Tenure Domain Model, Zambia. [Video] You Tube. Available online: https://www.youtube.com/watch?v=3Jwjdvlx3q0 (accessed on 6 September 2017).
49. United Nations Human Settlements Programme UN-Habitat. *Stories of Change: Global Land Tool Network Phase 2*; UN-Habitat: Nairobi, Kenya, 2019.
50. Global Land Tool Network. Securing Land Tenure for a Young Fish Farmer in Chamuka, Zambia. 2018. Available online: https://gltn.net/2018/10/05/fish-farming-for-a-living-chamuka-chiefdom-zambia (accessed on 17 May 2021).

 land

Article

Fit-For-Purpose Upscaling Land Administration—A Case Study from Benin

Steven Mekking [1,*], Dossa Victorien Kougblenou [2] and Fabrice Gilles Kossou [2]

1 Kadaster International, Land Registry and Mapping Agency of the Netherlands, P.O. Box 9046, 7300 GH Apeldoorn, The Netherlands
2 Agence Nationale du Domaine et du Foncier, Sainte Rita, Cotonou, Benin; kougblen@gmail.com (D.V.K.); gil_kossou@yahoo.fr (F.G.K.)
* Correspondence: steven.mekking@kadaster.nl; Tel.: +31-6524-817-32

Abstract: The government of Benin in 2013 decided upon a centralized land administration, with the purpose of recording the entire national territory in one land administration system to promote durable economic development by increasing legal certainty in real estate transactions. This is a major challenge, given that currently, of the estimated 5 million cadastral parcels, less than 60,000 parcels have a land title and are registered in the national land administration agency's central database. This case study describes how a transition to a fit-for-purpose approach in land administration makes it possible to realize the Benin government policy. In the context of Benin, the core of this approach is the introduction of a tenure system based on presumed ownership parallel to the existing title system with state-guaranteed ownership. From a quality perspective, this meant a shift in priorities from "good but slow" to "good enough and fast". A field test has proven that this new approach is necessary to realize the governmental purpose but puts pressure on the quality aspect and the related interests of established parties such as private surveyors. In the Benin case, this pressure is reduced by designing a land information system based on the Land Administration Domain Model (LADM) that makes it possible to include and keep track of both cadastral parcels with state-guaranteed ownership and cadastral parcels with presumed ownership in the database. Both ways of tenure security can therefore coexist, allowing landowners to choose between the level of legal security that best fits their needs and means.

Keywords: fit-for-purpose land administration; case study; Benin; land administration; cadaster; land registration; land administration domain model; LADM

Citation: Mekking, S.; Kougblenou, D.V.; Kossou, F.G. Fit-For-Purpose Upscaling Land Administration—A Case Study from Benin. *Land* **2021**, *10*, 440. https://doi.org/10.3390/land10050440

Academic Editors: Stig Enemark, Robin McLaren and Christiaan Lemmen

Received: 1 April 2021
Accepted: 18 April 2021
Published: 21 April 2021

1. Introduction

Benin, officially the Republic of Benin (French: République du Bénin) is a former colony of France which gained independence in 1960. Benin became in 1991 a democratic republic, in which the President of Benin is both head of state and head of government, within a multi-party system.

Benin is a relatively small West African country, elongated in latitude, and covers an area of 114,763 km^2. It is bordered to the south by the Atlantic Ocean, to the west by Togo, to the north by Burkina Faso and Niger and to the east by Nigeria. Benin's relief does not present great differences in altitude. The average altitude is 200 m [1] (p. 9). According to the results of the Fourth General Census of Population and Housing of May 2013, Benin has 10 million inhabitants (almost 12 million according to the 2019 projection). The country has a predominantly rural population, with 5.5 million inhabitants in rural areas and 4.5 million in urban areas and a crude birth rate equal to 36%, of which 32% is in urban areas against 39% in rural areas. Life expectancy at birth in Benin is 63.84 years [1] (p. 78). The capital of Benin is Porto-Novo, but the seat of government is in Cotonou, the economic capital). Benin is a developing French speaking country. In terms of the Human Development Index 2020, its position is 158 of the 189 [2].

The legal foundation under the land administration in Benin is the 2013 Land Administration Law [3] that together with the 2017 addendum [4] replaced different previous land laws. The execution of the land administration in Benin is assigned to the National Land Registry and Agency, l'Agence Nationale du Domaine et du Foncier (abbreviated as ANDF), founded in 2016 (Figure 1). ANDF is also responsible for the national cadaster.

Figure 1. Head office of l'Agence Nationale du Domaine et du Foncier in Cotonou.

With the introduction of the 2013 Land Administration Law and the establishment of ANDF, Benin has decided upon a centralized land administration, with the objective of recording the entire national territory in one digital central land administration system ("le cadastre national numérique"). Benin has also opted for a very high level of legal certainty, a Torrens-based system [5], with the provision of land titles (titres fonciers) that grant a practically indisputable property right to the natural person or legal entity stated on the title. This property right is guaranteed by the state and can only be lost through the expropriation procedure described in the 2013 Land Administration Law [3] (art. 210). The property right established by a land title remains even intact if it turns out later that the land title had been acquired fraudulently. In that case, damage compensation will be paid to the victim [4] (art. 147 nouveau). Each transfer of the property right, for example, upon purchase or inheritance, is only legally valid when the parcel concerned has a land title [3] (art. 17) and when the transaction is confirmed by means of a notarial deed registered in the national land administration [3] (art.157). The Land Administration Law also establishes the basic structure of the work process for the application for and issuance of a title, the establishment of additional rights and restrictions and the registration of transactions. This is subject to legal time periods. For example, ANDF is required to issue a title within 120 days after an accepted application, provided no formal objections have been submitted against the title application. If this happens, ANDF is no longer bound by the 120 days deadline (the deadline is then undefined) [6].

With the introduction of a state guaranteed property right and a legally established, uniform, issuance and maintenance process, the Benin government wants to end the prevailing uncertainty regarding real estate transactions that arose due to a wide variety of unclear procedures that were susceptible to official randomness and fraudulent practices, such as multiple sales of the same parcel and sales by parties other than the legal owner [7]. To facilitate the transition to the land title system, the Land Administration Act established a transition period. This period originally ran from 2013 through 2018, but the

2017 addendum of the Land Administration Law extended this period to 2023 [4] (art. 516 nouveau).

2. The Land Administration Problem

By setting a legal framework and establishing a centralized government body, Benin has decided upon a formal legal approach in order to realize a uniform land administration system that covers the entire country. Except for the legal transition period, no further specific measures have been taken to ease the transition to the legally prescribed title system. The following assumptions implicitly provide the support that this approach will deliver the envisioned legal certainty to Benin:

1. That the Land Administration Act compels all landowners, of their own accord, to apply for a land title within a foreseeable period;
2. That these title applications and resulting transactions generate a stream of revenue such that ANDF can provide its services in each municipality of Benin.

Seven years after the introduction of the Land Law in 2013, it appears these implicit assumptions will not be satisfied. The legal certainty has increased, but its effect is limited, since the transition to titles is progressing slower than expected. In April 2017, there were about 40,000 land titles issued [8]. By January 2021, this number had risen to about 60,000 [9]. This is only a very small part of the total number of Benin parcels to be registered, which ANDF estimates at about 5 million. It is obvious that this process will not succeed in registering land rights for a substantial portion of the Benin territory by 2023, the end of the transition period. The majority of the estimated 5 million parcels will then still be without title.

The reason for the slow increase in the number of titles is associated with the costs. The reference point for the determination of the price for the new land title was the situation before the introduction of the 2013 Land Law, when it cost a lot of time and money to obtain an official confirmation of landownership. The application process could take many years, where it was uncertain for the applicant all that time whether the desired certainty would be obtained [10] (pp. 8–10). With the introduction of the guaranteed lead time of 120 days and a fixed price of 100,000 West African CFA franc (180 US dollars) for the application of a land title at ANDF, the government of Benin provided a significant improvement. This improvement, however, is not enough. In addition to this amount, there are also the costs of the (municipal) documents to be obtained before the application and the mandatory use of a certified private surveyor. These costs vary and depend among other things on the parcel size, but ANDF estimates, on the basis of internal calculations, the total cost for obtaining a land title for a parcel of 500 m^2 at 300,000 CFA (540 US dollars). With a national headcount poverty rate estimated at 40.1% in 2015 based on the international poverty threshold set at $1.90 per person per day in purchasing power parity [11], the price for a land title is still far too high for most of the Benin population.

The lag in the volume of titles can lead to a negative spiral. Before the introduction of the Land Law, it was possible to formalize real estate transactions at municipal level by recording them in municipal property registers. This decentralized system was inefficient and susceptible to fraud and corruption, but it was often cheaper for the landowners and provided a stable source of income to the municipalities involved [10] (p. 41). In order to give them the opportunity to adapt, the 2013 Land Law provides in art. 516 a five-year transition period. This period was extended until 2023 by the 2017 adaptation of art. 516. Because the vulnerable groups that have the most interest in the legal certainty of the new titles cannot afford them and municipalities benefit from the continuation of the old method (and they are also allowed to do so during the transition period), there are, in practice, few incentives to comply with the new law. Parties are also not yet convinced of the enforceability of the title system. Benin has a history of unsuccessful land reforms [10] (p. 2) and a large informal economy of almost 60% of Gross Domestic Product [12] because rules are often not enforced. These past experiences lead to a wait and see attitude. The fact that the transition period has already been extended once and a new extension appears necessary, feeds the sepsis.

The lagging title flow has the negative effect that less money is coming in than expected. This leads to delays in the plans to open an ANDF office in each municipality. As a result, ANDF is less visible and increases the distance to the potential users, which reinforces the negative spiral and increases the likelihood that the increase in the necessary legal certainty envisioned by the legislator will not be achieved.

3. Addressing the Problem

3.1. Hypothesis

To support the Benin government, the Netherlands Embassy in Benin has funded a four-year project: le Projet de Modernisation de l'Administration Foncière (abbreviated as PMAF). In this project, staff from ANDF and Benin and Dutch land administration experts are working together in order to speed up the introduction of a sustainable national land administration.

The approach of the PMAF project is based on a closer analysis of the existing practice of land administration that assumes state-guaranteed property rights, with a land title (titre foncier) as tangible evidence. The major benefits of this system are the high degree of legal certainty and a sound recording through the mandatory registration of notarial deeds. The Benin government has also prepared well for the implementation of this system. The procedures and methods for issuing titles and the registration of transactions have been legally secured and translated into implementation rules and work procedures. The ANDF personnel has been trained to apply the instructions and an Information Technology system (Système d'Information Foncière) has been developed that supports the work process. ANDF and PMAF both considered this to be a solid foundation, provided that it would succeed in increasing the number of titles by lowering the threshold for obtaining the first title. The system could then maintain itself because each buyer understands that he/she runs too great risk if he/she does not have the title put in his/her name by AND, and there are also fewer financial thresholds, since there is no unit price for settling most transactions, but a rate based on the value of the parcel (currently this is usually 0.3% of the current market value). In this context, there is a need for an approach that is cost and time efficient for the initial establishment of the land administration. Such an approach was developed in the so-called Fit-For-Purpose Land Administration. This approach entails a land administration that has been designed to achieve tenure security for all within a relatively short time, with relatively low costs and applied within the legal, spatial, and institutional framework. Fit-For-Purpose Land Administration was jointly developed by the Fédération Internationale des Géomètres and the World Bank. The concepts and guidelines were first published in 2014 [13] and 2015 [14].

The analysis of the PMAF project led to the following hypothesis to have the envisioned system operate:

that a Fit-For-Purpose Land Administration approach
makes it possible to provide a land title (Titre Foncier) to all of the parcels
in a village or neighbourhood in a limited time for little money.

This opens up the possibility for each landowner to acquire a first land title affordably, or possibly even free of charge, if this approach is adopted in a National Program paid for by the Benin government or donors.

To validate the hypothesis, the PMAF project planned a field test. For this purpose, four geographically dispersed municipalities (Bohicon, Tori Bossito, Sakété, and N'Dali) were selected, consisting of urban, rural, and mixed areas. Figure 2 gives an impression of the differences between the urban and rural test sites.

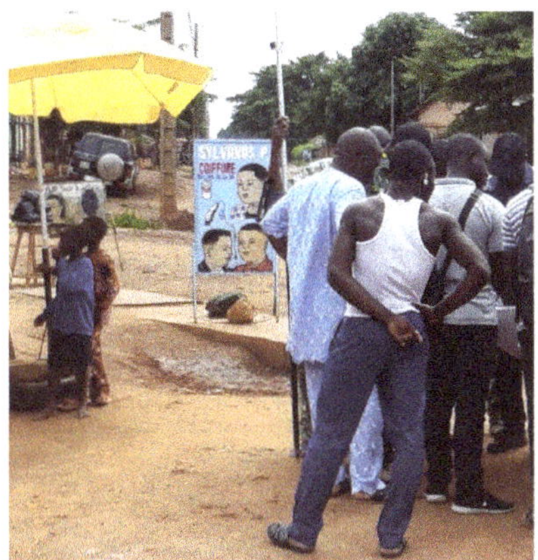

Figure 2. Test site in Bohicon (urban) and in Tori Bossito (rural).

This mix was chosen to gain insight into potential cultural and area-specific issues in elaborating a fit-for-purpose approach in the Benin context. The tested approach included using para surveyors from the area with knowledge of the local language and using the local land committees (Section Villageoise/Urbaine de Gestion Foncière) to inform the residents. Over the course of the test, this approach was improved by accepting testimony for claiming property rights, using members of the local land committees as mediators for problem solving and accepting a relatively low accuracy for surveying parcel boundaries.

3.2. Conceptual Framework

For the elaboration of the proposed approach two widely accepted frameworks have been used:

The earlier introduced Fit-For-Purpose Land Administration (FFP LA) approach combined with a field test for verifying if FFP LA will be a solution for the Benin challenges.

The Land Administration Domain Model (LADM) [15] as solution for using the results of the Fit-For-Purpose land administration approach in a future proof land information system [16,17].

3.3. Result of the Field Test of the Fit-For-Purpose Land Administration Approach

In the three months of the test period, spatial and administrative data for a total of 2349 parcels with a total surface area of 3500 hectares were collected in the four selected municipalities. The test proved that a fit-for-purpose approach can achieve fast production at low cost. The field teams produced data for an average of 11 parcels per working day with an average cost of 18 US dollars per parcel. This amount includes all costs directly related to data collection (the labor costs of the field teams, the cost of the collection equipment, the cost of the boundary stones, the cost of guidance and mediation by the local land committees, the cost of communication and publication and the cost of post-processing and quality control). One-off costs such as the development of methodologies and procedures and the construction of the central land administration information system (the digital cadaster) are not included in the costs per parcel. These are covered from the PMAF project budget.

However, the test also showed that the fast production and low price are only feasible if the original hypothesis of a title as end result is relinquished and a lower level of legal certainty is tolerated where the ownership is not established as absolute right but as a verified presumption of ownership.

In contrast to the property right documented by a title, this presumed ownership is not absolute but can be disputed. It is applicable until a legal procedure decides otherwise. The introduction of the possibility for later correction provides space for a less formal, faster, and cheaper approach that better aligns with the political goal of quickly achieving a land registration that covers the entire country. This modified assumption made it possible to apply cost-reducing and speed-increasing measures as part of replacing the existing way of registering land rights by a fit-for-purpose approach tailored to the Benin context. These measures were partly derived from the method for building a communal cadaster in rural areas of Benin, the creation of a Plan Foncier Rural [18], and consisted of:

(1) Allowing testimonies as a basis for establishing the presumed ownership that will be confirmed by means of a one-time inspection open to the public in the village. This is a break with the current practice where owners must always present a document when applying for a title in order to demonstrate their presumed ownership. Working with testimonies is efficient but could lead to institutionalized land grabbing in specific cases, such as with unresolved estates. This risk is manageable because any victims can still rectify it later in court. This rectification is not possible with a title;

(2) Engaging members of the local land administration committees to resolve disputes through mediation. The Land law in Benin regulates that each village and each municipality must have a land administration committee, une Section Villageoise de Gestion Foncière [3] (art. 428). This committee consists of residents with authority who have been elected by the population to represent a particular interest group (such as women and young people). The land administration committee members are familiar with the local situation and can therefore mediate when there are conflicts regarding ownership or the parcel borders. To fulfil this task, the committee members must be trained and receive remuneration for their effort, but this is cheaper than hiring external mediators or legal experts. This grass roots mediation is not possible in the title process, where conflicts are directed straight to the court and a title parcel is recorded in the registration only when the court has passed judgment. In the alternative approach a parcel is registered as a conflict parcel when the mediation was unsuccessful, and a court ruling proves necessary. This does not delay the registration process and allows potential buyers and other parties with an interest in a parcel to see that there is a conflict as long as the case is still in court;

(3) Accelerating the measurement process by allowing the landowners to play a more active role. This additional participatory step allows the involved landowners themselves marking the parcel boundaries using locally produced boundary stones;

(4) Accepting a lower measurement accuracy of one meter for all cadastral parcels (urban and rural), rather than aiming for precision in terms of centimeters. This lower accuracy allows the use of simple measurement equipment and software that can be operated after limited training. As a result, relatively expensive certified land surveyors are only needed for specialized tasks like supervision and quality control. When issuing a title, such inaccuracy would be unacceptable because the stated surface area of the parcel has a legal meaning and therefore cannot change later on. That is why, with a title, the highest surveying standards must be satisfied immediately.

In the methods section, the last two changes are explained in more detail.

3.4. LADM-Based IT Architecture

After the field test, ANDF and PMAF agreed that adding the registration of presumed ownership is necessary in order to meet the goals of the Benin government. To realize this, not only the legal framework and the processes and procedures need to be adapted, but also the IT architecture. Important architectural challenges were to make the legal status of registered parcels as soon as possible visible to the public and to show both the parcels collected with the existing title process and the "fit-for-purpose" collected parcels. For this IT innovation, ANDF wanted to benefit from the knowledge and experience of PMAF.

A specific point of attention in the ANDF-PMAF partnership in terms of data and innovation was the development of a land administration information system based on international standards and best practices. The most important innovation was the introduction of a data-centric architecture (Figure 3), where data and processes are decoupled [19]. This decoupling enables the database to feed the central information data store, the digital cadaster, from different sources. The use of the LADM as an overarching standard, with a profile tailored to Benin's needs [20], ensures that parcel data from these different sources and with different levels of quality can be linked so that the user can always understand which level of legal certainty is provided for the cadastral parcel he or she is interested in.

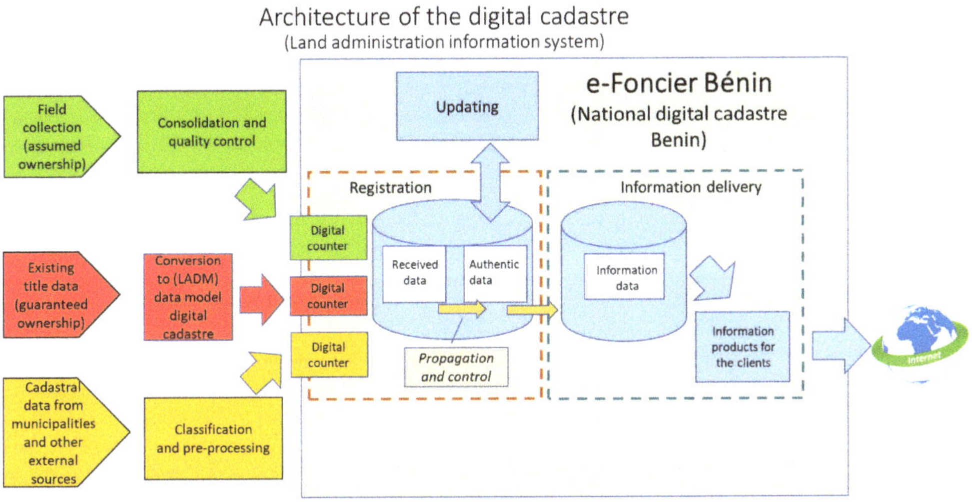

Figure 3. Architecture of the digital cadaster.

The following figure shows how this architecture is constructed.

The green flow consists of owner data and parcel boundaries that are collected by field project teams. This parcel data is, at the village level, consolidated by the providers responsible for a particular work area (the software components used for this process are explained in more detail in Section 4.3.1). The provider creates an input file from this bundled data and sends it electronically to the central IT system for validation and further processing under the control of the national land administration agency (ANDF). The central IT system then provides the services required for public viewing of the collected and consolidated parcel data (such as a draft cadastral map of the village) and the other process steps required to give the field data legal status as authentic cadastral data. This dataflow regards parcels collected in accordance with the agreed upon Fit-For-Purpose approach, hence with a presumed ownership based on a participatory approach that includes room for testimonial-based ownership claims. This flow is a new arrangement, enabling the imposition of procedural and technical standards on the parties that are contracted for the field data collection. Of these, the most important requirement is that the data collection software they use must provide data that is compliant with the Benin LADM profile. This means that the delivered data does not have to be converted but only checked before it can be incorporated in the central database. As the input data is standardized, these checks can also be largely automated.

The red flow regards data from the current titling process. For this purpose, an information system with its own data model already exists. A conversion module allows the core data from this system to be converted and duplicated automatically into a LADM-compliant data so that it can also be incorporated into the central database of the national digital cadaster. In doing so, the existing IT system does not need to be adjusted. Thus, the investment made continues to pay off and it relieves the ANDF organization because the current work method for processing title applications and linking rights and restrictions to titles can continue. Furthermore, the decoupling also ensures that any delay in the construction of the digital cadaster does not adversely affect the realization of the legal time periods for issuing the titles. After all, the IT system required for this purpose functions autonomously.

The orange flow regards the reading in of cadastral data from external parties. This is mainly data from existing municipal cadastral administrations. These data are very diverse, which sets specific requirements for the conversion and control. For example, it may require that paper plans and registers can be provided, which must first be digitized before they can be processed. It may also be necessary to adjust them geometrically again because, for example, they were collected in a local coordinate system. Because this conversion can cost a lot of time and money, an intake is always done first to determine whether the conversion of an external data source is worth the effort. In certain cases, it is more efficient not to use the data source and to collect the data once again in the relevant area (to use the green input flow). In the future, this input stream may also be used to record usage rights, as registered by the municipalities, in the national land register.

The update process has also been decoupled. A "Fit-For-Purpose" update process (the blue "updating" block in Figure 4) is being developed to keep track of parcels of land with assumed ownership, allowing frequent transactions such as sales and heritage to be carried out more easily and with less costs. An important element in this new update process is the use of forms from the cadastral database instead of the mandatory use of notarial deeds. In line with the Fit-For-Purpose approach, this means that the maintenance process for parcels with a presumed ownership is limited to common transactions such as sales and inheritance. Relatively rare transactions requiring a high degree of legal certainty, such as the provision of a mortgage, will not be supported. However, the existing title information system does support this kind of transaction and by means of the "red" stream they can also be processed in the cadastral database. In this case also, the use of the LADM profile ensures that these different maintenance transaction flows are displayed in a coherent manner.

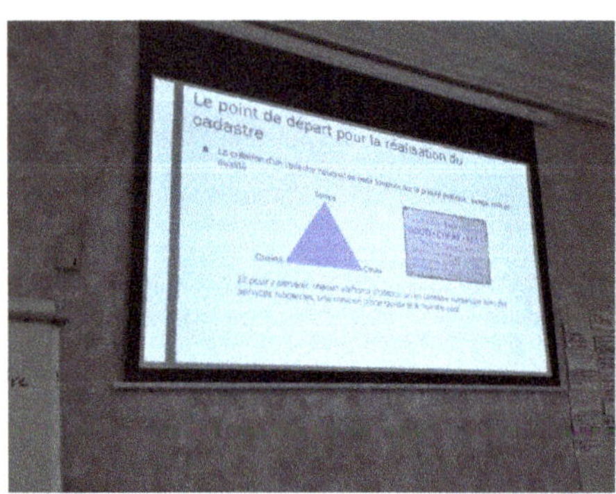

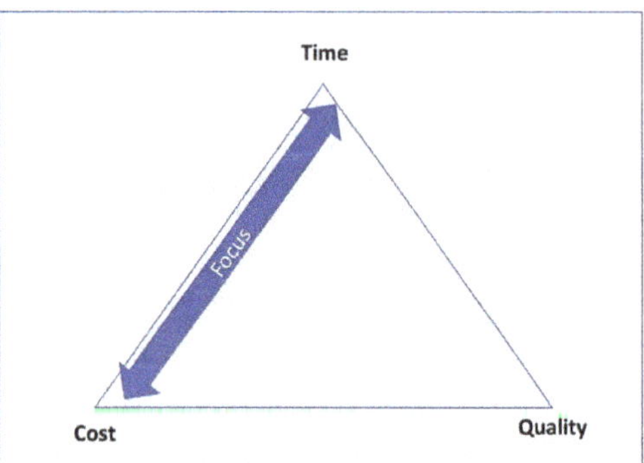

Figure 4. Purpose workshop and the iron triangle with the chosen focus.

Another important element in the "Fit-For-Purpose" maintenance process and the Benin LADM profile is that mutation transactions are based on consulting the current parcel status in the national digital cadaster. The linking of persons to assumed property rights as registered there is considered to be "the truth". Owners therefore do not need to submit paper documents to prove their ownership rights in the case of property transactions. After processing a transaction, the (new) owner does receive a printed notification, so that he can check whether the transaction was processed correctly, but this document plays no formal role in subsequent transactions. This digital working is in line with the Benin government's aspiration to become a smart government [21] (p. 38).

4. Materials and Methods

4.1. Method for Defining the Purpose

The most important step in the approach to establish a national Benin land register was to define the goal. The original emphasis in Benin, as in many former French colonies in West Africa, was on the accurate administrative registration of the land combined with the issuance of titles that provided the title holder with a high degree of legal protection [9] (p. 3). Traditionally, therefore, this was the purpose of land registration. As explained in Section 2 of this paper, this high degree of legal security was accompanied by high costs and a slow pace of registration. This raised the question for the PMAF project whether there should not be a reorientation of the objective. To clarify this, the "iron triangle / project management triangle" [22] was used.

In workshops with the ANDF management and stakeholders, the focus for the realization of the national land register was discussed using the axes of the iron triangle (Figure 4). Because the application of the iron triangle stimulates the parties involved to speak out about priorities, since of the three possible aspects: time, cost, and quality, only two can be given priority, a shared picture emerged about the desirable focus. In these workshops it soon became clear that time was the most important factor. Because the realization of the national land register is part of the priority list of the president of Benin, as elaborated in the National Government Action Plan, le Programme d'actions du gouvernement 2016–2021 [21] (p. 76), ANDF was under heavy pressure from politicians who wanted to see visible results. Although the workshop participants recognized that a project that had only started in 2019 could not be expected to complete the national land register by 2021 (the year of presidential elections), it was necessary to prove in practice that a rapid nationwide coverage of a national land register was feasible before the plan ended.

In addition to speed, cost also played an important role in the political context. The National Government Action Plan is very ambitious and covers a large number of areas in which Benin wants to develop. For this development, the Benin government is heavily dependent on external funding: 2829 billion CFA (5.14 billion US dollars), compared to 700 billion CFA (1.27 billion US dollars) from its own resources [21] (p. 82–85). Because the development of the national cadaster must compete with other projects in attracting external funding, it was important that it could not only be shown to be quick to implement, but that it could also be done efficiently, at a low price per parcel. Thus, in addition to time, cost was the second key element.

The discussions about focus and priority based on the triangle thus provided a different strategic orientation than the legal administrative perspective of the current land administration process, which focuses on quality.

The fact that ANDF and the stakeholders agreed that it was not quality but time and costs that should be central, was a big change that was very decisive for being able to apply the Fit-For-Purpose Land Administration approach in Benin.

4.2. Method for Identifying the Parcel Boundaries

A specific part of the Benin approach is that the residents themselves mark their property boundaries before the arrival of the surveyors. This approach consists of several steps. The process starts with the strengthening of the land management committee of the village or neighborhood (la Section Villageoise de Gestion Foncière) so that their members can help explain and prepare the process of participatory demarcation. The next step is that the PMAF project provides materials and engages local craftsmen so that the boundary stones can be made locally. When the boundary stones are ready, the local land committee distributes them to the landowners and instructs them on their proper use. Finally, the landowners use them to demarcate the common parcel boundary in consultation with their neighbors (Figure 5). Members of the land management committee oversee this process and are available to mediate if the residents cannot immediately agree.

Figure 5. Monumentation of a boundary by an owner.

This method promotes the active participation of the landowners to a great extent. After all, they determine and mark the boundaries of the parcels themselves. Moreover, making the boundary stones on the spot is a visible event that helps to involve the village population and creates goodwill because local craftsmen can earn some money with it. The method is also efficient. If all the parcels are demarcated, the parcel boundaries are clear and the (para)surveyor only needs to record the coordinates of the stones and verify that both neighbors agree. This verification can take place asynchronously. If both neighbors separately indicate the same stone as the boundary, it can be assumed that they are in agreement [23]. This preparatory step avoids that the measuring teams must wait until both owners are on site to indicate the location of the parcel boundary. The latter can take a long time in rural areas.

The use of boundary stones in combination with the acquisition of boundary coordinates in the field differs from the application of the fit-for-purpose approach, as described by Enemark et al., where boundaries are drawn on aerial photographs or satellite imagery. This approach was also tested in Benin but proved not to be feasible, due to the poor quality of the national aerial photograph set and because the boundaries in the terrain are often not clearly visible, due to vegetation. Field tests showed that the owners and their neighbors could only indicate the boundary of their parcel on the aerial photograph on site, and sometimes only with the assistance of a project team member, as can be seen in Figure 6. Identification of the boundaries using this imagery at a central place in the village was difficult for the rightsholders.

Due to these issues, the usage of aerial photography did not save any time compared to other methods, such as the designation and measurement of (pre-positioned) boundary points. However, the procedures are designed in such a way that this option can still be added at a later stage, for example for the registration of very large parcels of land, such as a national park.

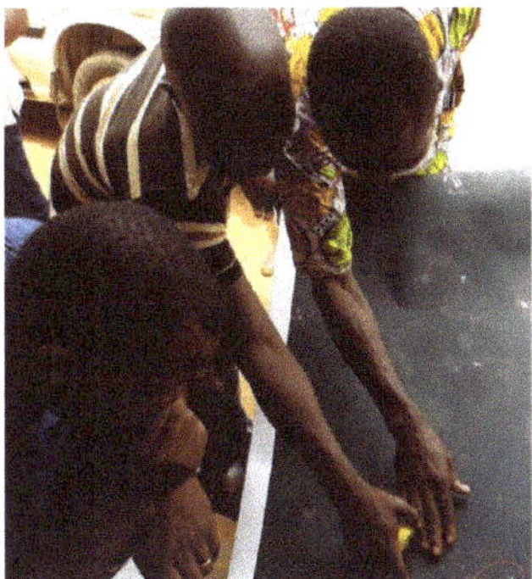

Figure 6. Field tests using aerial photography to identify parcels.

4.3. Approach for Choosing Materials for the Collection of Field Data

4.3.1. GNSS Receiver

The main criterion for choosing the hardware and software for field data acquisition was the cost in relation to the minimum quality requirements. When the field test showed that the use of satellite images was not saving time in the Benin landscape, the project moved to applying Global Navigation Satellite System (GNSS) receivers. A GNSS receiver calculates a position on earth by measuring the "time" that a radio signal travels between the satellite and the receiver. They are widely used for navigation ("GPS"), but they can also be used to determine the coordinates of parcel boundaries. Three categories of GNSS receivers were tested to gain insight into the quality of the measured boundaries under real-life conditions. These tests were conducted by ANDF and representatives of the Institut Geographique National du Bénin (IGN). IGN is responsible for the certification of land surveying work in Benin. ANDF and IGN were particularly interested in the accuracy with which boundaries could be measured. The following categories of solutions were tested:

1. Built-in receiver Android tablet (Galaxy Tab2);
2. Budget professional GNSS receiver (Emlid Reach RS);
3. High end professional GNSS receiver (Navcom S-3040).

In the test, the Galaxy Tab2 Android tablet achieved a varying accuracy between 2 and 6 m. The Emlid achieved an accuracy of at least 30 cm and the Navcom achieved an accuracy of at least 15 cm. When discussing the test results with ANDF, it was concluded that budget professional GNSS receivers offer the optimal balance between quality and cost. Based on this, ANDF in consultation with its stakeholders determined that the boundaries of parcels of land with a presumed ownership right should be surveyed with a minimum accuracy of one meter. ANDF estimates that this provides sufficient margin to ensure that budget GNSS equipment can be used everywhere in Benin.

The choice of the one-meter tolerance is not based on scientific research but is the result of the search for a compromise between accuracy and costs, considering the context of Benin, where parcels of land normally are surveyed by certified surveyors. According to ANDF and IGN, these surveyors apply a professional standard whereby for parcels with a land title a precision of at least 15 cm must be achieved, due to the legal status of the registered surface. With this as a starting point, several scenarios were developed in which precision was linked to possible functions of a cadaster. The scenario with boundary coordinates with 1 m standard deviation for all cadastral parcels, regardless of whether they are located in urban or rural areas, appeared to offer the best balance between efficiency and fulfilling the legal, fiscal, and technical functions of the cadaster, as described in the Benin Land Registration Act [2] (art. 453).

The average parcel price mentioned in Section 3.3 is based on the use of budget GNSS receivers (with the price of a Emlid kit as a reference), assuming a price of 1,260,000 CFA (2290 US dollars) per unit and a depreciation period of three years. Since equipment in

Benin must endure a lot because of the tropical climate and rough terrain, a relatively short depreciation period was chosen. The Emlid Reach should be considered as an example. This equipment was used in the test, but for larger scale data collection, the PMAF project will again choose from "budget" GNSS receivers available in Benin. This will involve applying modified selection criteria that incorporate the experiences from the test.

4.3.2. Data Collection Software

Cost reduction was also a key consideration in the selection of software for the recording of parcels and rights. To save on personnel costs, the first preference was for software packages that could process the geographical and administrative/social parcel data in an integrated manner, so that one person could register a parcel. However, a study conducted by PMAF showed that such integrated software is relatively expensive [24]. At the time of the research (July–December 2019), USAID's Mobile Applications to Support Tenure (MAST) appeared to be the only available free open-source solution that could meet the stated functionality requirements. At the time, however, PMAF did not succeed in getting MAST working because the publicly available software libraries were outdated. It also became clear at that time that there were not any consultants specialized in MAST available who could support PMAF, which meant that this solution was discarded. Besides MAST, only a limited number of commercial suites were found that integrated the required geo- and socio-functionality. The licensing costs for these were so high, however, that they were also discarded. The fact that labor costs in Benin are very low played an important role in this decision. Based on the personnel costs for the field test [25], data collection by one person would save only 160,000 CFA (290 US dollars) per month, whereby this saving would be partly cancelled out by extra intake time because actions such as registering and photographing identity documents can no longer be performed in parallel.

For these reasons, PMAF has chosen to split the functionality for collecting administrative parcel data and the functionality for collecting parcel boundaries. For the collection of administrative data, the PMAF project has contracted a local software developer to develop an open-source Android socio-app adapted to the LADM profile for Benin. A low-cost commercial Android geodata collection app, such as Locus GIS, will be selected for on-site capture and verification of the parcel perimeter. This app will process the coordinates of the measured boundary points (such as the placed boundary stones) provided by the budget GNSS receiver. PMAF will also have the developer of the socio-app develop a consolidation application for the integration of the data from both applications. This integration is done on the basis of assigning temporary (project) parcel numbers, with team members exchanging them so that they each enter the same number for the same parcel in their application. When the field data are processed in the national digital cadaster, this temporary project number is replaced by a definitive unique identifying parcel number, with the original project number remaining available as a reference. The "consolidation" application also takes care of bundling parcel data at village level in data files that meet the interface specifications of the central information system (the digital cadaster).

In the calculation of the average parcel price mentioned in Section 3.3, a one-off investment of 1,950,000 CFA (3545 US dollars) was assumed for building the socio-application and the consolidation application, on the basis of previous experience with Beninese software development. For the use of the geo-application, an amount of 36,000 CFA (65 US dollars) per device per year was assumed, on the basis of the offers available in the Google play store.

4.4. Methods of Promoting Participation by Women and Youth

An important goal of the PMAF project is the active participation of women and youth in building the national land register and claiming their property rights. This is given a lot of attention in the practical elaboration of the data collection. An important element is the use of role models. In the selection of the parties who collect the data, the proportion of young people and women in the measuring teams is a criterion. In addition, attention has been paid to ensuring that women and young people play an active role in the visual language of the information materials, as can be seen in Figure 7.

Figure 7. Example of information material.

Another measure to ensure that the position of women is strengthened during the collection and registration of cadastral parcels was the hiring of an NGO, named "Cercle d'Action pour le Développement Durable et la Paix" (abbreviated CADDUP). Its task is to ensure that women actively stand up for their rights and to ensure that their interests are not consciously or unconsciously violated when procedures and working methods are implemented. This NGO has its own budget from PMAF, is independent from the parties that provide the data collection, and has no ties with ANDF and other cadastral institutions. An important activity of CADDUP is organizing, training, and facilitating local vigilance committees that help women stand up for their rights and that can act as representatives of women who cannot demand these rights themselves because they fear reprisals from their social group.

5. Discussion

The case study demonstrates how a traditional high-quality land administration system can be combined with another, fit-for-purpose, land administration system in order to speed up the implementation of parcellation with nation-wide coverage. However, this promise is based on a limited test and has not yet been validated at larger scale. Whether it will become a reality depends on several assumptions that need to be further investigated. Part of that research will be carried out by the PMAF project when, in the period 2021–2023, data for approximately 200,000 parcels of land have been collected with the presented Fit-For-Purpose approach will be made accessible in the newly created digital cadaster.

The assumptions that will be further investigated are listed below, including a short description of the measures that will be taken by the PMAF project to fulfil the expectations or to limit the damage when that is not possible:

(1) *A legal basis is realized in a timely manner.*

The current law considers a digital cadaster, and this is positioned as an essential guarantee for certainty about the ownership of land. The legislator has given this digital national cadaster a broader scope than just the disclosure of titles, where the exact content can be arranged by decree. The ANDF legal experts expect that the creation of this decree can introduce the alternative of a "presumption of ownership", by registering it as such in "the cadaster". This expectation is based on an interpretation of chapter 5 of the Benin land law which outlines the organization and functions of the cadaster. Because a decree can be established relatively easily in Benin, it is possible to quickly give a legal basis to the collected fit for purpose parcel data. Because this is such an important assumption, the project is providing funds for external specialists to carry out a legal analysis to independently determine what is needed to legally anchor the use of the Fit-For-Purpose procedures for the construction and maintenance of the national digital cadaster. By giving this priority, ANDF and PMAF will know what else needs to occur in the legal area and whether there is sufficient time available to arrange it. With this knowledge, additional resources can be organized, if necessary, to adjust the legal framework or in the worst case, the data collection can be delayed until there is an adequate legal basis for use and maintenance;

(2) *The people of Benin have confidence in the national digital cadaster and its "Fit-For-Purpose" construction and maintenance.*

The approach described creates a new legal, institutional, and technical ecosystem for the registration and mutation of property rights. This system will only produce the intended effects if users understand its importance and have confidence in its proper functioning. ANDF and the PMAF project are aware of the importance of creating trust and this is also included in the outcomes the project must achieve. These outcomes are operationalized by actively involving representatives of chain partners and population groups, and by basing the development of procedures and products on the interests and needs of both the population (the demand side) and the agencies involved (the supply side). However, much will depend on the way the Benin government is going to apply the system. If the latter uses the cadaster mainly to levy taxes and not as a tool for a fairer and better use of land, the population will avoid the land register and fall back on the current informal practices. The same will happen if there is no sustained investment in the knowledge and capacity of the actors involved. In that case, bureaucracy will return, and vulnerable groups will continue to face impregnable barriers. By having important guiding principles embedded in strategic government plans and by developing customer oriented sustainable business models, PMAF is trying to mitigate these risks (see also the discussion points below), but whether this is adequate will have to be seen in the years following the project's completion;

(3) *A pricing model is introduced that offers affordable legal certainty for everyone.*

In order to not exclude any parties, the price for transactions on parcels with presumed property must be adjusted to the ability of the Benin population to pay and be in proportion to the value of the cadastral parcel involved in the transaction. Various solutions are available for this purpose, such as a rate based on the sale price or sliding scales based on the location and the surface area of the cadastral parcel. It is mainly the role of PMAF to provide alternative business models and to show with good communication to the partners in the chain that a low rate combined with a large volume will benefit everyone.

(4) *Property transactions can be carried out easily.*

In addition to making transactions affordable, the users must also be able to perform them easily. This regards very practical things such as taking illiteracy into account and preventing long travel times because motorized transport is scarce and relatively expensive in rural areas. An important task here for the project is to provide efficient and simple maintenance procedures and to ensure that no process steps are added that make it unnecessarily more complicated for the users. In addition, ANDF must mobilize the political support that is needed to move parties, which have not yet embraced thinking from the user standpoint, in the right direction;

(5) *Certified surveyors can be used without driving up the cost.*

Currently, the intention is to grant the first tranche of the field data collection to established surveyor agencies. This well-organized professional group of certified surveyors views the project with much interest and obviously sees this as an opportunity to increase its revenue. Because the surveyor agencies have a lot of knowledge and experience and can mobilize a lot of "opposing force" if they should be passed over, it has been decided to first outsource the collection of data to them. The proposal request will provide the tested work method and the corresponding cost as a framework to stimulate sharp pricing. As the collection of data is marketed in different tranches, other parties may also be involved at a later stage, if subcontracting to certified surveyors' agencies proves too expensive;

(6) *A separate GNSS antenna is required to register parcel boundaries.*

The choice made in Benin to base the registration of all cadastral parcels on an accuracy of at least 1 m may makes it possible to register parcel boundaries with the help of a smart phone's internal GNSS antenna. Currently, smart phones that can reliably deliver this precision are not yet common, but it is possible that they will be by the end of 2022, when the PMAF project ends. PMAF will follow developments in this area as the use of standard smart phones in combination with user-friendly collection applications may save significant costs compared to the current situation where professional surveying equipment is used.

(7) *The tenure security of women and other vulnerable groups is reinforced.*

When recording property rights, it is important that this occurs with specific attention to vulnerable groups, including women and young people. This approach has been evaluated and developed further using the gender evaluation criteria for large-scale land tools [26]. The approach registers the ownership of men as well as women. In many cases, a usage right is granted for women's access to land. To accelerate the data collection, the decision has been made to only record the property right and refrain from registering secondary usage rights. This issue was one of the difficult considerations between speed and quality. This choice was determined to be acceptable based on the assumption that when arranging the property right, landowners are more willing to grant long-term usage right because they then no longer run the risk that a claim on the land can be placed through this use. This would still indirectly reinforce the legal certainty of women and young people, with the added effect that they are more willing to invest more in working the land, which will increase the food security in time. Moreover, the LADM profile and the system architecture of the national digital land registry are designed to allow for the later addition of secondary usage rights, for example by granting specific updating rights to municipalities, so that they can populate the cadastral database on the basis of agreements drawn up at local level that define specific land use agreements.

(8) *The entire population of Benin will benefit from the project results.*

As explained earlier, Benin has a long history of failed land reform and, at the moment, land rights are still mainly something for the rich. However, the PMAF project's motto is "Sécurité foncière pour tous" (land rights for all), and ANDF and PMAF truly believe that by applying "Fit-For-Purpose the Benin way", it will be different this time. Because food security and inclusiveness are key policy priorities of the Dutch embassy (the formal sponsor of the project), it is important that it can be proven that the faith in the benefits of introducing a national "Fit-For-Purpose" digital cadaster is justified and that the position of vulnerable groups actually improves. Performance indicators have been established and a monitoring system has been set up to verify this.

To verify that the presented benefits are realized in a sustainable manner, it may be useful to repeat this case study when the PMAF project is completed.

6. Conclusions

The Benin case shows that the hypothesis that a Fit-For-Purpose approach could speed up the realization of a nationwide land administration is valid and that it is possible and advantageous to combine this with maintaining existing, more traditional, land administration systems.

The Benin example also shows that combining two different approaches for registering land rights with the development of a Fit-For-Purpose information system based on a situation-specific LADM profile, offers the possibility to include both in the information provision. By doing so, stakeholders can easily see what level of legal certainty applies to a cadastral parcel and thus what procedure they need to follow to carry out transactions. The the IT architecture presented also shows how existing information systems can be exploited so that no complex migration process is required, and existing IT investments continue to pay off.

The case demonstrates that there are several ways of applying a Fit-For-Purpose approach in land administration and that it can be necessary to adapt it to better match the context of the country in which it is applied. After all, the political officials of a country are the ones who determine the "purpose" in a fit for purpose land administration approach. This is a reality that a project must be able to handle. The responsible parties, ANDF and PMAF, believe that a "Fit-For-Purpose approach in the Benin way" will be up to the challenge. Whether that is actually the case will have to be proven in the remaining lifetime of the project.

Author Contributions: Conceptualization, S.M. and F.G.K.; methodology, S.M. and F.G.K.; writing—original draft preparation, S.M.; writing—review and editing, D.V.K. and F.G.K. All authors have read and agreed to the published version of the manuscript.

Funding: This research received no external funding. The funding of the publication costs for this article has kindly been provided by the School of Land Administration Studies, University of Twente, in combination with Kadaster International, The Netherlands.

Acknowledgments: The case study is based on the activities of the Projet de la Modernisation de l'Administration Foncière (PMAF, www.projectpmaf.com, accessed on 25 March 2021) The PMAF project is executed by a consortium of three parties: MDF Training and Consultancy, VNG International, and Kadaster International and is carried out in close cooperation with l'Agence Nationale du Domaine et du Foncier, République du Bénin (www.andf.bj, accessed on 25 March 2021). The PMAF-project is funded by the Embassy of the Kingdom of the Netherlands in Cotonou, Benin.

Conflicts of Interest: The authors declare no conflict of interest.

References

1. Institut National de la Statistique et de l'Analyse Economique. *Annuaire Statistique 2019*. Available online: https://insae.bj/publications/publications-annuelles#annuaire-statistique (accessed on 17 April 2021).
2. United Nations Development Programme. *Human Development Report 2020. The Next Frontier: Human Development and the Anthropocene*; United Nations Development Programme: New York, NY, USA, 2020. Available online: http://hdr.undp.org/en/content/human-development-report-2020 (accessed on 25 March 2021).
3. Présidence de la République du Bénin. *LOI N02013-01 DU 14 AOÛT 2013 Portant Code Foncier et Domanial en République du Bénin*. Available online: https://www.andf.bj/index.php/le-foncier-au-benin/textes-sur-le-foncier/loi-2013-01-du-14-aout-2013/download (accessed on 25 March 2021).
4. Présidence de la République du Bénin. *LOI 2017-15 DU 10 AOÛT 2017 Modifiant et Complétant la Loi n° 2013-01 du 14 août 2013 Portant Code Foncier et Domanial en République du Bénin*. Available online: https://www.andf.bj/index.php/le-foncier-au-benin/textes-sur-le-foncier/loi-2017-15-du-10-aout-2017-modifiant-et-completant-la-loi-2013-01/download (accessed on 25 March 2021).
5. Simpson, S.R. *Land Law and Registration*; Cambridge University Press: Cambridge, UK, 1976.
6. Agence National du Domaine et du Foncier—Confirmation de Droits Fonciers. Available online: https://www.andf.bj/index.php/procedures-et-demarches/confirmation-des-droits-fonciers (accessed on 25 March 2021).
7. Agence National du Domaine et du Foncier—Accès au Foncier. Available online: https://www.andf.bj/index.php/le-foncier-au-benin/acceder-au-foncier (accessed on 25 March 2021).

8. Agence National du Domaine et du Foncier—Le Foncier au Bénin en Chiffres. Available online: https://www.andf.bj/index.php/le-foncier-au-benin/le-foncier-en-chiffres (accessed on 25 March 2021).
9. Land Administration Modernisation Project (PMAF). Available online: https://www.projectpmaf.com/home (accessed on 25 March 2021).
10. Philippe Lavigne Delville. Chapter 7: History and Political Economy of Land Administration Reform in Benin. In *Benin Institutional Diagnostic*; Economic Development & Institutions, 2019. Available online: https://edi.opml.co.uk/resource/benin-land-rights-and-reforms (accessed on 25 March 2021).
11. The World Bank in Benin—Overview. Available online: https://www.worldbank.org/en/country/benin/overview (accessed on 25 March 2021).
12. International Monetary Fund. *The Structure of the Beninese Economy*; IMF Country Report No. 18/02; International Monetary Fund: Washington, DC, USA, 2017. Available online: https://www.imf.org/~{}/media/Files/Publications/CR/2018/cr1802.ashx (accessed on 25 March 2021).
13. Enemark, S.; Bell, K.C.; Lemmen, J.; McLaren, R. *Fit-For-Purpose Land Administration*; Joint FIG/Worldbank Publication, International Federation of Surveyors (FIG): Copenhagen, Denmark, 2014.
14. Enemark, S.; McLaren, R.; Lemmen, C. *Fit-For-Purpose Land Administration, Guiding Principles for Country Implementation*; United Nations Human Settlements Programme UN-Habitat: Nairobi, Kenya, 2016.
15. *Technical Committee: ISO/TC 211 Geographic Information/Geomatics. ISO 19152:2012 Geographic Information—Land Administration Domain Model (LADM)*; International Organization for Standardization: Geneva, Switzerland, 2012.
16. Lemmen, C.; Van Oosterom, P.; Bennett, R. The land administration domain model. *Land Use Policy* **2015**, *49*, 535–545. [CrossRef]
17. Van Oosterom, P.; Lemmen, C. The Land Administration Domain Model (LADM): Motivation, standardisation, application and further development. *Land Use Policy* **2015**, *49*, 527–534. [CrossRef]
18. République du Bénin, Ministère de l'Agriculture, de la Pêche et de l'Élevange. Resume de la Procedure du Plan Foncier Rural. Available online: https://rmportal.net/library/content/frame/resume-de-la-procedure-du-plan-foncier-rural.pdf/at_download/file (accessed on 25 March 2021).
19. Berdys, H. *SIF Architecture and Recommandations, Internal Project Documentation*; Projet de Modernisation de l'Administration Foncière: Cotonou, Bénin, Unpublished Work; 2020.
20. Van Staalduinen, J.; Lemmen, C.; Berdys, H.; Mekking, S.; Kossou, F. *Land Administration Domain Model (LADM)—Country Profile for Benin, Internal Project Documentation*; Projet de Modernisation de l'Administration Foncière: Cotonou, Bénin, Unpublished Work; 2020.
21. Présidence de la République du Bénin. Programme d'actions du Gouvernement 2016–2021. Available online: https://revealingbenin.com/telechargements/francais/ (accessed on 25 March 2021).
22. Barnes, M. Construction project management. *Int. J. Proj. Manag.* **1988**, *6*, 69–79. [CrossRef]
23. Morales Guarin, J.M.; Lemmen, C.H.J.; de By, R.; Molendijk, M.; Oosterbroek, E.-P.; Ortiz Davila, A.E. On the Design of a Modern and Generic Approach to Land Registration: The Colombia Experience. Presented at 8th Land Administration Domain Model Workshop 2019, Kuala Lumpur, Malaysia, 2019; pp. 217–243. Available online: https://wiki.tudelft.nl/pub/Research/ISO19152/LADM2019Workshop/LADM2019_paper_D1_F.pdf (accessed on 25 March 2021).
24. Oosterbroek, E.P.; Vranken, M. *Applications et Equipements GNSS Pour l'Acquisition Massive de Données de Terrain au Bénin, Internal Project Documentation*; Projet de Modernisation de l'Administration Foncière: Cotonou, Bénin, Unpublished Work; 2020.
25. Dossoumou, H.; Mekking, S. *Synthèse de Traitement des Données TOP Calculations, Internal Project Documentation*; Projet de Modernisation de l'Administration Foncière: Cotonou, Bénin, Unpublished Work; 2020.
26. The Global Land Tool Network Gender Evaluation Criteria for Large-Scale Land Tools. GLTN Secretariat Facilitated by UN-HABITAT, Nairobi, Kenya. Available online: https://gltn.net/download/gender-evaluation-criteria-for-large-scale-land-tools/ (accessed on 25 March 2021).

 land

MDPI

Article

The Fit for Purpose Land Administration Approach-Connecting People, Processes and Technology in Mozambique

Marisa Balas [1,*], João Carrilho [1] and Christiaan Lemmen [2,3]

1 Departamento de Ciências Sociais e de Gestão, Universidade Aberta, 1250-100 Lisboa, Portugal; jcarrilhoster@gmail.com
2 Faculty of Geo-Information Science and Earth Observation (ITC), University of Twente, 7514 AE Enschede, The Netherlands; chrit.lemmen@kadaster.nl
3 Kadaster International, Cadastre, Land Registry and Mapping Agency of the Netherlands, 7311 KZ Apeldoorn, The Netherlands
* Correspondence: marisa.balas@exi.co.mz; Tel.: +258-82-317-4880

Abstract: Mozambique started a massive land registration program to register five million parcels and delimitate four thousand communities. The results of the first two years of this program illustrated that the conventional methods utilized for the land tenure registration were too expensive and time-consuming and faced several data quality problems. The purpose of this research was to conceptualize, develop and test a country-specific Fit For Purpose Land Administration (FFPLA) approach for Mozambique, denominated as FFPLA-MOZ, intertwining three pillars: people, processes, and technology, to solve the constraints faced in systematic registrations. Such a contextualized approach needed to be: (i) in line with legislation; (ii) appropriate to the circumstances and needs of the systematic registration; (iii) cost-effective; (iv) based on available technology; and (v) fit to establish a sound and sustainable land administration system. By connecting people, processes, and technology, the FFPLA-MOZ approach achieved several benefits, including cost and time reduction, increased community satisfaction, and improved quality of work and data. The FFPLA-MOZ approach also supported a more robust community engagement through a more participatory land registration, denominated community-based crowdsourcing. Initial observations indicated that strong leadership and commitment were of extreme importance to ensure change management, capacity development, and project delivery for the success of these initiatives. The research only focused on the registration of land under good faith and customary occupations, as well as community delimitations. The next stages should focus on other land management activities and integrate other cadastres.

Keywords: fit-for-purpose land administration; land tenure security; community-based crowdsourcing; pro poor land recordation; innovative technology; SiGIT

Citation: Balas, M.; Carrilho, J.; Lemmen, C. The Fit for Purpose Land Administration Approach-Connecting People, Processes and Technology in Mozambique. *Land* **2021**, *10*, 818. https://doi.org/10.3390/land10080818

Academic Editor: Volker Beckmann

Received: 31 May 2021
Accepted: 28 July 2021
Published: 4 August 2021

1. Introduction

Access to land is a crucial element that allows women and men to play their full role in building peaceful, stable, and prosperous societies [1]. However, about 75 percent of the world's population cannot safeguard their land rights, a critical situation, especially for the poor and the most vulnerable in society [2]. The most recent data indicate that worldwide less than 15 percent of the land belongs to women [3], which constitutes a fundamental obstacle to women's economic and social empowerment, and consequently, an obstacle to sustainable development [4–8]. While some authors indicate that formal titling is needed to achieve market integration [9], others argue that existing customary systems often offer enough security of tenure and most initiatives to register customary land do not provide for the promised benefits [10]. Nevertheless, experience has proven that formal registration of land rights increases land tenure security [11].

The 2030 global agenda [12] recognized the need to build affordable, equitable, and sustainable land administration systems to identify the way land is occupied and used, an urgency reflected in several targets of the 17 sustainable development goals (SDGs) [13]. The International Land Coalition (ILC) stated that "the solution to reach all the SDGs is a people-centered land governance, in which land rights are an obligation and are at the heart of all SDGs: without land rights, there is no sustainable development!" [14].

The 2030 agenda is, therefore, a key driver for countries throughout the world, especially developing countries, to develop adequate and accountable land policies and regulatory frameworks for meeting the SDGs. This can be achieved by designing a land administration that meets the needs of people and their relationship to land, supporting the security of tenure for all [2], including women, as they serve as a basis for greater food production and food security [15], contributing for better welfare of families [7,16–22].

The solutions for a better land administration require attention to the specific context in which the registration occurs, ensuring that existing rights, both individual and communal, are recognized [11]. The Fit For Purpose Land Administration (FFPLA) approach has been conceptualized in reaction to the challenges set by the 2030 agenda [23], requiring cost-effective, time-efficient, transparent, scalable, and participatory land administration, including participatory surveying, volunteered land administration and crowdsourcing [2,24]. The principle of the FFPLA approach is that the spatial, legal, and institutional frameworks for land administration are in balance in such a way that tenure security can be established and maintained in a timely and affordable way, always aiming at the local, regional or national needs [25].

The FFPLA was applied in several countries such as Indonesia, Colombia, Nepal, Benin [13,25–27] and is recommended by organizations such as the International Federation of Surveyors and the World Bank that have published jointly the FFPLA reference guide [2], as well as the Global Land Tool Network (GLTN), UN-Habitat and Kadaster that published the guiding principles for country implementation [28].

This research used the FFPLA approach as a reference framework to enhance the security of tenure for all in Mozambique. It supports solutions for affordable land administration and can be quickly developed and incrementally improved over time [2,24,28–31]. The FFPLA indicates that efforts should be taken to achieve complete coverage and a complete overview first, and then improving incrementally over time where and when needed, by enhancing spatial accuracy, legal requirements, and institutional processes.

1.1. *Context of Mozambique*

Mozambique is a Southern African country with an area of approximately 800,000 km^2, a population of 30.8 million inhabitants, 66 percent of which live in rural areas, engaged in agricultural activities [32].

Mozambique is considered one of the poorest countries in the world. The latest report on the incidence of poverty based on the household budget survey, conducted in 2014 and 2015, showed that 46.1 percent of the population in Mozambique lives below the national poverty line, with poverty being highest in rural areas (53.1 percent) [33]. The UNDP[1] 2020 Human Development report placed Mozambique in 181st place out of 189 countries, with an HDI[2] of 0.456, a life expectancy at birth of 60.93 years, and an inequality coefficient[3] of 54 [34]. The UNDP report also indicated that the percentage of the population living below the internationally accepted poverty line is 62.9 percent, a value higher than the national poverty line.

The legal and regulatory framework determined that the land belongs to the State, and it cannot be sold nor alienated. The land law is generally accepted as favorable to reduce inequality in access to land and security of tenure, to protect the poor and the communities through community delimitations and registration of good-faith and customary occupations [35]. However, as more than 80 percent of the land under good faith or customary occupations is not registered [36], there have been complaints of land grabbing, causing dissatisfaction and insecurity among those more vulnerable who cannot defend themselves. This results from increased pressure on the land, mainly due to the aftermath of a 16-year civil war, climate changes, weak soil management policies, large-scale project investments including the extractive industry, and the need for food and fuel [37]. This situation is particularly relevant for women and other vulnerable groups since that the percentage of land registered for these segments of society is less than 20 percent [38].

Although land-use rights based on good faith and customary occupations are considered in the Mozambican legislation, their formalization through registration was not performed regularly, nor systematically, due to lack of skilled people and resources, weak land administration practices, as well as due to the high costs involved in the first land registration process, resulting in just a tiny percentage of land rights being formally registered.

Another impediment was the lack of an efficient Land Information Management System (LIMS) to support the registration [39], as these systems are usually costly [40]. Not having an appropriate LIMS posed severe limitations in the country's decision to promote equality in access to land and security of tenure and weakened the Government's ability to make appropriate reforms resulting in unsustainable development, weak productivity, and diminished investments on land. Although the establishment and maintenance of a LIMS could be expensive, the Government understood that refusal to establish such a system would be even more expensive, as stated by Dale & McLaughlin [41]. The Government also recognized that the LIMS needed to be appropriate to the circumstances and needs of the country; otherwise, there could be more harm than good, as Williamson [42] alerts.

To address these issues, two significant initiatives took place. The first one, in 2010, named "Acesso Seguro à Terra", aimed at performing systematic individual land registrations in the four northern provinces of Mozambique and designing, developing, and implementing a LIMS for Mozambique (denominated SiGIT[4]). This initiative was supported by the Government of Mozambique and the USA-funded Millennium Challenge Corporation. In 2015, another initiative, known as "Terra Segura", continued the registration effort further, aiming to regularize five million land parcels and delimitate four thousand communities, consolidating the land administration system. This last program was a flagship program of the Government of Mozambique, with the support of the World Bank [43].

1.2. Problem

These systematic land registration programs adopted approaches that were consistently time-consuming and costly. The "Acesso Seguro à Terra" program lasted four years and only registered 200 thousand parcels, mainly in urban areas. The "Terra Segura" program results by the end of 2016, almost two years after the start of the program, illustrated that only 220 thousand additional parcels were registered and 400 communities were delimitated, out of the goal of five million parcels and four thousand communities, respectively. Additionally, the average costs of such first registration were close to 50 USD per parcel and 10.000 USD per community [43]. The registration process used paper forms resulting in high rejection rates due to poor quality of data such as invalid locations, invalid cross-reference information of land uses and type of occupation, and null data in mandatory data fields. Another constraint was that community delimitations and individual registrations were performed separately, which increased the costs as there were duplicate activities and dissatisfaction from communities. The registration was not systematic and was not delivering the required security of tenure [37].

It became clear that conventional methods of land tenure regularization were both too expensive and time-consuming, facing several data quality problems. At that pace and with those costs, the targets could not be met, and more importantly, security of tenure would not be achieved for all. There was, therefore, a need to adopt an approach that would solve the inefficiencies, data quality problems, and high costs involved in the registration process.

1.3. Objective

The purpose of this research was to conceptualize, develop and test a country-specific land administration approach, the FFPLA-MOZ approach, to ensure that any systematic land registration program could be run affordably and efficiently, securing land tenure for all, rich and poor, women and men, and young and old, providing accurate, complete, and up to date information, enhancing a participatory engagement of communities, and reducing gender inequalities.

2. Materials and Methods

2.1. Research Design

The "research onion" proposed by Mark Saunders, Philip Lewis, and Adrian Thornhill [44] was utilized as a reference framework to design this research (Figure 1). The highlighted words in Figure 1 were the choices made for this research design.

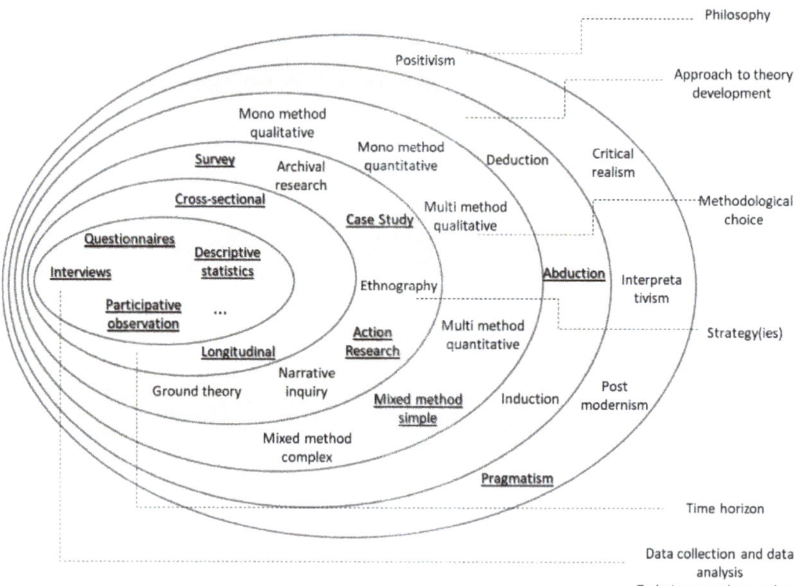

Figure 1. The "research onion" as a reference framework for the design of this research. Source: Adapted from ©2018 Mark Saunders, Philip Lewis and Adrian Thornhill [44].

The research followed a pragmatic philosophy, value-driven, characterized by a focus on practical solutions and outcomes, improving future practice, and taking the problem identified as a starting point [44,45].

The approach to theory development was abductive, moving from theory to data (as in deduction) and from data to theory (as in induction) [44,46]. The first step was to select the reference frameworks that supported the design of the conceptual model, which then supported the development of the FFPLA-MOZ. The tests and improvements to the FFPLA-MOZ were performed through data collection, analysis, and interpretation, for which multi-methods, qualitative and quantitative, were used to ensure a richer approach [45]. Qualitative methods allowed for observing, describing, and interpreting the existing land registration processes, their actors and their interpretation of the reality, and the improvements required, as well as for evaluating the interest of the participants in the registration process. Quantitative methods allowed for collecting observable and quantifiable data, such as time and cost spent per registration, essential to improve the FFPLA-MOZ. All improvements were tested through subsequent data collection and so forth [44].

The research adopted different but complementary strategies, namely, action research and case studies:

- The purpose of the action research strategy was to promote organizational learning and to produce practical outcomes through identifying issues, planning action, taking action, and evaluating action [44] in a PDCA[5] quality improvement cycle [47,48]. Action research was fundamental to improve all the components of the FFPLA-MOZ approach, where one step, a specific case study, provided inputs to the next, based on the data collected, observations and lessons learned. The time horizon for this part of the research was longitudinal, over four years (January 2016–January 2020).
- The *case studies* were defined to test the FFPLA-MOZ in specific scenarios as a way to obtain a better understating of the complex reality through contextual description and analysis [49], in a real-life context [50], delimited in time and space [51].
- Three case studies were defined with an incremental level of difficulty: (i) a peri-urban area, where parcels followed an urbanization plan, and the limits of the parcels were visible; (ii) a community where parcel boundaries were not clear; and (iii) a cluster of communities with no clear boundaries both for individual parcels and communities.
- The first case study involved 49 parcels, over a week, with a team of four people. The second case study involved 2500 parcels from two communities in two different provinces, during three months, with four teams of three people, two for community delimitations, and two for individual registrations. This was a complete coverage of the territories of these communities. The third case study involved a cluster of eight communities. In this case, the communities were delimited, but land registrations were performed in the center of the cluster, involving 820 parcels from three communities. One team of four people was dedicated to community delimitations, and three teams of three people were performing individual land registrations. The time horizon for this part of the research was cross-sectional [44,45], meaning that data from each of the case studies were collected quantitatively and qualitatively at a given moment, not over a long period. The patterns and differences between the different case studies were analyzed, providing improvements from one case study to the following [45].

Several techniques and procedures to collect and analyze data were utilized in this research, namely:

- Semi-structured interviews surveys—to collect valuable information from experts of different thematic areas in an exploratory and descriptive manner. Interviews helped deepen the knowledge regarding the FFPLA approach and to answer "what", "who", "where", "when", "how" and "how much" questions [52] related to land administration processes. This part of the research supported the design of the conceptual model and guided the development of the FFPLA-MOZ approach.
- Participative observation—from the perspective of both an insider, as a participant, and an outsider, as an observer, to experience each of the case studies settings and further improve the approach based on reflections and learning process [53] throughout the active research period.
- Questionnaire surveys to 3369 land tenants and 10 community councils—as part of the case study strategy—to collect data from a large number of respondents in a standardized and economical way, allowing easy comparison and validation of data [44]. Questionnaires were designed to collect data related to individual parcels and communal lands and data related to titleholders. These were implemented into a mobile application with embedded data quality and time processing controls. The data collected were inserted into a database which then provided reports and dashboards of the most crucial performance measurements such as rejection rates; the number of parcels registered per project, per day, per team; the number of parcels registered by gender and land use; and the number of hectares registered.
- Questionnaire surveys to 24 end-users and 5 team managers of the mobile and cloud applications—to assess end-user satisfaction regarding the overall data quality, ease of use of the tools, and teams' productivity. Qualitative data collected from these questionnaires were analyzed through descriptive statistical analysis.

- The research also utilized data from secondary sources [44,45], such as online satellite imagery and maps from Google maps, to obtain images to support the registration process. The national land cadastre database was also consulted to check for existing parcels to avoid overlapping registered parcels.

2.2. Reference Frameworks

The research used the Fit For Purpose Land Administration (FFPLA) approach [24,30,31] as a reference framework and took into account the World Bank Land Governance Assessment Framework (LGAF) [54]. These frameworks were utilized in line with the country's land legal and regulatory framework and having in mind the country's context: a relatively large country with more than 80 percent of unregistered land occupations, specifically in rural areas; different typologies, cultures, and languages; difficult access to some remote areas; statutory laws, norms and customary practices, not well known, and sometimes in conflict with each other; high level of illiteracy; scarcity of skills and resources; limited access to internet connectivity; and limited access to high-resolution images [37,55].

The FFPLA guiding principles for country implementation [28] provided the principles and criteria relevant in the Mozambican context, primarily to design the spatial, legal, and institutional requirements. These guidelines also enhanced the need to include change management, capacity development, and project delivery [30] in the design of the managerial processes that would support the activities to "recognize", "record", and "review" land rights. "Recognize" involves a procedure for recognizing, classifying, and developing a typology in land rights based on an assessment of existing legitimate rights in the countryside. "Record" means collecting data on evidence of land rights. "Review" means assessing the evidence of rights and any possible outstanding claims so that, when conditions are met, the security level of the rights will be increased [28].

The recent LGAF assessment [35] was helpful to provide insights from the current land governance practices, particularly the assessments made to land rights recognition, individual and communal lands regulations, public land management, land cadastre and public information, dispute resolution mechanisms, and institutional arrangements and policies.

2.3. Conceptual Model

The conceptual model for the FFPLA-MOZ approach was established through three intertwined pillars: People, Processes, and Technology (Figure 2):

1. *people* or interested parties—to promote knowledge, awareness, and active participation of individuals and communities in the land registration. Apart from the participation of government agencies and services providers, active community engagement was advocated by several studies and frameworks [24,54,56–58]. Participatory land-use planning created a higher impact within the communities, increased ownership and satisfaction, and promoted a more sustainable land administration [59]. Crowdsourcing was also an essential approach for first registrations and maintenance of the land cadastres [58,60]. In this research, crowdsourcing was community-based in order to avoid land grabbing.
2. *processes*—to implement the rules of the game in a standardized manner. A comprehensive land registration set of processes and procedures, combining both managerial and operational processes (end-to-end), integrating the individual land registration (RDUAT[6]) and community delimitation (DELCOM[7]) processes. This was helpful to achieve several benefits, including (i) increased quality of work; (ii) reduction of costs, (iii) reduction of time; and (iv) more accurate and up-to-date information. Processes design respected the key principles of the spatial, legal, and institutional frameworks proposed by Enemark et al. [24].
3. *innovative technology*—to facilitate data acquisition and validation while working in the field, improve data quality, reduce time, and reduce work, errors, and rework costs. Innovative solutions had to be adequate to "recognize", "record", and "review"

land rights cost-effectively. Solutions implemented in similar country contexts [61] proved to be effective to reduce costs [27,62–64] were also analyzed. The guiding principles for building land information systems for developing countries [65] were considered in the design and development of all technological innovations.

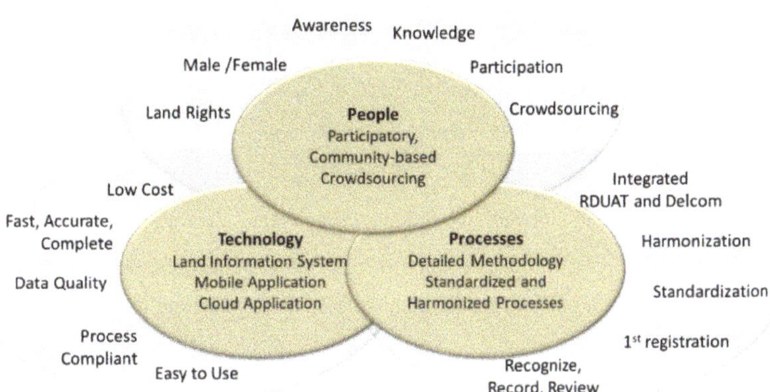

Figure 2. The Conceptual Model for developing the FFPLA-MOZ approach based on three intertwined pillars: People, Processes and Technology.

The FFPLA-MOZ approach was conceptualized, developed, and tested through several steps. First, the principles and criteria to be met were defined based on the conceptual model. Second, the processes, their respective activities and tasks, were defined and described in detail, integrating both RDUAT and DELCOM, harmonizing all processes and sub-processes, and creating a standard of work for any systematic land registration program. Third, the necessary tools were developed, respecting the designed processes. These tools were integrated with the existing LIMS (SiGIT). Fourth, tests were performed to validate the processes and the developed tools through three case studies in a continuous improvement cycle.

3. Results

3.1. The Mozambican Fit for Purpose Land Administration Approach

3.1.1. Principles and Criteria

The FFPLA-MOZ approach adopted the following criteria [66]:

Participation of citizens—Ensure active community participation. The registration process started with a public meeting, with all groups and communities being represented. Together with the local community council, the objectives of the work and the way the work was to be executed were explained to the community. Because women and other vulnerable groups were usually the ones suffering the most insecurity of tenure [8], specific meetings were organized with these segments of the community so that everyone understood that their rights were to be recognized and recorded, following the motto "land rights for all, rich/poor, old/young, male/female" (Figures 3–5). Sensitization sessions to promote gender equality and advocate land rights were held throughout the community, utilizing role-play sessions with different scenarios of land tenure inequalities. The training was given to community members selected to be part of the registration teams, including women, the elderly, and young people.

Figure 3. Meeting the local community council—Chizavane cluster.

Figure 4. Meeting with women separately to ensure they understand their rights.

Figure 5. Teaching a "Queen" (female community leader) to read the map.

Inclusiveness—Registration must be systematic, covering the selected area, respecting existing registered parcels, contemplating only parcels under good-faith occupation or customary practices that comply with legitimacy criteria [67], *ensuring that no one is left behind.* The registration took into consideration the fact that there were existing parcels in the land cadastre. Illegal land grabbing or conflicts were excluded and were signaled as such. Parcels within protected areas were recorded for further action. Since the FFPLA-MOZ approach was tested through specific case studies, it was not possible to cover the entire area of a given community, with exception for the second case study that had specifically that purpose. The first and the third case studies only contemplated parts of the community land.

Adequate accuracy for public and community lands—Adopt flexibility based on the demands for accuracy depending on tenure types such as public lands, community lands, and protected areas. Ensure that scale and accuracy is sufficient for securing the various kinds of legal rights. Have different tenure forms recognized through the legal framework. Monumentation is not to be done. Where boundaries were clear, these were delineated over a map and later validated with the measurements taken with the GNSS[8] system at known ground control points. This illustrated an accuracy of ±20 m, less than the ±30 m required by government regulations. In locations where the boundaries were fuzzy or dynamic (e.g., rivers), the proposal was to define a buffer [68]. Where the boundaries were not clear or were difficult to access, the proposal was to start with an initial sketch of the limits and wait for the results of the individual land registrations to delineate either a final limit or define a buffer. This was possible through an additional feature added to the mobile application, where each parcel received a name tag of the community to which it belonged. With this functionality, instead of geometrical straight lines defining the boundaries of the communities (a common practice being utilized before), boundaries could be more accurate respecting the parcels' geographic location (Figure 6).

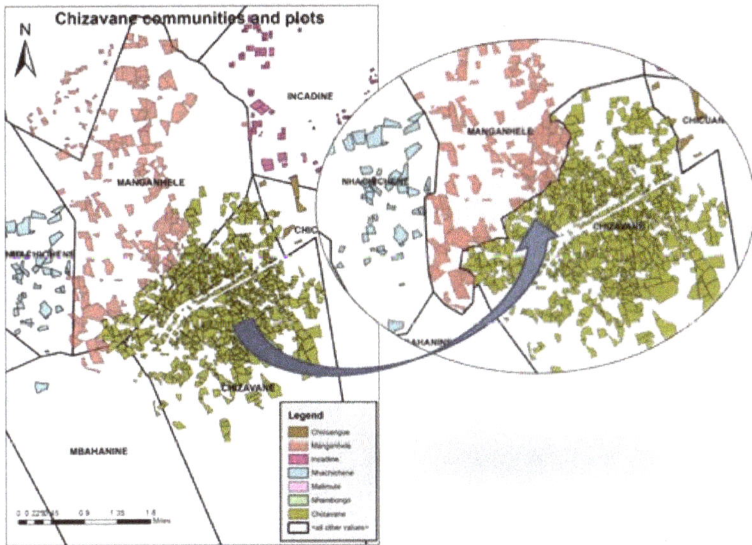

Figure 6. Example of the adjustments made to the community boundaries based on the individual parcels' registrations.

Adequate precision for individual land parcels—The purpose is to bring tenure security to customary and good-faith occupations. Accommodate a range of methods to measure and record parcel boundaries, including identifying visible boundaries on imagery. Individual parcels under good faith or customary occupations were recorded with precision under ±1.5 m. Beacons were not utilized to formalize the boundaries as monumentation was time-consuming, not consistently effective or respected, and it delayed the process of land recordation and registration. Community leaders recommended utilizing "live" monuments such as trees planted at the vertices of the parcel.

Reliability—Ensure technology captures spatial and administrative data with quality controls capturing people to land relationships. Ensure community participation, with solid social preparation, to increase the reliability of the entire process. People to land relationship was collected through the mobile application, seeking alphanumeric data of tenants, administrative data of land uses, as well as geographic data of the parcel (Figure 7). The mobile application was adjusted to collect alphanumeric and geographic information in the same form. The application included quality controls, image capturing of documents, and information regarding the local authority's validation (Figure 8). Only complete and validated forms were uploaded to the cloud. Tenants' data could be shared between teams using a feature created in the mobile application. Parcels that were rejected afterwards were sent back to the team, indicating the error to be corrected.

Figure 7. Using the mobile application to collect alphanumeric and geographic information.

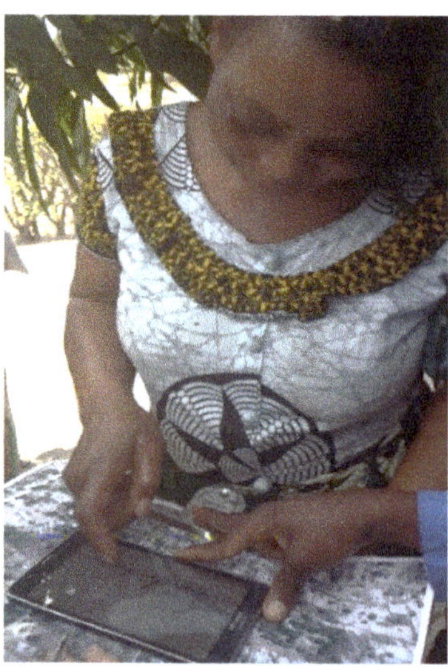

Figure 8. All parcels were validated by the local authority.

Update or ability to be updated—Consider opportunities for ongoing updating, sporadic upgrading, and incremental improvement of the land information whenever relevant or necessary for fulfilling land policy aims and objectives. In the last case study, apart from the new registrations, there was a need to update 150 parcels registered previously in a different initiative. These updates were needed for the following reasons: (i) inclusion of additional tenants, after sensitization of the community with regards to gender equity; (ii) adjustment of boundaries to resolve detected conflicts; (iii) adjustment of boundaries where the parcels infringed protected areas; and (iv) disaggregation of a parcel to contemplate several other parcels.

Digital services: flexibility in the technological solutions—Adopt an effective, scalable, sustainable, and secure land information system, in an incremental approach, as this is more sustainable than other more ambitious, faster implementations [30,31], *primarily oriented to more developed economies.* Although the ultimate technological solutions were sophisticated and supported the most innovative features, the FFPLA-MOZ solutions were relatively simple to accommodate the limitations of the Information and Communication Technology (ICT) infrastructure and shortage of skills. Flexibility in the technological solutions was implemented through a series of innovations, described in detail in Sections 3.1.3 and 3.2.2.

3.1.2. Integrated and Harmonized Processes

The processes were defined with the participation of different stakeholders from the Government, civil society, NGOs[9], service providers, and experts from the different subject areas. All processes, their respective activities and tasks, were harmonized, integrating both individual registrations and community delimitations, resulting in a set of guidelines and norms. Efforts were made to ensure that these processes would also be Fit For the Future. This component of the FFPLA-MOZ was comprised of a Process Wheel, a Process Orchestration Diagram, and several Context Diagrams:

- The Process Wheel (Figure 9) consisted of two main cycles: a managerial cycle in an outer circle and an execution cycle in an inner circle. The managerial cycle comprised three management processes: planning the period, monitoring and controlling all RDUAT/DELCOM projects, and evaluating the period. This outer cycle ran within a specific period, generally defined as a year. The execution process was performed in a cycle of a shorter duration for every RDUAT/DELCOM project. The inner cycle was comprised of five subprocesses: Prepare the Field Work, Prepare Communities, Collect Data in the Field, Process Data, Deliver Documents, and Close Project.

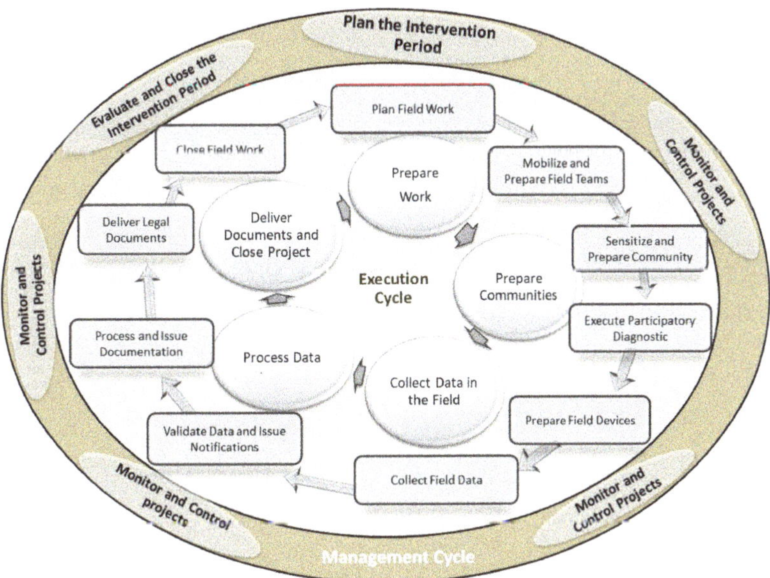

Figure 9. The processes of the FFPLA-MOZ approach. Outer ring: Managerial processes (Plan, Monitor and Control, and Evaluate and Close). Inner ring: Execution processes (Prepare the Field Work, Prepare Communities; Collect Data in the Field, Process Data, Deliver Documents and Close Project). Source: Developed from [37].

- The Process Orchestration Diagrams (Figure 10) illustrated all the processes, subprocesses, respective activities, and their sequence of execution, which supported the configuration of the workflows within the applications. It was an important tool to visualize those specific or common activities both to individual registrations and community delimitations, and therefore could be harmonized. In Figure 10, for example, the activities with white text font were only performed for community delimitations. These were to be completed in a later stage of the registration, outside the scope of this research.

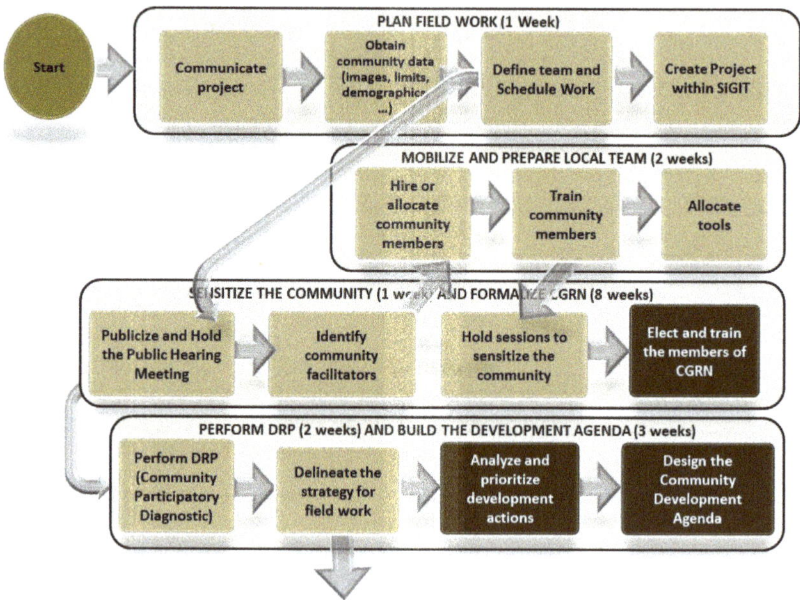

Figure 10. Example of part of the orchestration of processes. A visualization of the workflow for each process or subprocess. Source: [37].

- The Context Diagram described each of the processes embedded in the FFPLA-MOZ approach in greater detail. The Context Diagrams were designed uniformly to create a norm for utilization, facilitating monitoring and control. Each process was described in terms of Goals and Objectives, Activities and Tasks, Techniques, Roles and Responsibilities, Inputs and Outputs, Metrics, as well as Risks and Mitigation measures.

3.1.3. Technology Innovation

Technology innovation was deemed essential to accelerate data collection, reduce costs and errors, increase data quality, and enhance community participation, provided it was accessible, cost-effective, and less skill demanding. Technology innovations were under the responsibility of the LIMS developer, a local company that had been supporting the operation of SiGIT since its implementation [69]. Innovation was achieved through three main activities:

First, the land information system—SiGIT—was adjusted to accommodate new and redesigned business processes, especially those related to good-faith and customary occupations and community delimitations. The system incorporated all land-related legislation, regulations, and policies into its workflows for all land administration processes from registration to taxation, expansion or reduction of area, transmission of rights, revocation or cancellation of rights (Figure 11). This adjustment also differentiated workflows for urban and rural areas. SiGIT was also changed to allow delimitations and individual registrations to co-exist within the same spatial project area and combine several activities within the same registration project. The interoperability between the SiGIT back-office application and the mobile and cloud applications (SiGIT mobile and SiGIT Cloud) was built, with several data quality controls in all data exchange interfaces. New analytics reports for the monitoring and control processes were created. The data model chosen for the SiGIT system was based on the ISO 19152:2012—Land Administration Domain Model (LADM) [70]. No changes were made to the systems architecture nor to the base technology. All the changes to the SiGIT system followed the change management procedure in place.

Figure 11. SiGIT Land Information System Business Processes. Source: [37].

Second, the mobile application—SiGIT mobile—was developed to contemplate the FFPLA-MOZ processes' requirements. It provided a user-friendly interface that would speed up collecting field data, easy to utilize both by surveyors and non-surveyors, including community members. The mobile application was developed with its database, which allowed to structure data and reuse objects that were previously collected in the field (parcel holders, documents, parcels), resulting in increased efficiency. The mobile application provided four methods of spatial data capture: (i) embedded mobile device GNSS (adequate for community delimitations); (ii) drawing the parcel on a map (adequate when images were clear and with good precision); (iii) pairing the mobile application with a GNSS device; and (iv) entering coordinates manually. It also included spatial data libraries to provide several spatial functions for data validation, manipulation, quality control, and the ability to work with offline maps, enhancing flexibility to work both alphanumeric and spatial data.

Finally, a cloud module—SiGIT Cloud—was developed to capture data from the fieldwork, pre-process data, and send data to the SiGIT application (or back to the mobile application in case of errors). This module was available to service providers and required the surveyor's expertise. It controlled the quality of work of specific teams and individuals through specific performance dashboards.

The development of the mobile and cloud applications utilized agile development methodologies, open-source technologies and libraries from the Open Geospatial Consortium and Open Street Maps, both for the alphanumeric and geographic components.

3.2. Improvements

3.2.1. Improvements through Integrated and Harmonized Processes

The harmonization between RDUAT and DELCOM allowed the combination and integration of processes and activities, which resulted in several benefits:

i. Reduced total time per project: initially, a complete delimitation took around nine months, whilst with the new integrated and harmonized processes, the work was completed in five months, including individual land registrations within the community.
ii. Reduced overall program budget: after integrating individual registrations and community delimitations, recalculations of the "Terra Segura" program budget illustrated a reduction by 30 percent, just by optimizing and combining resources and activities.
iii. Reduced malpractices: adopting a uniform way to perform the work reinforced each participating role in the land administration and increased clarity on what to do, how to do, when to do it, thus promoting a more sustainable process-based approach to land administration.

3.2.2. Improvements through Technology Innovation

Data captured in the field through the mobile application were uploaded to the cloud and downloaded to the SiGIT application from the cloud. A team dedicated to quality control performed all the necessary checks to determine data rejection rates and team productivity. Comparisons were made between analogue (paper-based forms) and digital mobile data capturing methods regarding the following: rejection rates; time to collect, validate and process field data; types of errors encountered; compliance of the tool with the data quality requirements. Comparisons were also made between the different geographic data capture methods. Questionnaires were given to field workers utilizing these tools to determine their level of satisfaction and their recommendations to improve the tools. The following results were achieved:

i. Increased performance: implementing the SiGIT mobile application and its integration with both the cloud and the back-office system resulted in several optimizations. Data collection, validation, and processing time for individual registrations reduced from 35 to 18 min (alphanumeric data) and from 21 to 6 min (geographic data). The amount of time varied according to the number of people that needed to be registered for each parcel and if walking around the parcel boundary was required. Rejection rates dropped from 44 percent to 1 percent, and possible types of errors reduced from 16 to 3.
ii. Reduced costs: the costs of community delimitations when in a cluster were reduced by 60 percent, and systematic individual parcel registrations were reduced by 70 percent. Within this research, the cost of registering the individual parcels was 15 USD, and to delimitate a community was 2.000 USD. These reductions were achieved through a combination of processes and technology improvements. These costs did not include imagery acquisition as free images were utilized and georeferenced. It also did not include the design of community development agendas to manage natural resources, as this activity was not part of this research.
iii. Flexibility and accuracy of geographic information: with the SiGIT mobile application, the flexibility of capturing geospatial information was increased as four methods were simultaneously available for geographic information data collection. The decision of which method to utilize significantly depended on the working conditions of the area and the availability of adequate imagery. Overall, the utilization of the android GNSS was simple. Still, the precision was not adequate (precision varied from ± 3 m to an average of ± 10 m but in some cases going over ± 18 m). The second method, the ability to draw the parcels over a map, proved to be a simpler and faster data collection procedure. Google Maps that were freely available were georeferenced using a GNSS and ground control points visible both on the map and in the field. This method had a drawback in situations where the boundaries were not clear or the quality of the image was poor. In the cases where good quality images were utilized, the method achieved ± 0.52 m precision. In situations where the boundaries of a parcel were not clear on the image, functionality was added to the mobile application that permitted the user to walk around the

parcel collecting geographic data. The third method utilized a differential GNSS connected via Bluetooth to the mobile application. This method achieved the best precision (±0.32 m) but was the most expensive method and required a professional surveyor to operate the GNSS. The fourth method was used when the differential GNSS could not interconnect with the mobile application.

iv. Increased control of team's performance: one of the problems from previous exercises was the lack of information regarding teams' performance during fieldwork and afterwards. Performance control was a requirement both from service providers and the Government. The cloud application provided a dashboard with several indicators that helped to correct several bottlenecks throughout the workflow.

v. Increased end-user satisfaction: all field users indicated that the mobile application was easy to use and navigate, with straightforward questions. They also indicated that the spatial component embedded in the mobile application was a substantial value-added. They complained that the mobile application mandated filling in information previously skipped in paper forms. All office users and quality controllers indicated their workload reduced considerably as all data were automatically uploaded and checked. They also indicated that the number of rejections and re-works was minimum and that the dashboards were helpful to control teamwork and project completion rates.

4. Discussion

The FFPLA-MOZ approach demonstrated that it is possible to reduce complexity, time, and costs, increase efficiency, as well as work quality and data quality. This was achieved through a series of improvements at each test interaction. In this section, the three pillars of the conceptual model of the FFPLA-MOZ approach are discussed: People, Processes, and Technology.

4.1. People: Awareness, Engagement, and Participation

The case studies illustrated that most people were not aware of their rights, and in some situations, the customary practices infringed the statutory law promoting gender inequalities.

The role-play training adopted for sensitization regarding gender equality and vulnerable groups' protection proved to be highly effective to illustrate land rights and land tenure security for all. Overall, in this research, the percentage of parcels belonging to women increased to 42 percent instead of 20 percent that was the average before the research started [38,71]. More recent statistics from the "Terra Segura "program indicated that the percentage of parcels belonging to women reached 56 percent in 2020 (Figure 12). However, this subject requires further investigation as land tenure equality is both for men and women, and therefore co-titling should also be promoted.

Figure 12. A group of women getting their land titles from the president of Mozambique.

It was clear that, in a more participatory engagement of the community (the third case study with community-based crowdsourcing), the work progressed somehow at a slower pace in the beginning but then achieved the same pace as if private surveyors would be doing the registration. Private surveyors still needed to be present to provide training, guide and supervise the work, control the quality of data collected, and help communities be familiar with the instruments of work, either the mobile application or the printed images of the community area. Communities responded well to the interpretation of printed maps, which helped define their boundaries in the cluster. The third case study was the case that observed more engagement and appropriation from community members, resulting in an increased level of confidence in the information collected and reduced overall costs. However, this more participatory engagement was only possible due to the utilization of mobile technology that was easy to use coupled with imagery with enough resolution to identify the spatial boundaries of community land and individual parcels.

To support the work of the community team, the community organized the payment of a small fee (around 1 USD per parcel), as a contribution from each household being registered, which ended up generating a lot of engagement from all selected team members and contributed to the smooth execution of the fieldwork. It also generated a precious income both to the older people and the young team members working in the field, usually unemployed [69]. This was a new approach as usually service providers contracted community members to participate in the registration process.

Some drawbacks were noted. During the public announcement meeting, it was explained to everyone the study would only include registrations in the center part of the cluster. Those that were not contemplated felt left behind and hoped that the process would continue afterwards. Additionally, some beneficiaries did not receive the individual parcels titles due to the unavailability of the SiGIT operations from the beginning of 2020, which constituted a setback in the engagement achieved during the fieldwork.

4.2. Processes: Integration and Harmonization

The research design was adequate to explore, describe and explain the working processes, and implement quality improvements at each new cycle of interactions.

By integrating individual registrations and community delimitations, it was possible to reduce redundant work, time and costs, and increase community satisfaction and willingness to participate in the registration process. Previously, before integrating these processes, communities complained about their individual lands being left out of the delimitation process or vice versa. The active participation enhanced overall land administration and produced more accurate and complete data. Additionally, communities understood the value of their land and how to manage it accordingly and understood the purpose of land registrations as a mechanism to reduce conflicts and avoid land grabbing.

Using clusters of communities helped define the boundaries of all communities belonging to the cluster, all at once, reducing time and costs. The amount of work was further reduced when the mobile application embedded functionality to tag each parcel with details of its community. Subsequently, when all parcels were uploaded into the cloud application, it was possible to utilize the individual parcel registrations located on the borderlines of the communities, utilizing different colors to represent each community, to delineate the boundaries in a more precise manner. This solution required that some activities change their execution order: community delimitations started with initial sketches, but final delimitations were only performed after all individual parcels had been registered. With this solution, it was possible to reduce the amount of work and time to delimitate communities.

One additional improvement to be implemented in the future is the creation of interoperability with other cadastres. Observations indicated that it took at least 5 min to capture information from each tenant (biography, pictures of documents), which was time-consuming, especially when several people were to be registered as co-tenants [72]. This could be solved through the interoperability with upstream cadastres such as the identity cadastres[10]. It was also recommended the interoperability with downstream cadastres that provide information to support the development agendas of communities such as mining concessions and forestry concessions.

Another important issue considered was the fact that cadastral information became quickly outdated, as illustrated in the third case study. Updating and maintaining cadastral data referred to the need for registers to be trustable and reflecting the actual spatial and legal/legitimate situation on the ground, while upgrading related to improving the accuracy for specific purposes, more generally to meeting societal needs. It became evident the need for appropriate processes and tools to be designed, developed, and implemented—preferably at the community level—to ignite the required updates when and where they occur. An initial solution to resolve this constraint was the implementation of a basic community-based land cadastre [73], providing communities with a dossier that included (i) all parcels registered within the community, with specific sections for updates (amendments); (ii) a map with all the registered parcels within the community; and (iii) a map with the community boundaries. This dossier was delivered to the community council and was available for public consultation. Subsequently, there was a recommendation to improve the mobile application with functionality to automate this dossier and to register requests for updates. Another recommendation, perhaps a more difficult one to implement, was to entitle the community with the authorization competence and power to formalize land rights for smaller parcels under good faith and customary occupations.

Within the "Terra Segura" program, the issue of monitoring and control proposed by the FFPLA-MOZ is still to be resolved. The Government did not yet manage to create a specific unit dedicated to this aspect due to the lack of funds to contract specialists and establish that function. Lack of funds also contributed to a reduced number of registration projects and the cancellation of others. Up to March 2020, a total of 826 community delimitations and 1,375,586 individual parcels had been processed by the SiGIT land cadastre. From then on, 220,805 parcels and 606 communities remain to be migrated to the system. This illustrates that, although there was an increased number of parcels registered and communities delimited, there were bottlenecks to be resolved, including funding. The involvement of the World Bank as a financer should resolve part of the funding problem.

4.3. Technology: Innovations to Improve Digital Services

The tools developed specifically for this research were based on the Mozambican context and the FFPLA-MOZ processes requirements. They were improved and fine-tuned through several adjustments that would not have been easy to implement if an off-the-shelf tool had been selected. The innovations were evaluated in light of the recommendations for a successful land information system implementation, namely: sustainability, security, and scalability [74]:

Sustainability: Achieved through user capacity and ability to implement new requirements, cultural changes, integration of data, provision of online services, increased revenue generation, stakeholder support. This was the dimension facing more threats during the research. The training was given to all end-users, contemplating government officials, service providers, and community members. There was recognized competence to utilize the tools throughout the country. Online supporting services and a service desk unit were created to support all internal and external users. Data integration within all stages of the registration process was accomplished through specific automated software, and those operating with the SiGIT mobile application could see their work immediately integrated into the land cadastre. However, there were insufficient expertise and ICT skills within the Government, which implicated that the ICT services, including the SiGIT application lifecycle management, were outsourced. Meanwhile, these contracts were cancelled due to insufficient funds. Similar constraints were observed in other countries [75]. For any systematic registration to succeed, the ICT infrastructure, including the application, cannot be the bottleneck. The Government must monitor and control aspects of the registration and leave both the registration projects and the ICT maintenance and support to specialized companies.

Scalability: Achieved through a phased approach implementation of business rules and processes, innovation to comply with legislative changes and new business requirements, integration, and interoperability between systems. SiGIT was initially built for the daily operations of a provincial land office and therefore did not contemplate the requirements for massive land registration programs, both in terms of software application and ICT infrastructure. The Government implemented an upgrade to the ICT infrastructure. Several new modules and additional functionality were developed into the SiGIT application, creating more robust and reliable services. These newer versions of SiGIT were released in a phased approach, based on the established priorities. All users agreed that the newer versions and upgrades delivered additional value and facilitated their activities. They also indicated that the interactive user manuals were a significant gain for speeding up the learning process. The incremental phased approach proved to work well. The existing solutions were improved throughout the research. Over time, they can be further enhanced to embed new technology and create greater functionality, given that more effective resources are available and maturity levels in utilizing these solutions have increased.

Security: Achieved through security access, security of data, ICT policies, and the guarantee of business continuity in case of disasters. The ICT infrastructure was upgraded to increase security. Synchronization mechanisms from mobile to cloud to provincial to the central database were created with increased security through encryption mechanisms. Secure access protocols and views to SiGIT data were implemented for service providers' external access. However, there were severe constraints in terms of communications as well as data center operations. There was a need to improve the security policy and the business continuity policy, and recommendations were given to the Government for that purpose.

5. Conclusions and Recommendations

By adopting the FFPLA-MOZ approach, three pillars—people, processes, and technology—were connected, resulting in several improvements in the systematic land registration. This was possible because (i) all processes were harmonized, integrating both individual registrations and community delimitations, resulting in a set of guidelines and norms which increased quality of work and reduced costs and time; (ii) community active participation enhanced overall land administration and produced more accurate and complete data; (iii) technology was affordable, easy to use, and was fit to accelerate data collection and processing, respecting all defined working processes.

Community-based crowdsourcing was effective for collecting and recording data so that the community itself organized the data collection, which was then submitted for subsequent data processing. This solution presented limitations unless (a) services providers were willing to prepare and engage the community in more tasks; (b) community land cadastres were integrated into the national land cadastre; (c) there were means to keep alive the interaction between the community and the government land offices; and (d) the tools that were developed for field data collection, updates, and consultation, were fit for community usage.

Gender equity is an essential aspect of land tenure security. Several strategies were tested to ensure tenure security for women and other vulnerable groups with positive outcomes, such as role-play training with real case scenarios of land tenure inequalities and selecting women to participate in the registration process. Meeting separately with women and other vulnerable groups also proved to work better to disseminate land rights. These approaches need to be further improved and included in the sensitization activities of any systematic registration.

Within this research, the FFPLA-MOZ provided both the Government, communities, and individuals with accurate and up-to-date information regarding land use and land rights of the communities. The approach can be launched nationwide to help create an effective and sustainable land administration, support security of tenure, reduce conflicts, and avoid land grabbing. This requires that all stakeholders must appropriate it at strategic, tactical, and operational levels. There is also a need to redistribute competencies and capabilities within the institutions in charge of the cadastre services. Since these land registration initiatives are usually complex, they require strong leadership, change management, project delivery, capacity building, and sufficient funds. Therefore, tangible and intangible costs and benefits need to be considered when evaluating the affordability and sustainability of such initiatives.

Author Contributions: Conceptualization M.B. and J.C. and C.L.; methodology, M.B.; software development, M.B.; fieldwork M.B.; analysis, M.B. and J.C. and C.L.; writing—original draft preparation, M.B.; writing—review and editing, M.B. and J.C. and C.L.; funding acquisition—fieldwork, M.B.; funding acquisition—publishing, C.L. All authors have read and agreed to the published version of the manuscript.

Funding: Funding for the development of the technology innovations was kindly provided by EXI Lda, a local private company responsible for the lifecycle management of the SiGIT, the Land Information Management System in Mozambique. Funding of the publication costs for this article has kindly been provided by the School of Land Administration Studies, University of Twente, in combination with Kadaster International, the Netherlands.

Institutional Review Board Statement: Not applicable.

Informed Consent Statement: Not applicable.

Data Availability Statement: All the data that was collected is part of the Mozambican Land Cadastre.

Acknowledgments: The authors would like to thank the communities involved in the case studies, namely the communities from Marracuene, Gurue, Malema and Chizavane. Special thanks to Simão Joaquim, José Almeirim de Carvalho, and Daniel Queface from the National Directorate of Land. The authors would also like to acknowledge the contributions of service providers such as ITC, Verde Azul, Topmap, and Atopocomé for their valuable contributions during the tests of the FFPLA-MOZ approach. Special thanks to Kadaster International for the valuable clarifications with regards to the FFPLA approach. Special thanks to Enzo Scarcella, from the Africa Leadership Initiative-South Africa, for providing the tablets for the fieldwork. Last but not least, special thanks to EXI and the SiGIT team for all the efforts to implement the required innovations.

Conflicts of Interest: The authors declare no conflict of interest. The funders had no role in the study's design; in the collection, analyses, or interpretation of data; in the writing of the manuscript, or in the decision to publish the results.

Notes

1. UNDP—United Nations Development Program
2. HDI—Human Development Index
3. Also known as GINI index, measures the extent to which the distribution of income (or, in some cases, consumption expenditure) among individuals or households within an economy deviates from a perfectly equal distribution.
4. SiGIT—Sistema de Gestão de Informação sobre Terras (Land Information Management System)
5. PDCA = Plan, Do, Check and Act.
6. RDUAT—Regularização do Direito de Uso e Aproveitamento da Terra (in English: Land Tenure Regularization). Good faith occupations and customary tenure are recognized by law, but do not require registration and titling. RDUAT is a formalization of this recognition, by recording and reviewing data regarding tenant's and parcel's data.
7. DELCOM—Delimitação Comunitária (in English: Community Delimitation). Community rights are recognized in the law. The process of delimitating communities envisages recognizing, recording and reviewing these rights, followed by a general community land use agenda.
8. GNSS—Global Navigation Satellite System
9. Non-Governmental Organizations
10. Identity cadastres such as the ID Card database, the electorate registration database, or the driving license database, contain biographic information such as name, date of birth, place of birth, filliation, marital status.

References

1. UN-Habitat. *Women and Land in the Muslim World: Pathways to Increase access to Land for the Realization of Development, Peace and Human Rights*; UNHabitat: Nairobi, Kenya, 2018.

2. Enemark, S.; Bell, K.C.; Lemmen, C.; McLaren, R. *Fit-For-Purpose Land Administration. FIG Publication No 60. Joint Publication;* International Federation of Surveyors (FIG), World Bank: Copenhagen, Denmark, 2014.
3. ILC. *Women's Land Rights;* International Land Coalition: Rome, Italy, 2019.
4. Singirankabo, U.A.; Ertsen, M.W. Relations between land tenure security and agricultural productivity: Exploring the effect of land registration. *Land* **2020**, *9*, 138. [CrossRef]
5. Daley, E.; Englert, B. Securing land rights for women. *J. East. Afr. Stud.* **2010**, *4*, 91–113. [CrossRef]
6. Daniel Ayalew, A.; Deininger, K.; Goldstein, M. *Environmental and Gender Impacts of Land Tenure Regularization in Africa: Pilot Evidence from Rwanda;* Africa Region Gender Practice Policy Brief, World Bank: Washington, DC, USA, 2011.
7. Odeny, M. Improving Access to Land and strengthening Women's land rights in Africa. In Proceedings of the Annual World Bank Conference on Land and Poverty, Washington, DC, USA, 8–11 April 2013; p. 16.
8. Yngstrom, I. Women, wives and land rights in Africa: Situating gender beyond the household in the debate over land policy and changing tenure systems. *Oxf. Dev. Stud.* **2002**, *30*, 21–40. [CrossRef]
9. De Soto, H. *The Mistery of Capital—Why Capitalism Triumphs in the West and Fails Everywhere Else;* Black Swan: London, UK, 2001.
10. Payne, G.; Durand-Lasserve, A.; Rakodi, C. The limits of land titling and home ownership. *Environ. Urban.* **2009**, *21*, 443–462. [CrossRef]
11. Deininger, K. Towards sustainable systems of land administration: Recent evidence and challenges for Africa. *Af. JARE* **2010**, *5*, 205–226.
12. UN General Assembly. Transforming our world: The 2030 Agenda for Sustainable Development. Available online: https://sdgs.un.org/2030agenda (accessed on 25 April 2021).
13. Musinguzi, M.; Enemark, S. A Fit-For-Purpose Approach to Land Administration in Africa—supporting the 2030 Global Agenda. *Int. J. Technoscience Dev.* **2019**, *4*, 69–89.
14. International Land Coalition. *The International Land Coalition and Sustainable Development Goals Working Together to Realise Land Governance;* International Land Coalition: Rome, Italy, 2018.
15. ActionAid International. *Securing Women's Right to Land and Livelihoods a Key to Ending Hunger and Fighting AIDS;* ActionAid International: Johannesburg, South Africa, 2008.
16. FAO. *Land Access in Rural Africa: Strategies to Fight Gender Inequality;* FAO: Rome, Italy, 2008.
17. Chigbu, U.E.; Paradza, G.; Dachaga, W. Differentiations in women's land tenure experiences: Implications for women's land access and tenure security in sub-saharan Africa. *Land* **2019**, *8*, 22. [CrossRef]
18. Haldrup, K. Mainstreaming Gender Issues in Land Administration: Awareness, Attention and Action. In Proceedings of the FIG XXII International Congress—Washington DC, Washington, DC, USA, 19–26 April 2002.
19. Agarwal, B. Gender equality, food security and the sustainable development goals. *Curr. Opin. Environ. Sustain.* **2018**, *34*, 26–32. [CrossRef]
20. Daley, E.; Flower, C.; Miggiano, L.; Pallas, S. *Women's Land Rights and Gender Justice in Land Governance: Pillars in the Promotion and Protection of Women's Human Rights in Rural Areas;* International Land Coalition: Rome, Italy, 2013.
21. FAO. Improving gender equity in access to land. In *FAO Land Tenure Notes 2;* FAO: Rome, Italy, 2006.
22. FAO. *Governing Land for Women and Men—A Technical Guide to Support the Achievement of Responsible Gender-Equitable Governance of Land Tenure. A technical Guide to Governance of Land Tenure;* FAO: Rome, Italy, 2013; p. 120.
23. McLaren, R.; Enemark, S. Fit-For-Purpose Land Administration: Developing Country Specific Strategies for Implementation. In Proceedings of the Responsible Land Governance: Towards an Evidence Based Approach: World Bank Conference on Land and Poverty, Washington, DC, USA, 20–24 March 2017.
24. Enemark, S.; McLaren, R.; Lemmen, C. *Fit-For-Purpose Land Administration Guiding Principles;* Global Land Tool Network (GLTN): Copenhagen, Denmark, 2015.
25. de Zeeuw, K.; Dijkstra, P.; Lemmen, C.; Unger, E.-M.; Molendijk, M.; Oosterbroek, E.-P.; van den Berg, C. Bridging the Security of Tenure Gap: Fit-For-Purpose Initiatives. In Proceedings of the FIG Working Week 2019 Geospatial Information for a Smarter Life and Environmental Resilience, Hanoi, Vietnam, 22–26 April 2019.
26. Morales, J.; Lemmen, C.; De By, R.; Molendijk, M.; Oosterbroek, E.-P.; Ortiz Dávila, A.E. On the Design of a Modern and Generic Approach to Land Registration: The Colombia Experience. In Proceedings of the 8th International FIG Workshop on the Land Administration Domain Model, Kuala Lumpur, Malaysia, 1–3 October 2019; pp. 217–242.
27. Mekking, S.; Kougblenou, D.V.; Kossou, F.G. Fit-For-Purpose Upscaling Land Administration—A Case Study from Benin. *Land* **2021**, *10*, 440. [CrossRef]
28. GLTN, UN-Habitat. *Fit-For-Purpose Land Administration Guiding Principles for Country Implementation;* Global Land Tool Network (GLTN): Copenhagen, Denmark, 2016.
29. Enemark, S. Fit-for-purpose land administration for sustainable development. *GIM Int.* **2016**, *30*, 16–19.
30. McLaren, R.; Enemark, S.; Lemmen, C. Guiding Principles For Building Fit-For-Purpose Land Administration Systems in Developing Countries: Capacity Development, Change Management and Project Delivery. In Proceedings of the 2016 World Bank Conference on Land and Poverty, Washington, DC, USA, 14–18 March 2016.
31. Lemmen, C.; Enemark, S.; McLaren, R.; Antonio, D.; Gitau, J.; Dijkstra, P.; De Zeeuw, C. Guiding Principles for Building Fit-For-Purpose Land Administration Systems in Less Developed Countries: Providing Secure Tenure for All. In Proceedings of the 2016 World Bank Conference on Land and Poverty, Washington, DC, USA, 14–18 March 2016.

32. INE. Annual Statistics. Maputo—Moçambique. Available online: http://www.ine.gov.mz/ (accessed on 31 May 2021).
33. Government of Mozambique. *Voluntary National Review of Agenda 2030 for Sustainable Development*; Government of Mozambique: Maputo, Mozambique, 2020.
34. UNDP. *Human Development Report 2020—The Next Frontier Human Development and the Anthropocene*; UNDP: New York, NY, USA, 2020.
35. DINAT. *Avaliação da Governação de Terras—Land Governance Assessment—Mozambique, Based on the WorldBank LGAF Framework*; DINAT: Maputo, Mozambique, 2017.
36. De Carvalho, J.A. *Balanço dos Registos de Terras Activos no SiGIT—Internal Report DINAT*; 2018; Unpublished.
37. Balas, M.; Carrilho, J.; Pascoal, T.; Queface, D.; Carimo, R.; Pinheiro, A. Secured Land tenure Program in Mozambique—A Fit For Purpose Approach. In Proceedings of the 2016 World Bank Conference on Land and Poverty—Scalling up Responsible Land Governance, Washington, DC, USA, 14–18 March 2016.
38. DNTF. *Estudo de Auditoria Social e de Género na Gestão e Administração de Terras*; DINAT: Maputo, Mozambique, 2014.
39. Larsson, G. *Land Registration and Cadastral Systems: Tools for Land Information Management*; John Wiley and Sons Inc.: Hoboken, NJ, USA, 1991; Available online: https://search.library.uq.edu.au/primo-explore/fulldisplay?vid=61UQ&docid=61UQ_ALMA218 0440060003131&lang=en_US&context=L&query=browse_subject,exact,environmental%20sciences%20%23fast&sortby=title& mode=browse (accessed on 25 April 2021).
40. Hanstad, T. Designing Land Registration Systems for Developing Countries. *Am. Univ. Int. Law Rev.* **1998**, *13*, 647–703.
41. Dale, P.; McLaughlin, J. *Providing a Detailed Discussion of a Third Type of Cadastre, the "Multi-Purpose Cadastre." In Land Information Management*; Oxford University Press: Oxford, UK, 1998; pp. 63–84. ISBN 978-0198584049.
42. Williamson, I. The Justification of Cadastral Systems in Developing Countries. *Geomatica* **1997**, *51*, 21–36.
43. World Bank. *Mozambique Land Administration Project (Terra Segura)*; The World Bank: Washington, DC, USA, 2018; p. 164551. Available online: https://documents1.worldbank.org/curated/en/933141621369593098/pdf/Disclosable-Restructuring-Paper-Mozambique-Land-Administration-Project-Terra-Segura-P164551.pdf (accessed on 25 April 2021).
44. Saunders, M.; Lewis, P.; Thornhill, A. *Research Methods for Business Students*, 8th ed.; Pearson Education Limited: New York, NY, USA, 2019.
45. Bryman, A. *Social Research Methods*, 4th ed.; Oxford University Press: Oxford, UK, 2012.
46. dos Santos, L.A.B.; do Vale Lima, J.M.M.; Garcia, F.M.G.P.P.; Monteiro, F.T.; da Silva, N.M.P.; dos Santos, R.J.R.P.; Afonso, C.F.N.L.D.; da Piedade, J.C.L.; Laranjo, P.J.F.; de Almeida Fachada, C.P. Orientações Metodológicas para a Elaboração de Trabalhos de Investigação. Cadernos do IUM N.8. 2019. Available online: https://www.ium.pt/s/wp-content/uploads/20190821_CAD-08 _Miolo_WEB-1.pdf (accessed on 12 January 2021).
47. Deming, W.E. *Out of the Crisis*; MIT Press: Cambridge, MA, USA, 1986.
48. Sokovic, M.; Pavletic, D.; Pipan, K.K. Quality Improvement Methodologies—PDCA Cycle, RADAR Matrix, DMAIC and DFSS Industrial management and organisation Industrial management and organisation. *J. Achiev. Mater. Manuf. Eng.* **2010**, *43*, 476–483.
49. Corcoran, P.B.; Walker, K.E.; Wals, A.E.J. Case studies, make-your-case studies, and case stories: A critique of case-study methodology in sustainability in higher education. *Environ. Educ. Res.* **2004**, *10*, 7–21. [CrossRef]
50. Yin, R.K. *Estudo de Caso: Planejamento e Métodos*, 2nd ed.; Bookman: Porto Alegre, Brazil, 2001.
51. Neuman, W.L. *Social Research Methods: Qualitative and Quantitative Approaches*; Pearson New International Edition: Essex, UK, 2014.
52. UN-GGIM. *Framework for Effective Land Administration: A Reference for Developing, Reforming, Renewing, Strengthening, Modernizing, and Monitoring Land Administration*; UN-GGIM: New York, NY, USA, 2020.
53. Spradley, J.P. *Participante Observation*; Waveland Press: Long Grove, IL, USA, 2016.
54. Deininger, K.; Selod, H.; Burns, A. *Areas Covered by the Land Governance Assessment Framework*; The World Bank: Washington DC, USA, 2012.
55. Balas, M.; Carrilho, J.; Murta, J.; Matlava, L.; Marques, M.R.; Joaquim, S.P.; Lemmen, C. A Fit-For-Purpose Land Cadastre in Mozambique. In Proceedings of the 2017 World Bank Conference on Land and Poverty—Responsible Land Governance: Towards an Evidence Based Approach, Washington, DC, USA, 20–24 March 2017.
56. Groenendijk, E.M.C.; Lemmen, C.; Bennett, R.M. Educational tools for fit for purpose land administration: Experiences and lessons from Mozambique. *GIM Int.* **2016**, *30*, 29–31.
57. Norfolk, S. *Aplicação da Cadeia de Valor da Terra Comunitária (CaVaTeCo) nos Processos de formalização da posse da terra e a Emissão de Certificados de DUATs pelas Comunidades Locais*; Terra Firma: Maputo, Mozambique, 2017.
58. RICS. Crowdsourcing Support of Land Administration: A New, Collaborative Partnership between Citizens and Land Professionals. RICS Research Report (November). 2011. Available online: https://www.clge.eu/wp-content/uploads/2012/05/16310_ RICS_Crowdsourcing_Report-final-WEB.pdf (accessed on 15 January 2021).
59. Dalupan, M.; Haywood, C.; Wardell, D.A.; Cordonnier-Segger, M.C.; Kibugi, R. *Building Enabling Legal Frameworks for Sustainable Land-Use Investments in Zambia, Tanzania and Mozambique: A Synthesis*; IDLO: Nairobi, Kenya; CIFOR: Nairobi, Kenya, 2015.
60. McLaren, R. Crowdsourcing support of land administration—A partnership approach. *Int. Fed. Surv. Artic. Mon.* **2011**. Available online: https://www.fig.net/resources/monthly_articles/2011/mclaren_december_2011.asp (accessed on 31 October 2020).
61. Deininger, K.; Augustinus, C.; Enemark, S.; Munro-faure, P. *Innovations in Land Rights Recognition, Administration, and Governance*; A World Bank Study: Washington, DC, USA, 2010.

62. Jones, B.; Lemmen, C.; Molendijk, M. Low Cost, Post Conflict Cadastre with Modern Technology. In Proceedings of the Responsible Land Governance: Towards and Evidence Based Approach, Washington, DC, USA, 20–24 March 2017; The World Bank: Washington, DC, USA, 2017.
63. Reydon, B.; Molendijk, M.; Lemmens, C. Land Titling Costs: Evidence from Literature and Cases Using FFP. 2020. Available online: https://research.utwente.nl/en/publications/land-titling-costs-evidence-from-literature-and-cases-using-ffp (accessed on 21 February 2021).
64. Antonio, D.; Njogu, S.; Nyamweru, H.; Gitau, J. Transforming Land Administration Practices through the Application of Fit-For-Purpose Technologies: Country Case Studies in Africa. *Land* **2021**, *10*, 538. [CrossRef]
65. Enemark, S.; Mclaren, R.; Lemmen, C.; Antonio, D.; Gitau, J. Scaling Up Responsible Governance: Guiding Principles for Building Fit-For-Purpose Land Administration Systems in Developing Countries. In Proceedings of the 2016 World Bank Conference on Land and Poverty, The World Bank, Washington, DC, USA, 14–18 March 2016.
66. Balas, M.; Carrilho, J.; Murta, J.; Joaquim, S.P.; Lemmen, C.; Matlava, L.; Marques, M.R. Mozambique Participatory Fit For Purpose Massive Land Registration. In Proceedings of the 2017 World Bank Conference on Land and Poverty—Responsible Land Governance: Towards an Evidence Based Approach, The World Bank, Washington, DC, USA, 20–24 March 2017.
67. Laarakker, P.; Zevenbergen, J.; Georgiadou, Y. Land Administration Crowds, Clouds, and the State. In *Advances in Responsible Land Administration*, 1st ed.; CRC Press: Boca Raton, FL, USA, 2015; pp. 91–114.
68. Carrilho, J.; Balas, M.; Marques, M.R.; Macate, Z. Addressing Fuzzy Boundaries in Community Delimitations for Systematic Cadastre in Mozambique. In Proceedings of the 2019 World Bank Conference on Land and Poverty—Catalyzing Innovation, Washington, DC, USA, 25–29 March 2019.
69. Balas, M.; Joaquim, S.; Almeirim, J. SiGIT Land Information System and the Challenges Imposed by the Fit for Purpose Approach to Land Administration. In *FIG Congress 2018 Embracing Our Smart World Where the Continents Connect: Enhancing the Geospatial Maturity of Societies*; FIG: Istambul, Turkey, 11 May 2018.
70. ISO. *ISO 19152:2012 Geographic Information—Land Administration Domain Model*, 1st ed.; ISO: Geneva, Switerzeland, 2012.
71. Hartlief, V.; Ntauazi, C.; Deus, N.R.; Santpoort, R.; Steel, G. *Documento de Trabalho 2: Assegurar os Direitos da Mulher à Terra no Continente Africano—Moçambique*; LANDac: Utrecht, The Netherlands, 2018.
72. Balas, M.; Carrilho, J.; Joaquim, S.P.; Murta, J.; de Carvalho, J.A.; Lemmen, C. Assisted Community-Led Systematic Land Tenure Regularization. In Proceedings of the 2018 World Bank Conference on Land and Poverty—Land Governance in an Interconnected World, Washington, DC, USA, 19–23 March 2018.
73. Balas, M.; Carrilho, J.; Vaz, K. The Role of Communities in Land Cadastre Maintenance. In *FIG Working Week 2019 Geospatial Information for a Smarter Life and Environmental Resilience*; FIG: Hanoi, Vietnam, 26 April 2019.
74. Lewis, L.K. *Vital Considerations for Software Implementation: How to Ensure Success of Your New Land Administration System*; Reuters, T., Ed.; Thomson Reuters Aumentum: Toronto, ON, Canada, 2009. Available online: https://tax.thomsonreuters.com/site/wp-content/pdf/aumentum/10-Considerations-3S-Whitepaper.pdf (accessed on 14 February 2016).
75. Land Equity International. *Land Administration Information and Transaction Systems: State of Practice and Decision Tools for Future Investment*; Report No. 95332419c00892020; Land Equity International: Wollongong, Australia, 2020.

 land

MDPI

Article

Applying the Fit-for-Purpose Land Administration Concept to South Africa †

Christopher Williams-Wynn

Surveyor-General, Eastern Cape, Department of Agriculture, Land Reform and Rural Development, East London, Pretoria 0001, South Africa; Christopher.WilliamsWynn@dalrrd.gov.za; Tel.: +27-82-577-5574

† Initial concepts of this paper were submitted to the Fédération Internationale des Géomètres (FIG) Working Week, Amsterdam, The Netherlands, 10–14 May 2020. Draft proceedings of the cancelled conference were published on 14 April 2020. Available online: https://www.fig.net/resources/proceedings/fig_proceedings/fig2020/papers/ts03h/TS03H_williams-wynn_10363.pdf.

Abstract: What potential will the fit-for-purpose land administration concept have of working in the Republic of South Africa? This question is asked against the existence of a high-quality cadastre covering most of the South African landmass. However, a large proportion of the people living in South Africa live outside of this secure land tenure system. Many citizens and immigrants reside on communal land, in informal settlements, in resettled communities, in off-register housing schemes, and as farm dwellers, labour tenants and other occupants of commercial farms. Reasonable estimates suggest that there are more than 5 million land occupations that exist outside the formal land tenure system and hence outside the formal land administration system. This paper looks at the current bifurcated system and considers how the application of the fit-for-purpose land administration system can expand the existing cadastral system and provide security of tenure that is beneficial and acceptable to all. It demonstrates that, not only could it work, but it is also considered to be necessary. This paper uses South Africa as a case study to demonstrate how adjustments to institutional, legal and spatial frameworks will develop a fully inclusive, sufficiently accurate land administration system that fits the purpose for which it is envisioned. These country-specific proposals may well be of international interest to assist with the formulation of fit-for-purpose land administration systems being developed in other countries.

Keywords: fit-for-purpose land administration; spatial; legal; and institutional frameworks; land tenure security; pro-poor land recordation; land governance reform; cost effectiveness; innovative technology

Citation: Williams-Wynn, C. Applying the Fit-for-Purpose Land Administration Concept to South Africa. *Land* **2021**, *10*, 602. https://doi.org/10.3390/land10060602

Academic Editors: Stig Enemark, Robin McLaren and Christiaan Lemmen

Received: 29 March 2021
Accepted: 1 June 2021
Published: 5 June 2021

1. Introduction

Land administration is defined in the Land Administration Domain Model as "the process of determining, recording and disseminating information about the relationship between people and land" [1]. The Framework for Effective Land Administration released by the United Nations Committee of Experts on Global Geospatial Information Management notes that "all people have the right to an adequate standard of living, regardless of whether underlying people-to-land relationships are formal, informal, statutory, customary, legal, legitimate, or otherwise in nature" [2], (p. 7). The Voluntary Guidelines on the Responsible Governance of Tenure of Land, Fisheries and Forests in the context of National Food Security (VGGT) notes that "access to land, fisheries and forests is defined and regulated by societies through systems of tenure. These tenure systems determine who can use which resources, for how long, and under what conditions" [3], (p. iv). Therefore, one of the guiding principles of responsible tenure governance (a key element of land administration) is to "promote and facilitate the enjoyment of legitimate tenure rights" [3], (p. 3), and equitable access to land, fisheries and forests. This should be applicable to all forms of land tenure, whether it be public, private, communal, indigenous, customary, or informal [3], (p. 7).

The International Land Measurement Standard describes "Land Tenure" as "the rules and arrangements connected with owning specified interests in the land. This can be defined as the relationship, whether legally or customarily defined, among people, as individuals or groups, with respect to land and associated natural resources (water, trees, minerals, wildlife, etc.). Rules of tenure define how property rights in land are to be allocated within societies. Land tenure systems determine who can use what resources for how long, and under what conditions" [4] (p. 24).

Enemark and McLaren highlight that, while there exists a wealth of literature that "emphasises the need for security of tenure and elaborates on its benefits, including the opportunities of significantly contributing to poverty reduction and sustainable development, the conventional approaches to land do not make this a reality" [5], (p. 3). Conventional land administration systems require high accuracy standards for identification, mapping and recordation of land rights. They are generally expensive and operate within a judicially oriented legal framework. As an alternative approach to conventional land administration, the fit-for-purpose land administration concept considers the cultural, social, economic and political context of a country and builds the components of land administration to benefit all people, regardless of their economic or social status [6], (p. 6). In recording land occupation and use, it recommends the use of visible features rather than invisible boundaries based on monumentation [5], (p. 21). It promotes the use of modern (advancing and affordable) technology such as geographic information system mapping technology (GIS), rectified imagery (computer software-generated true-scale aerial photographs) and Global Navigation Satellite System position fixing (GNSS) [6], (pp. 16–17) and [5], (p. 32). It advocates that adjudication, recordation and dispute resolution should be handled through transparent, flexible and simple administrative procedures [5] (p. 27), utilising a human rights approach with all interested and affected parties participating.

The VGGT emphasises that a secure tenure system supports the recognition and respect of all legitimate (formal and social) tenure right holders and their rights, promotes the safeguarding of their rights against threats and infringements and promotes access to justice, thereby minimising tenure disputes, violent conflicts and corruption [3], (p. 3). All people who legitimately occupy land should be provided with a form of secure land tenure that is affordable resulting from highly participatory, quick and efficient methods of recordation, and which can be incrementally improved whenever desired [5], (pp. 3, 5). The fit-for-purpose land administration concept with its three key pillars (institutional, legal and spatial frameworks) supports all these goals by maximising the documenting and recording of people-to-land relationships, thereby facilitating their recognition and inclusion [7], (p. 13).

2. Research Problem and Methodology

The publication "Fit-for-purpose Land Administration: Guiding Principles for Country Implementation" sets out certain principles for consideration. Firstly, the pro-poor fit-for-purpose approach "will lead to social inclusion, increased equity and better recognition of human rights" [7], (p. 5). Secondly, land administration functions "include the areas of: land tenure (securing and transferring rights in land and natural resources); land value (valuation and taxation of land and properties); land use (planning and control of the use of land and natural resources); and land development (implementing utilities, infrastructure, construction works, and urban and rural developments)" [7], (p. 9). Thirdly, "there is a consensus that governing the people-to-land relationship is at the heart of the global agenda and that there is an urgent need to build appropriate and basic systems using a flexible and affordable approach to identify the way land is occupied and used by all, whether these land rights are legal or locally legitimate" [7], (p. 13).

The Fédération Internationale des Géomètres (FIG) guide on Fit-for-purpose Land Administration (Publication No. 60) adds that "Fit-for-purpose means that the land administration systems—and especially the underlying spatial framework of large-scale mapping—should be designed for the purpose of managing current land issues within a specific country or region" [8] (p. 6). However, comparing the current land administration system of the case study area, i.e., the Republic of South Africa, with the fit-for-purpose land administration concept, reveals two major problem areas that require consideration.

- The existing land administration system of South Africa is constructed on the foundation of the official cadastral records, many of which are old, outdated and prepared before standards existed. Many documented land parcels are therefore inaccurate, especially where the boundaries are based on topographical features that are dynamic in nature. A solution needs to be found to improve existing boundary records through utilising innovative technology and datasets that are readily available.
- Many legitimate land occupations are excluded from the formal land administration system, especially those undocumented rights on communal land, in informal settlements, in resettled communities, in off-register housing schemes and housing of farm dwellers, labour tenants and other occupants of commercial farms. The land administration records are therefore incomplete. A solution needs to be found to bring existing legitimate land occupations into the country's land administration system.

Recognising that South Africa is looking to overcome past racially based inequity of land distribution to achieve socioeconomic stability and inclusive economic growth for all South Africans, the fit-for-purpose land administration concept is being considered to close these gaps. Officials from the Tenure Reform and Spatial Planning components of the South African government have proposed the development of an integrated Land Administration System that:

- Includes a legally secure tenure for those with insecure tenure;
- Promotes socioeconomic stability and growth;
- Develops an efficient land management system that is relevant to all;
- Effects a unitary, non-racial and flexible land tenure system that supports an equitable redistribution of land resources; and
- Links all people to their *indawo*—as it is called in isiZulu (i.e., the land they occupy, use or to which they have rights).

Almost all the landmass of South Africa has been covered by land parcels delineated on diagrams kept in the Offices of the Surveyors-General [9] and registered in deeds effected in the Deeds Registries [10]. These are the two key components of its cadastral system and the foundation of the land administration system. The methodology of this research is, therefore, to analyse the current South African Land Administration System in relation to the institutional, legal and spatial frameworks as constituted by the fit-for-purpose land administration concept [7], (pp. viii and 17). These analyses require an understanding of the history and evolution of the South African land administration system (Section 3) and the resulting current land administration system (Section 4). Analyses and solutions are provided within each of the three frameworks (Sections 5–7), and recommendations on the way forward are provided (Section 8), leading to some conclusions (Section 9). It is posited that the fit-for-purpose land administration concept [7,8] can address the identified problems through careful consideration and standardised application. To implement the South African government's vision, it is argued that the fit-for-purpose land administration concept is highly suited to provide a mechanism to bring all legitimate land occupations, currently excluded, into a unitary land administration system.

3. A Brief History of the Land Administration System in the Case Study Area: South Africa

The Republic of South Africa uses a cadastral system introduced during the Dutch occupation of the Cape, based on Roman-Dutch law and brought to the Cape in 1652 by the *Vereenigde Oostindische Compagnie* (VOC) [11], (p. 42). The first land parcels were granted by the Dutch authorities to former employees of the VOC who originated from Europe but chose to remain in the Cape on conclusion of their term of service. They were known as "free *burghers*" and were given land on condition they assisted in the production of fresh food to supply passing trade ships. The grants were recorded on diagrams (identifying position and extent) and in deeds (text describing the link of the land parcel to the *burgher*). When the British took over the Dutch colony in 1806, all the land within the colony, excluding the existing registered land parcels belonging to the *burghers*, was proclaimed as belonging to the (British) Crown.

Before the establishment of a fully inclusive government in South Africa in 1994, the right to an adequate standard of living resulting from the use and enjoyment of a land parcel within a secure land tenure system excluded most of the indigenous people and enforced separation of race groups. This land administration system was dominated by European (initially Dutch and later British) colonial and ultimately apartheid policies, laws and practices. Fisher and Whittal note that: "Indigenous, native, Asian, ethnically-mixed and freed-slave descendent South Africans have borne the brunt of cruel systems of governance that used land administration as a tool to engineer society along racial lines" [11], (p. 347). Only between 1895 and 1923 were attempts made to extend land rights to a few indigenous peoples of South Africa, in recognition of faithful service. Apart from this brief extension of the land administration system to include people of colour, most occupation by those who did not have European ancestry was restricted to communal land (Figure 1 and Figure S1 [12] in the Supplementary Materials), reserves and dormitory townships.

Figure 1. Extract from 2018 rectified imagery [12], (Ref: 2531 AC 10), showing part of a rural village known as Buyelani on communal land in the Mpumalanga Province, at scale: 1:5000 (approximate).

Roughly 16 million hectares (or 13% of the South African landmass) was set aside as communal land, an umbrella term for customary land, land under traditional authorities, tribal land and any land that was part of the South African homeland system [13], (p. 13). Even though communities occupy it, communal land comprises land registered as state land, land proclaimed as state land, land assumed to be state land (land that has never been registered), land resumed by the state and "Trust" land. Records of any rights of individual occupation within communal areas are mostly informal and insecure.

Areas of communal land were not self-sustainable and by as early as 1961, they were "overcrowded, and not one of them could feed its own population except for food purchased by wages from outside" [14], (p. 60). As a result, many people migrated to the cities, towns, mines and farms in search of work and wages to feed their families. The migration has not only occurred from communal areas within the South African borders. Migrants from neighbouring countries, in search of a better life, number in the millions; many have relinquished their original nationality and assimilated into local communities to obtain South African citizenship; others have obtained the required work permits for foreign nationals. Thus, in addition to the insecure tenure pervasive in areas of communal land, insecure tenure is also prevalent in:

- Informal settlements (not only shack dwellers, because informal settlements can include some substantial residences), mostly on the outskirts of urban areas;
- Communities that have been resettled on state-acquired commercial farms as part of the government's redistribution policy where, if the land was transferred to the community, it was transferred to a communal property association;
- Government-initiated housing schemes that were developed in an organised fashion (Figure 2 and Figure S2 [12] in the Supplementary Materials), but where title deeds could not be issued to beneficiaries because of administrative hindrances;
- People that reside on commercial farms as farm workers or labour tenants (i.e., people who may occupy a portion of a farmer's land in exchange for labour); and
- "Ingonyama Trust land", which is an anomaly of communal land that many in government still consider to be state land. Approximately 2.8 million hectares of land is held in trust by the Zulu monarch for the benefit of the Zulu nation. Some leasehold titles have been granted by the Ingonyama Trust Board to residents and business sites.

Figure 2. Extract from 2018 rectified imagery [12], (Ref: 2230 BC 22), showing part of a rural village known as Tshiungani, a government-initiated settlement scheme in the Limpopo Province, at scale: 1:5000 (approximate).

Figure 3 gives a good indication of population densities, where vast rural areas of the country have occupation densities of 30–300 people per square kilometre, but economically sustainable commercial farms are often larger than three square kilometres (300 hectares). Most of the darker-washed areas to the north and east of the figure is communal land and contains an estimated 2–3 million existing homesteads and other rights. The actual numbers of informal dwellings are unknown, but estimates suggest that similar numbers of dwellings as have been calculated in communal areas now also exist in informal settlements. This means that any proposals to bring the occupants of insecure tenure into a land administration system would possibly need to consider identifying more than 5 million land occupations. The numbers continue to increase.

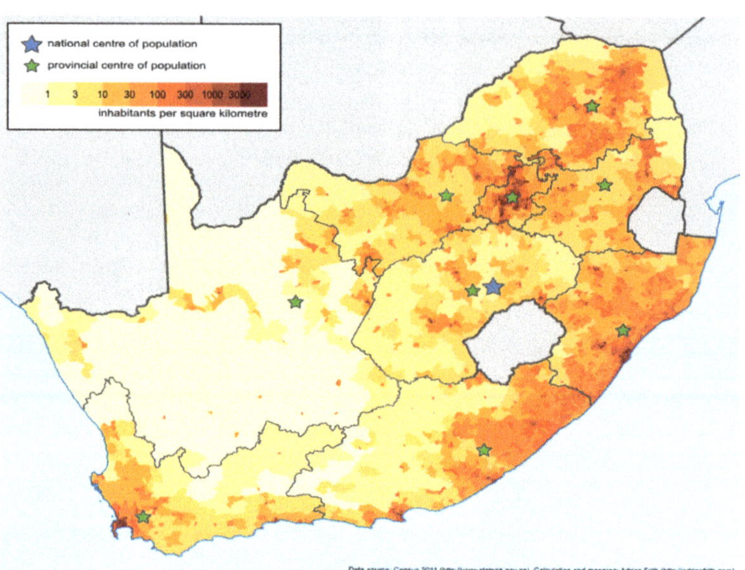

Figure 3. Map by Adrian Frith [15], indicating the number of inhabitants per square kilometre in South Africa as recorded in the last census in 2011.

4. The Resultant Land Administration System of the Case Study Area: South Africa

The biggest hurdle in South Africa is the multiplicity of different forms of land tenure (formal and less formal) that exist, often overlapping in areas where people-to-land relationships of a community are to be formalised.

- There are rights of the original title holder and successors in title, which were originally granted to people of European descent (known as "whites").
- There are instances of quitrent title, many of which are still recorded in the original Indigenous owner's name (known as "non-white"). Quitrent title was land granted by the Crown to loyal subjects—both European and Indigenous—on condition that the holder would supply the colonial administration a specified annual contribution, either in cash or in farm produce. Although no new quitrent titles were created after 1923 [16], the system was maintained by the state until 1934, when all quitrents belonging to "whites" were converted to freehold title and the remaining quitrent titles held by "non-whites" were, thereafter until recently, ignored.
- Less secure individual rights were granted to indigenous peoples by a Resident Magistrate under the formal "Permission to Occupy" (PTO) system. This system became the jurisdiction of the "*Bantu* Administration" system as in the example in Figure 4. In this example, the arable allotment (No. 50) referred to thereon has a quitrent diagram in the office of the Surveyor-General [16], (Reference S.G. No. 9154/1901) but was issued as a PTO right by an organ of state that no longer exists. It is not registered in the Deeds Registry.
- From 1939, policy succeeding quitrent title and running concurrently with the PTO policy was the "Betterment Scheme" policy, where state officials (usually Agricultural Extension Officers) relocated indigenous people from their smallholdings of arable land (often in extent of between three and seven hectares, with a further right to communal grazing), onto one acre (roughly 2000 square metre) sites in "Villages" (Figure 5) in an attempt to increase the efficiency of agricultural production through state-run cultivation of arable land and consolidation of grazing land.

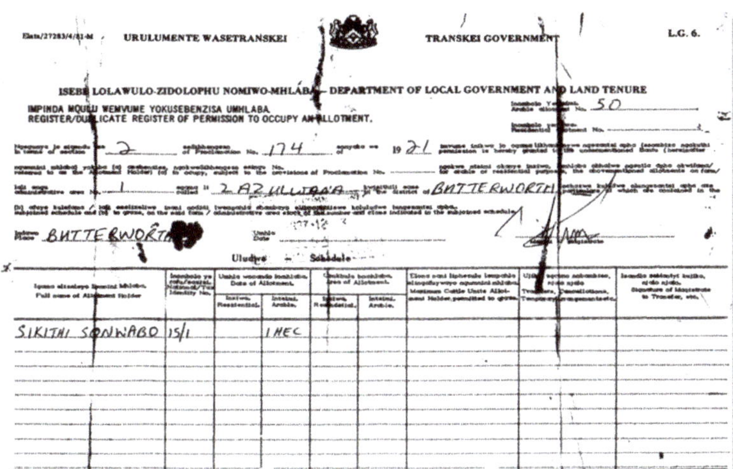
URULUMENTE WASETRANSKEI TRANSKEI GOVERNMENT L.G. 6.

ISEBE LOLAWULO-ZIDOLOPHU NOMIWO-MHLABA — DEPARTMENT OF LOCAL GOVERNMENT AND LAND TENURE

IMPINDA MQULU WEMVUME YOKUSEBENZISA UMHLABA
REGISTER/DUPLICATE REGISTER OF PERMISSION TO OCCUPY AN ALLOTMENT.

Arable Allotment No. 50

Residential Allotment No.

In terms of section 2 of Proclamation No. 174 of 1921 permission is hereby granted to the undermentioned Bantu (hereinafter referred to as the Allotment Holder) (a) to occupy, subject to the provisions of Proclamation No. ____ for arable or residential purposes, the abovementioned allotment on farm/administrative area No. 1 called "ZAZULWANA" in the district of BUTTERWORTH ... the subjoined schedule and (b) to graze, on the said farm / administrative area stock of the number and class indicated in the subjoined schedule.

Place BUTTERWORTH Date

Magistrate

Uludwe — Schedule

Full name of Allotment Holder	Identity No.	Date of Allotment: Residential	Date of Allotment: Arable	Area of Allotment: Residential	Area of Allotment: Arable	Maximum Cattle Units Allotment Holder permitted to graze	Transfers, Cancellations, Temporary arrangements etc.	Signature of Magistrate to Transfer, etc.
SIKITHI SONWABO	15/1		1 HEC					

Figure 4. A duplicate copy of a Permission to Occupy (PTO) obtained from the owner, Mr Sonwabo Sikithi, on the 26 October 2020. His PTO of 1 hectare is situated in the Administrative area of Zazulwana, District of Butterworth, Eastern Cape Province.

- The government-run PTO system floundered due to lack of maintenance [17], (p. 21), mainly because deaths and succession were seldom reported. There are suggestions that unwillingness to report changes was because "non-white" holders objected to having to continue with their annual contributions, when their "white" counterparts did not, due to the conversion of their rights to freehold. As a result, the African traditional leadership hierarchy (king, chief, headman or council) assumed the responsibility to apportion land to their subjects as they saw fit and were recognised only for as long as the recipient was considered a faithful subject of that traditional authority. These PTOs were usually not recorded in any government-administered system and hence were allocated without reference to existing demarcated rights [17], (pp. 18–19).

Figure 5. Photograph of a "Betterment Scheme" settlement known as Brook's Nek, on communal land in the former Transkei area of the Eastern Cape Province [18].

- Allocations were also determined by civil society and political structures, including "people farmers" and "land grabbers", especially where the traditional leadership was inefficient, or the settlement was outside the communal areas. These land allocations were often in recognition of membership or allegiance to a specific structure and were often orchestrated as mass invasions of a preidentified land area.

- Informal rights to land would extend to every person currently residing within the area of the community, no matter how they got there; whether by birth, voluntary migration, job seekers, forced resettlement, assimilation or sworn allegiance.
- Those rights may well extend beyond every person residing on the land. Traditional communities recognise the association of all descendants willing to maintain allegiance and therefore as having rights in the area. For example, Bishop Emeritus Andile Mbete detailed to the author that he has a "house" in a suburb of East London, a city in the Eastern Cape Province, but his "home" is near Willowvale, in a communal area in the former Transkei "homeland", from whence he and his ancestors came.

Over the years, all existing deeds (to the exclusion of most other land rights), issued under any administration, have been incorporated into the current deeds registration system, which is now governed by the Deeds Registries Act [10]. This deeds registration system is dependent on the preparation of diagrams that detail the dimensions, area and position of approximately nine million land parcels, with the support of figures delineating the shape in a prescribed format [9,15] (*vide* Figure S6 [16] in the Supplementary Material and extract in Figure 6).

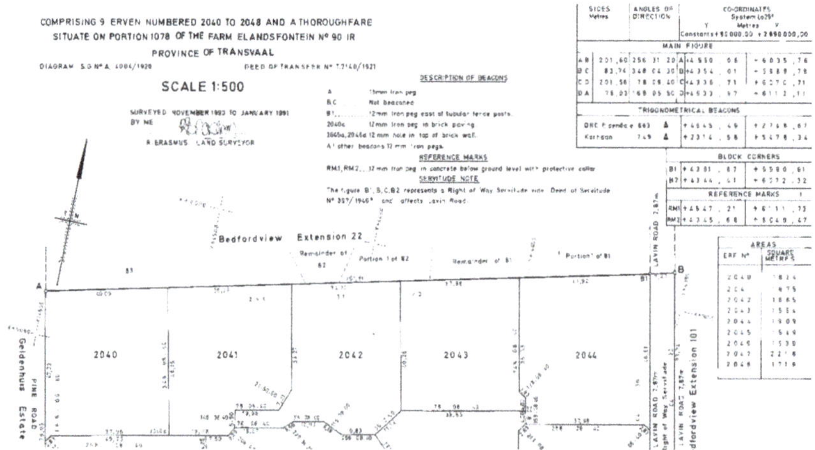

Figure 6. An example of a general plan indicating 9 erven in the formal cadastral system in a suburb known as Bedfordview, City of Johannesburg, in the Gauteng Province. Documents such as this are readily obtainable from the Cadastral Spatial Information dataset [16] (Ref: S.G. No. A. 1282/1991).

All diagrams of the South African cadastral system must now be based on points monumented (beaconed) in the ground and coordinated in the National Control Survey System (NCSS) [9,19] that was established and is maintained by the Chief Directorate: National Geospatial Information (NGI) based in Mowbray, Cape Town [17,18]. The NCSS evolved from its origins in about 1860 to its current state, which provides a complete National Reference Framework of base stations. Base stations include Trigonometrical stations (pillar beacons), Town Survey Marks (submerged under an inspection cover, usually at road intersections in urban areas) and, more recently, Continually Operating Reference Stations (CORS) that provide real-time and post-processing reference positioning data derived from GNSS transmissions [20]. This system has also been adopted by many of South Africa's neighbours.

The surveying of land parcels is governed by the Land Survey Act [9]. This Act defines a diagram as "a document containing geometrical, numerical and verbal representations of a piece of land, line, feature or area forming the basis for registration of a real right" [9].

Section 11 of the Act instructs that a land surveyor shall "carry out every survey undertaken by him or her in accordance with this Act, and in a manner that will ensure accurate results and be responsible to the Surveyor-General for the correctness of every survey carried out by him or her" [9]. The Act continues in Section 14 by stating that "no diagram of any piece of land shall be accepted in any deeds registry in connection with any registration therein of that land, unless the diagram has been approved by the Surveyor-General" [9]. Section 16 states that "no diagram shall be approved by the Surveyor-General unless it is prepared under the direction of and signed by a land surveyor" [9].

The system permits rectilinear, curvilinear and ambulatory boundaries between physical beacons. Most of the boundaries delineated in the Cadastral Spatial Information datasets are rectilinear boundaries, consisting of straight lines between the monumented points. (An extract of an area of rectilinear boundaries of the Cadastral Spatial Information database is shown in Figure 7 superimposed over rectified imagery supplied by NGI [12,20].) The line between the monumented points may also be curvilinear—a mathematical curve, a permanent, static topographical feature such as a wall or a fence, or even a dynamic topographical feature such as the middle of a river or the high-water mark of the coast.

Figure 7. 1map Online GIS [21] overlay over rectified imagery of a suburb known as Vincent Heights in a formal cadastre area in East London, Eastern Cape Province, at scale 1:2000 (approximate).

Cadastral Spatial Information has been created from a collection of the delineated boundaries surveyed over 300+ years as documented on the approved diagrams and survey records preserved in the Offices of the Surveyors-General. Each boundary line contained in the Cadastral Spatial Information database is only as accurate as the survey of the original diagram, or any subsequent resurvey of that boundary. Before 1929, standards for the survey of boundary lines were not specifically legislated. Therefore:

- Some early diagrams show boundary lines with no recorded mathematical data—these diagrams only indicate a drawn figure, complemented with the intended area of the land parcel;
- Boundaries may have been shown in relation to topographical features, such as the top of a hill, or following a river or the coast;

- Some of the data defining boundary lines are inaccurate due to poor survey practice and substandard equipment, or were simply paced, ridden, sketched by eye or drawn from memory;
- Many monuments (beacons) defining each end of the boundaries have disappeared completely, resulting in uncertainty of the legal position; and
- Occasionally, land surveyors or the Surveyor-General discover errors in mathematical data or overlaps of diagrams, which must be corrected.

The "general plan" (which is a composite diagram of several land parcels) shown in Figure S6 [16] in the Supplementary Materials indicates the position and extent of each land parcel recorded in relation to the NCSS and adjoining properties to an accuracy of a few centimetres. Once approved, these records are filed in the offices of the Surveyors-General. Each land parcel is then linked to a legal person or entity through the registration of a deed. Section 2 of the Deeds Registries Act [10] gives the Registrar of Deeds the powers and responsibilities to preserve all records of ownership or rights in land, based on the land parcel information held by the Surveyor-General. Further, the Registrar of Deeds oversees the examination and registration of any new deed lawfully submitted to his or her registry for transaction. The deed inextricably links land parcels to people by adding the "who" and the "how" of the rights to a land parcel. A land parcel only becomes a legal object once it has been registered in the Deeds Registry. All such legal objects form the foundation of the South African Land Administration system.

5. Analysis in Terms of the Institutional Framework

It cannot be emphasised enough that communities are more likely to preserve, protect and manage their rights when such rights in land are recognised. The current disparate land administration system does not make this a reality for all. The occupants of settlements such as the ones shown in Figures 8 and 9 (Figures S8 and S9 [12] in the Supplementary Materials) remain excluded from the current system. The author has demonstrated in previous research [18] that South Africans with less formal and insecure tenure want their legitimate occupation recorded and recognised. In almost every community that the author has worked for, with and in, the communities have been exemplary in their commitment to the exercise of recordation of their land rights, when the exercise is preceded by community focused and protocol-based communication. People want to participate in the process to ensure an equitable distribution of land rights.

Figure 8. Extract from 2018 rectified imagery [12] (Ref: 2531 AD 16), showing an informal settlement in an area known as Luphisi, in Mpumalanga Province, at scale: 1:10,000 (approximate).

Figure 9. Extract from 2019 rectified imagery [12] (Ref: 2732 AB 15), showing part of a rural settlement in an area known as Siphondweni, on communal land in the KwaZulu-Natal Province, at scale: 1:5000 (approximate).

During the 20th century and notwithstanding its bifurcated application, the South African Cadastral System established itself as a high-quality land information system [11], (pp. 225–327). It is currently in decline. Most land administration institutions in South Africa are inadequately resourced; many professional and technical officials trained in spatial information systems could not be retained within the administration. Nevertheless, the cadastral survey adjudication process and the Cadastral Spatial Information databases are still being run remarkably well on outdated systems, by willing but largely inexperienced and ill-equipped officials. Attempts to apply technological innovation, modern systems and updates to operations to improve efficiency, quality and accuracy have encountered many challenges and costly delays. The fit-for-purpose land administration system being proposed will include a prescribed institutional methodology of eight key protocols, some of which can run concurrently. These are set out in Table 1.

Table 1. A proposed institutional methodology of eight key protocols.

No.	Title	Description
1	Initial engagement with the community	Initial engagement with the community will determine the status of the settlement. Is the occupation a traditional community where their land is held in trust by the state or the Ingonyama Trust? Does the community have traditional, civic or social leadership? Does the community have a formal or recognisable identity? This will lead to the determination of position and extent of the settlement and how it is represented on official documents (filed in the Office of the Surveyor-General) of the underlying land parcels as recorded in the existing cadastre.
2	Determination of current formal land records	Determination must be made of who the land owners of the identified land parcels are by scrutinising official deeds records filed with the Registrar of Deeds. Dispossessed and successors in title of deceased land owners may require legal expropriation as prescribed in the relevant legislation. Where there is no record, the land will probably be unalienated state land (i.e., never registered) as decreed in the colonial annexation of the Southern African territories mentioned earlier.
3	Determination of current informal land records	Much preparation is necessary to ensure that the relationships, rights, protocols and culture of the community are understood and upheld. Community participation will assist with the determination of any additional rights overlapping formal ownership, whether registered, recorded, social or recognised. These rights could be filed in many places, such as the Magistrate's Office, in offices of the Department of Agriculture, or kept by the Traditional Council. There are also existing rights held by members that may not be recorded in any official archive. (It is always a marvel when, on inquiry, a member of the community produces a well-preserved document that had been issued to a forefather many generations before. Figure 4 is such an example.)
4	Project planning and funding	It is not only the members of the community that are to be included in the process: relevant government institutions have a vested interest. As part of the preparatory work, those organs of state empowered with developmental functions must determine the availability of funding. This will also have to be presented to the community for agreement together with a detailed, defined project plan. Some funds may be allocated to a communal trust fund to assist the community in supporting the implementation. In certain instances, compensation may need to be considered to acquire land under existing land rights for the provision of services.

Table 1. *Cont.*

No.	Title	Description
5	Stakeholder engagement	Information sessions must be widely advertised (in writing and by word of mouth) to ensure maximum participation and involvement of the community. Any source of information on community membership should be sought. This may range from records of heads of households held by the leadership, to verbal nomination from participants from the community. It is noted here, that the members present should be requested to consider those not present, including those members not in permanent residence. On state land, the Minister (as the nominal owner) ultimately approves the transfer of land, or the issuing of rights. The Minister must therefore be convinced that all processes have been followed and that the community has been adequately consulted. A generally accepted principle in government circles is that 80% of the community must support or agree to the proposal and such decisions must be recorded in official community resolution documents.
6	Identification of land occupation	Thereafter, the names of community members can be linked to the position of each homestead and the extent of any land occupation. Using readily available rectified imagery [12] based on the NCSS [19,20], the limits of occupation, use and other rights attached to each homestead can be identified in conjunction with inputs from the community. It is always important to ensure the community's participation in the identification of all boundaries and not to assume that what is visible is the recognised extent of any right. While applying this step, recognition must also be given to non-allocated areas within the settlement and questions should be asked as to who holds the rights to those areas?
7	Institutional capacity	The institutional framework must include an investigation into the capacity of the administration to process and maintain the increased numbers of land rights and land transactions [5], (p. 7) that will result from the implementation of this inclusive land administration system.
8	Ongoing maintenance of the system adopted	Lastly, the institutional arrangements must include processes by which people are able to communicate transactions of any land right with the responsible authority. Any legitimate change to the people-to-land relationships must follow a simple process to ensure that the system administrators are able to ensure all land records are current, correct and complete.

6. Analysis in Terms of the Legal Framework

Any fit-for-purpose land administration implementation strategy must include a formal legal framework, which satisfies the political mandate of the ruling party, and which should ideally integrate all land information into a single system to reduce cost and improve access to information. New systems and policies are being considered but will take time to implement. The one key area that is still being debated is whether there should be a differentiation (in terms of the continuum of land rights) between recordation and registration of land rights. This is particularly sensitive in South Africa because of the discriminatory processes imposed on previously disadvantaged groups in the country as described in Section 3. For example, in Figure 9 (Figure S9 [12] in the Supplementary Material), there may well be a diagram and deed for the school site visible on the upper left, but it is highly unlikely that the occupations visible over the rest of the image (some surrounded by hedges) would currently have any form of secure tenure. Similarly, in Figure 10 (Figure S10 [12] in the Supplementary Materials), the visible agricultural units may have some form of recognised right, but there are no formal records of those rights in the offices of either the Surveyor-General or the Registrar of Deeds.

Figure 10. Extract from 2015 rectified imagery [12] (Ref: 3130 AA 11), showing small-scale farming in a rural area known as Mahana, on communal land in the Eastern Cape Province, at scale: 1:5000 (approximate).

Current thinking is to separate recordation from registration:

- Recordation would document the reality of land occupation, land use, rights in or to land as it is on the ground. For example, it could include social tenure as is conceptualised in the Social Tenure Domain Model [22]. It would recognise the identity of a person where their right to a specified piece of land is undisputed, recognising that person as a participant with individual, community, communal or informal rights. Recordation (possibly in the form of a Certificate of Land Right to provide a legally secure tenure) may not necessarily proceed to registration as currently prescribed in law.
- Registration (title deed or other forms of legal tenure [10]) would be retained for currently registered land parcels, formal rights, state land, external boundaries of land for identified communities and any applicable legislation that improves rights of insecure tenure (e.g., Upgrading of Land and Tenure Rights Act [23], and Interim Protection of Informal Land Rights Act [24]).

Most important in the South African context is that, while the recordation process will be cheaper and quicker than the requirements of the existing cadastral system, it must not be inferior or less secure. Current legislation will require only minor adjustments to apply the principles of fit-for-purpose land administration. These minor adjustments are primarily because legislation has not kept pace with innovation, neither has it maintained its efficiency or simplicity. Such minor adjustments will also improve the quality and accuracy of the existing cadastre.

Beyond the legislation controlling the cadastral system, other legislation affecting land rights and land development will need to be assessed. For example, Section 6 (1) (b) of the Land Survey Act requires the Surveyor-General to ensure any diagram or general plan to be approved by him or her is "in accordance with any statutory consent in so far as the layout is concerned" [9]. The requirement of a "statutory consent in so far as the layout is concerned" frequently causes much hardship (time and cost) to developers and prospective beneficiaries of enhancements to land tenure. Scores of pieces of planning legislation remain on the statute books, and are administered by inexperienced and ill-equipped officials, insufficient in number and skills to assess any development of land in terms of the relevant statutes. New legislation, such as the Spatial Planning and Land Use Management Act [25], has been promulgated without the necessary resources to implement the legislation, especially in rural areas, and without repealing old order legislation. This has resulted in frequent duplication of requirements.

Statutory consent applications require extensive research, even if the development is an in-situ upgrade of a fully functional community. Planning and application fees are costly, prescribed circulation through multiple organs of state supplying services is lengthy and often inordinately delaying and the conditions ultimately attached to the statutory consent are often very onerous. Burdensome standards and expectations are frequently imposed on an existing, functional settlement, where all that the members of that community want are their rights recognised and recorded. Much can still be done through the repeal of outdated legislation and amendment and updating of useful legislation to accommodate technological advances and prevent duplication.

7. Analysis in Terms of the Spatial Framework

With extraordinary advances in recording methods using GNSS, rectified imagery (such as the extract shown in Figure 11 and Figure S11 [12] in the Supplementary Materials) and GIS, it is suggested that anything produced using a combination of these innovations will be more accurate than many of the existing land records existing in the Offices of the Surveyors-General.

Figure 11. Extract from 2019 rectified imagery [12] (Ref: 2723 AB 7), showing an informal settlement known as Molomowapitsana, in the Northern Cape Province, at scale: 1:5000 (approximate).

Strides must be made to replace antiquated systems and equipment with their highly efficient modern successors. NGI, as the South African National Mapping Agency, has extensive aerial photography coverage of the whole country [12]. The denser the population, the larger the scale of the photography and more frequently updates are undertaken. Much of this aerial photography has been adjusted into rectified imagery at an accuracy better than most of the spatial data recorded in the Offices of the Surveyors-General. As can be seen from the recent rectified imagery of which extracts are recorded in Figures 1, 2 and 8, Figures 9–11 (Figures S1, S2 and S4–S7 [12] in the Supplementary Materials), many of the unregistered land occupations are defined by fences, hedges or walls. Therefore, the country has usable rectified imagery from which all visible boundaries can be determined.

Further, there are also many instances where the original position was unclearly defined, either in terms of description or position. It is suggested that the fit-for-purpose land administration concept can assist in developing standards to guide the identification of positions of many existing boundaries using currently available resources, technology and datasets. The result will be more accurate cadastral records at much lower cost. For example, it is proposed that, where topographical features are used as boundaries, rectified imagery could be used to update those visible boundary positions, especially where they are of a dynamic nature, such as along rivers or the coast. The rectified imagery provided by NGI [12] would satisfy all the prescribed standards of land parcel delineation [9,16,19,20].

8. The Way Forward

As with any transformative policy implementation, there are major issues to be considered. Fit-for-purpose land administration recommends the possibility of an incremental approach, where the initial recording tenure rights "using simple and low-cost approaches … should be upgraded when need arises" [7], (p. 6), and is therefore built around the flexible recognition of different forms of land tenure.

Much preparatory work is necessary in identifying existing rights on any piece of land, including ownership, unrecorded succession of title, permissions, allocations, occupations and descendants. It is not as onerous as it sounds, as there are generally only two sources of information, firstly the official records held by the current administration (notably the Offices of the Surveyor-General [9,16] and Registrar of Deeds [10]) and, secondly, the institutional knowledge and preserved evidence of the community. It is strongly recommended that a more pervasive Land Rights Enquiry system is developed and implemented within state structures to resolve the plethora of overlapping and conflicting land rights. Already, some of the state-owned enterprises, such as the Electricity Supply Commission (ESKOM) and the South African National Roads Agency Limited (SANRAL), have internal structures engaging with the Surveyors-General and Registrars of Deeds to resolve land issues due to their need to negotiate acquisition of land for their infrastructure that traverses the communal areas.

Another matter to consider is that, during the 1980s, the state issued contracts to land surveyors to document many of the previously un-surveyed dormitory townships that had been laid out some distance away from the formal "white" towns, primarily as accommodation for African labour who worked in the nearby towns. A few hundred land surveyors surveyed hundreds of thousands of land parcels in a very short time and completed the project well ahead of the anticipated schedule. The townships had been laid out, houses built, and yards fenced. In most instances, land surveyors were able to survey the fence corner posts, i.e., boundaries were defined by physical features approximating rectangular sites and blocks as originally demarcated. Figure 12 is an extract of a survey in which the author was involved (Figure S12 [16] in the Supplementary Materials).

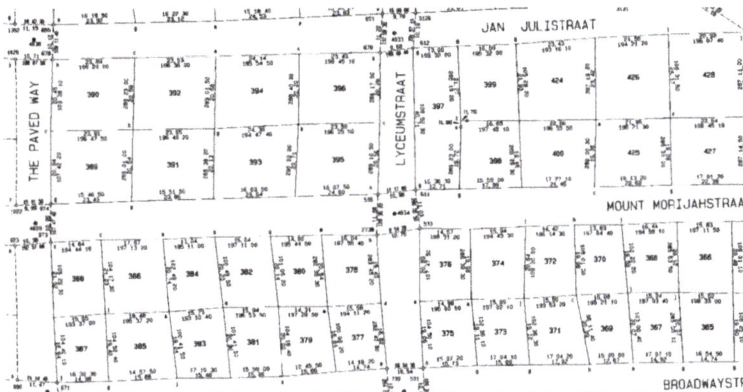

Figure 12. Extract from a 1985 general plan of a township with 1092 erven known as Makeleketla, Administrative District of Winburg, Free State Province [16] (Reference: S.G. L. No. 854/1985). The line intersections were determined by fixing the positions of the fence corner posts.

The land surveying team completed the fieldwork of all 1092 erven in a little over two weeks, using the technology available then (single-second theodolites, electronic distance measuring equipment and tape measures). Using more efficient modern technology, such as GNSS equipment, rectified imagery, drones (UAVs) and GIS software, the land surveying profession is confident that the six million land parcels still to be documented can be surveyed to the prescribed accuracies and standards set out in the Land Survey Act [8] and Regulations in an even quicker rate.

Regulation 3, appended to the Land Survey Act [9] standardising field measurements and observations, prescribes that "a land surveyor shall determine the positions of all stations and beacons within the limits of accuracy prescribed in regulation 5 and shall check every part of his or her survey". With regard to determining positions by photogrammetric methods, the standard prescribed in the Regulation is based on out-dated methods. Nevertheless, the facility to use aerial photography is already written into the South African spatial legislation and can be simply updated. The profession is ready and willing to prepare the necessary land parcel documentation, starting with the mark-up of rectified imagery and the creation of any prescribed cadastral plans.

With the quality of the available rectified imagery from NGI [12], many of the land parcels can be delineated straight from the imagery. For example, settlements such as those of Buyelani (Figure 1), Tshiungani (Figure 2), Luphisi (Figure 8) and Molomowapitsana (Figure 11) would require only cursory ground truthing (community participation), as the boundaries of the land parcels are sufficiently recognisable for recordation purposes from the imagery. On the other hand, settlements such as those of Siphondweni (Figure 9) and Mahana (Figure 10) may require more extensive verification on the ground, because hedges and edges of cultivation may not define the full extent of the occupants' rights. Experience of the author has shown that the members of the community will readily provide much of the additional information whenever trust is established.

The buy-in from the community members will also ensure that land administration records are maintained in the future. Already, government has considered "Community Information Centres" to be placed within every municipal area to facilitate ease of access by the communities. Even though it was acknowledged in Section 5 that the Cadastral System is in decline, resources are available to overcome this through state facilitation. Current "Information and Communication Technology" will make it possible for beneficiaries to participate in a web-based fit-for-purpose land administration system.

There are three issues that still require attention.

8.1. Institutional

A participatory and transparent method of ascertaining the link of people (both those who reside on the land parcel and those who have rights thereto) to the land parcel must be instituted. A community-based process is essential to ensure maximum participation and reduce the risk of excluding any rightful beneficiary. Any such recordation needs to be accepted into a fully capacitated Land Administration system and any subsequent transaction must be easily updated in that system.

8.2. Legal

The state needs to create the legal mechanism of recording the people-to-land relationships, whether it be the current registration system or another form of recording. In addition, the requirements of the plethora of national and provincial planning legislation and development controls, much of which is obstructive to any system of recording of existing land rights, would need to be addressed and overcome.

8.3. Spatial

Datasets, innovative technology and human resources are readily available, all of which will facilitate the implementation of the appropriate consultation, delineation and recordation processes. Controlling authorities must still give the go-ahead to utilise them.

9. Conclusions

The standard accuracy requirements and the requirement to use registered land surveyors to oversee the delineation of land occupation is not an expensive option. Extensive surveys based on streamlined processes, modern technology, participation by the communities and identification of existing visible features can and will be performed at very little cost and at great speed and precision—"cheap, accurate and fast". The legal requirements of existing cadastral surveying protocol can easily be achieved at very little cost per land parcel.

Government recognises the slow pace at which land registration is implemented in the current system and acknowledges that the cost of acquiring a title deed is too high for most. The government is therefore investigating a new, more inclusive, Integrated Land Administration System that will facilitate the recordation of the large number of new land rights.

In South Africa, there is generally political will to make it happen. Minimal amendments to legislation controlling the cadastral system are required. An additional form of deeds registration (or recordation) is a possibility. Rationalisation of impeding planning legislation is essential. It has been proven that people want their rights documented. Many boundaries are visible on current aerial imagery. Technology exists and is available to provide a high level of accuracy at minimal cost. The land surveying profession is well-established. The resultant land rights would easily be upgradable to formal title. There are therefore many positive aspects already in place and no outstanding issues are insurmountable. A fit-for-purpose land administration system should be implemented in South Africa so that land administration will be "designed to meet the needs of the people and their relationship to land, to support security of tenure for all and to sustainably manage land use and natural resources" [6] (p. 5).

The fit-for-purpose land administration system considered in this paper may assist to formulate recommendations in countries or regions with similar issues. Solutions proposed for South Africa may well find application in many other countries.

Supplementary Materials: Rectified imagery is available online on the link http://www.cdngiportal.co.za/cdngiportal/ [12]. Larger images and hence better quality copies of the following rectified imagery extracts are available as Supplementary Materials online at https://www.mdpi.com/article/10.3390/land10060602/s1: Figure S1: Buyelani; Figure S2: Tshiungani; Figure S4: Luphisi; Figure S5: Siphondweni; Figure S6: Mahana; Figure S7: Molomowapitsana. Scanned images of diagrams and general plans approved in the Offices of the Surveyors-General are available online on the link: http://csg.dla.gov.za/ [16]. Images of the complete document from which the following extracts are available as Supplementary Materials online at https://www.mdpi.com/article/10.3390/land10060602/s1: Figure S3: Bedfordview; Figure S8: Makeleketla.

Funding: Funding of the publication costs for this article has kindly been provided by the School of Land Administration Studies, University of Twente, in combination with Kadaster International, The Netherlands.

Institutional Review Board Statement: Not applicable.

Informed Consent Statement: Not applicable.

Data Availability Statement: Aerial photographs are extracted from the rectified imagery dataset of the Chief Directorate: National Geospatial Information, to whom Copyright is acknowledged. The extracts are reproduced under Government Printer's authorisation (Authorisation No. 11841 dated 30 November 2020.

Acknowledgments: The author acknowledges much support given by work colleagues and friends in academia in providing a platform to discuss and argue the concepts contained in this paper. In addition, the author acknowledges delegates from member states participating in the Fédération Internationale des Géomètres (International Federation of Surveyors') gatherings over the last seven years who have given the author the original ideas, research material, encouragement and guidance. Some of these are the guest editors of this Journal.

Conflicts of Interest: The author declares no conflict of interest. The funders had no role in the design of the study; in the collection, analyses, or interpretation of data; in the writing of the manuscript, or in the decision to publish the results.

References

1. The Land Administration Domain Model. LADM: ISO 19152. 2012.
2. The United Nations Committee of Experts on Global Geospatial Information Management (UN-GGIM). Global Consultation on the Framework for Effective Land Administration. UN-GGIM. 2020.
3. UN Food and Agriculture Organization (FAO). *Voluntary Guidelines on the Responsible Governance of Tenure of Land, Fisheries and Forests in the context of National Food Security*; The Food and Agriculture Organisation of the United Nations: Rome, Italy, 2012.
4. International Land Measurement Standard Coalition. *International Land Measurement Standard: Due Diligence for Land and Real Property Surveying*; International Land Measurement Standard Coalition: London, UK, 2019.
5. Enemark, S.; McLaren, R. Secure Tenure for All Starts to Emerge: New Experiences of Countries Implementing a Fit-for-Purpose Approach to Land Administration. In Proceedings of the World Bank Conference on Land and Poverty, Washington, DC, USA, 25–29 March 2019; Available online: https://vbn.aau.dk/ws/portalfiles/portal/311735227/Washington_2019._Enemark_v5_Paper_ID_294.pdf (accessed on 22 March 2021).
6. Musinguzi, M.; Enemark, S.; Mwesigye, S.P. *Fit-for-purpose Land Administration—Country Implementation Strategy: UN Habitat, Global Land Tenure Network publication*; Republic of Uganda, Ministry of Land, Housing and Urban Development: Kampala, Uganda, 2018.
7. Enemark, S.; McLaren, R.; Lemmen, C. *Fit-for-purpose Land Administration: Guiding Principles for Country Implementation*; United Nations Human Settlements Programme UN Habitat: Nairobi, Kenya, 2016.
8. Enemark, S.; Bell, K.C.; Lemmen, C.; McLaren, R. *FIG Guide to Fit-for-Purpose Land Administration*; Fédération Internationale des Géomètres (FIG) Publication No. 60; A joint FIG/World Bank Publication: Washington, DC, USA, 2014.
9. The Land Survey Act. Act No. 8 of 1997. 1997.
10. The Deeds Registries Act. Act No. 47 of 1937. 1937.
11. Fisher, R.; Whittal, J. *Cadastre: Principles and Practice*; South African Geomatics Institute: South Hills, South Africa, 2020.
12. Chief Directorate: National Geospatial Information. Geospatial Portal. Available online: http://www.cdngiportal.co.za/cdngiportal/ (accessed on 4 December 2020).
13. Republic of South Africa, Department of Agriculture, Land Reform and Rural Development. Land Audit. 2013. Available online: https://s3-eu-west-1.amazonaws.com/s3.sourceafrica.net/documents/119320/South-Africa-Land-Audit-Report-2013.pdf (accessed on 30 May 2021).
14. Brookes, E.H.; Webb, C.D.B. *A History of Natal*; University of Natal Press: Pietermaritzburg, South African, 1965.
15. Frith, A. Centres of Population of South Africa. Available online: https://adrian.frith.dev/images/mean-centres-150dpi.png (accessed on 2 March 2021).
16. Chief Surveyor-General. Cadastral Spatial Information Dataset. Available online: http://csg.dla.gov.za/ (accessed on 2 June 2021).
17. Republic of South Africa, Public Service Commission. *Report on the Evaluation of Land Administration in the Eastern Cape*; Republic of South Africa, Public Service Commission: Pietermaritzburg, South African, 2003.

18. Williams-Wynn, C.D. What People Want. In Proceedings of the Fédération Internationale des Géomètres (FIG) Working Week, Sophia, Bulgaria, 17–21 May 2015; ISBN 978-87-92853-35-6. ISSN 2307-4086. Available online: https://fig.net/resources/proceedings/fig_proceedings/fig2015/papers/ts01b/TS01B_williams-wynn_7684.pdf (accessed on 25 February 2021).
19. Clarke, D.G. *Standard for the National Control Survey System, Chief Directorate*, 3rd ed.; National Geospatial Information, Department of Rural Development and Land Reform: Cape Town, South Africa, 2018.
20. Chief Directorate: National Geospatial Information. Geodetic and Control Survey Services. Available online: http://www.ngi.gov.za/index.php/what-we-do/geodetic-and-control-survey-services (accessed on 4 March 2021).
21. Klopper, N. 1map Online GIS. Available online: https://www.1map.co.za/apps/onemap2020 (accessed on 2 March 2021).
22. Lemmen, C. The Social Tenure Domain Model—A Pro-Poor Land Tool. FIG Publication No. 52. May 2013.
23. The Upgrading of Land and Tenure Rights Act. Act No. 112 of 1991. 1991.
24. The Interim Protection of Informal Land Rights Act. Act No. 31 of 1996. 1996.
25. The Spatial Planning and Land Use Management Act. Act No. 16 of 2013. 2013.

 land

MDPI

Article

Application of FFPLA to Achieve Economically Beneficial Outcomes Post Disaster in the Caribbean

Charisse Griffith-Charles

Department of Geomatics Engineering and Land Management, The University of the West Indies, St. Augustine, Trinidad and Tobago; Charisse.Griffith-Charles@sta.uwi.edu; Tel.: +868-682-8730

Abstract: Fit-for-purpose mechanisms for developing land administration systems have been posited to be especially effective in resource strapped economies since these mechanisms quickly create the settings for economic as well as social and environmental development. Competition for depleted resources in the face of recent deleterious events such as climate change, Covid-19, hurricanes and other natural hazard impacts, and global economic crises, among other challenges, should nudge many developing countries toward the application of Fit for Purpose Land Administration (FFPLA) as opposed to costly and lengthy standard methods. Problems arise in convincing states of the benefits of applying the FFPLA. This paper explores how fit-for-purpose methods for establishing and upgrading land administration infrastructures have become increasingly imperative to developing countries, particularly small island developing states (SIDS) of the Caribbean, in light of declining economies. The experiences of Caribbean countries, with a focus on Trinidad and Tobago, Barbados, Saint Lucia, and Jamaica, in implementing adjudication and titling for their land administration, are compared to FFPLA guidelines in terms of major objectives, supportive legislation, and method of application. Based on the outcomes of the evaluation, it is suggested that including more facets of the FFPLA, primarily for progressing the process toward economically beneficial success, would be an advantage.

Keywords: fit-for-purpose land administration; cadastre; FFPLA

 check for updates

Citation: Griffith-Charles, C. Application of FFPLA to Achieve Economically Beneficial Outcomes Post Disaster in the Caribbean. *Land* **2021**, *10*, 475. https://doi.org/10.3390/land10050475

Academic Editors: Stig Enemark, Robin McLaren and Christiaan Lemmen

Received: 1 April 2021
Accepted: 29 April 2021
Published: 2 May 2021

1. Introduction

Systematic adjudication and titling are theoretically the ideal method for establishing a comprehensive cadastral component of the land administration in a jurisdiction [1–4]. A comprehensive cadastre is posited to result in benefits to the country such as land market growth, and invigoration, reduction in conflict over land rights, increased well-being of the society, and increases in agricultural productivity, as well as benefits to individual landholders such as security of tenure, and access to credit [3,4]. The veracity of the economic benefits and return on the significant expenditure accruing to a specific jurisdiction, however, is highly dependent on a multiplicity of localized factors including market opportunities, credit availability, and cultural approaches to credit risks as has been previously examined [1,2,4] including recently in Rwanda, and Ethiopia [5,6]. Individual studies have tracked the outcomes of these programmes over time with varying results as contextual factors, and stated objectives may differ. If economic, social, or environmental benefits are to be experienced, however, the parcel fabric and attribute ownership information should be comprehensive, or nearly so, over the jurisdiction so that transparency and access and thus confidence in the database is achieved and land governance and management decision-making strategies can be employed.

The assumption may be made that social benefits may accrue more readily if the coverage is complete and focused on formerly informal or tenure insecure communities while economic benefits may more readily result from an emphasis on capturing areas with more marketable properties. Decentralisation of the process of establishing a land administration system has been explored by, for example, Ho et al. [7]. It is therefore conceivable that a comprehensive cadastre may be established for a region or community for a specific purpose. Given the need for comprehensive coverage over a jurisdiction or community, different methods, as proposed in the Fit for Purpose (FFP) approach and Fit for Purpose Land Administration (FFPLA) [8,9], are appropriate for quickly and cheaply capturing land administration data. This paper concentrates on the characteristics of FFPLA [8,9] that would lead to acquisition of comprehensive parcel location and tenure data for a jurisdiction with a focus on maximizing economic, timely, and cost-efficient outcomes.

The main issue restricting the application of FFPLA for many developing countries, including those of the Caribbean, is achieving the acceptance of the state, including institutions, technocrats, professionals, and politicians, and also the beneficiary population, that the merits of conducting an FFPLA development process outweigh the cost, time, and labour implications of the standard high-precision procedures previously promoted as the ideal. States may sometimes believe that the quality of an FFPLA product is inferior to that produced by the more rigorous processes. This evaluation examines the experiences with adjudication and titling being undertaken by Caribbean countries and determines how closely the methods being currently used fit the FFPLA tenets and whether there is correlation between the methods used and the current status of the countries in development. In the following sections, the paper first describes the current status of land registration in the Caribbean, then presents the evaluation methodology, and, after analysis of the extant available data, suggests the way forward for land administration to achieve an economically beneficial outcome both from the anticipated outcomes of the process and the minimising of the cost of the registration activity.

2. Land Registration in the Caribbean

The FFPLA approach, as proposed by FIG and UN-Habitat [7,8] promotes the achievement of security of tenure but in a cost-effective and time sensitive way. FFPLA focuses on flexibility, inclusiveness, participation, affordability, reliability, attainability, and upgradeability. To achieve all of these requirements, the process must be performed to attain efficiency by using as few resources as possible to obtain only as much as is absolutely necessary for achieving a specific beneficial goal.

There are other cost-conscious tools available such as the Pro-Poor Land Recordation Tool (PPLRT) [10,11] that are largely applied in rural contexts and in large informal community settings where civil governance is strong. These can also be applied together with and as a complement to the FFPLA. However, the countries of the Caribbean are small with very little distinction between rural and urban areas and with small informal communities interspersed among the formal areas, making it difficult to apply community managed tenure systems without a lot of support and intervention from the state and state institutions. The Social Tenure Domain Model (STDM), for example, has been successfully piloted in Saint Lucia and in St. Vincent and the Grenadines, but has not been scaled up to other communities in those countries [12].

The 'land administration' in FFPLA is defined variously but generally centres around systems, institutions and databases that organize, manage, acquire, and disseminate tenure, value and use data related to individual land units [8,9,13]. For most countries, and certainly for Caribbean countries, the land administration is taken to be the cadastre for the spatial boundary information and the legal registry for the formal tenure information. With few exceptions, the English-speaking Caribbean countries, in recognition of their similar, and fairly recent, British colonial history, still retain constitutional monarchies, with the Queen as the titular head of state and the privy council in England as the final court of appeal. This may have implications for the reluctance to adopt more flexible and appropriate tenure laws that deviate from the legal regimes with which the countries are most familiar.

Some of the Caribbean English-Speaking countries (Anguilla, Antigua, British Virgin Islands, Cayman Islands, Montserrat, Saint Lucia and the Turks and Caicos Islands), as shown in Figure 1 and Table 1, have successfully registered 100% of their parcels in relatively quick and efficient three-year long compulsory systematic adjudication and titling projects to install English-type titling land registration which involves low precision demarcation maps, and unique parcel IDs. The others have generally attempted to continue voluntary sporadic Torrens type titling alongside their largely voluntary deed registration systems. This is as shown in Table 1 [14] where the percentage of parcels registered in deed or title or dual systems is provided.

Figure 1. The Caribbean countries in context.

As indicated by the question marks, many countries can only estimate the number of parcels that are recorded, as the systems are not structured to be parcel based nor are they integrated land information systems that can be queried for this type of information. It should be noted that these estimates do not define the term 'parcel' so it is not stated whether legitimate, community or state acknowledged, contested or adversely possessed land is also included in the total estimated parcel count. Some of the systems are majority deeds based with the registration of deeds not being mandatory but serving as evidence of priority in the presence of any contesting claims. Deeds registration in these countries serve the purpose of supporting the majority of transactions and, because of the confidence imposed in the systems through use over time, it is difficult to convince landowners to voluntarily convert to the registration of title.

Table 1. Land registration in the English-speaking Caribbean adapted from study by Johnson [14].

Country	% Registered
Anguilla	100
Antigua	100
Bahamas	?
Barbados	~10
Belize	~60
Bermuda	?
British Virgin Islands	100
Cayman Islands	100
Dominica	50
Grenada	?
Guyana	?
Jamaica	45
Montserrat	100
St. Kitts-Nevis	50
Saint Lucia	100
St. Vincent and the Grenadines	?
Trinidad and Tobago	50
Turks and Caicos Islands	100

The FFPLA approach, as applied to several countries with completed parcel databases, even when this term was not solidified in use, has therefore been shown to result in a timely and comprehensive land database. This fact begs the question of why the other countries have not followed the same or a similar process.

3. Materials and Methods

This evaluation examines the experiences with adjudication and titling being undertaken by countries of the Caribbean in general with specific examples from Trinidad and Tobago, Barbados, and Jamaica, and others where a comprehensive land administration has been proposed and initiated but has not yet been achieved. Evaluation criteria used were: firstly, the presence and applicability to FFP of defined objectives, secondly, how supportive the legislative structure and therefore the process of implementation is to the FFP, and thirdly, experiences in application in relation to FFP guidelines. Data was compiled from published documents, legislation, and press statements made by officials responsible for the land administration in Caribbean countries. Some of the data emanates from the various land management, land administration, and several other land related projects that have been embarked on over several years in the Caribbean as funded by international aid agencies or regional governments to address the significant challenges that the countries have experienced while grappling with climate change, including the effects on the intensification of natural hazard impacts such as hurricanes. Global economic crises including energy market upheavals surrounding oil and gas and the most recent Covid-19 pandemic have also intensified the need for examining where the limited resources need to be applied.

Land policy and land policy related documents in the Caribbean, accessed and reviewed for compatibility with the vision of the FFPLA, include:

- The National Land Policy of Saint Lucia of 2007 [15];
- The Saint Lucia Land Policy Issues Paper of 2013 [16];
- Land Policy and Management in the Caribbean [17]
- The St. Kitts/Nevis Land Policy Issues Paper of 2013 [18];
- The National Land Policy of Jamaica of 1997 [19];
- A Methodological Framework for Comparative Land Governance Research in Latin America and the Caribbean [20]; and
- The Barbados Growth and Development Strategy 2013–2020 of 2013 [21].

These documents give an indication of the status and background of individual Caribbean countries and their specific direction and vision regarding land development. From these, the ability of the FFPLA to allow the countries to achieve their land related goals can be discerned.

Land titling and regularisation legislation, accessed and reviewed for compatibility with the suggested procedures of the FFPLA, include:

- The State Land Regularisation of Tenure. Act 25 of 1998 of Trinidad and Tobago [22];
- The Registration of Title to Land Act 2000 of Trinidad and Tobago [23];
- The Land Adjudication Act 2000 of Trinidad and Tobago [24];
- The Land Tribunal Act 2000 of Trinidad and Tobago [25];
- The Land Adjudication Amendment Act No. 10 of 2018 of Trinidad and Tobago [26];
- The Registration of Titles Cadastral Mapping and Tenure Clarification (Special Provisions) (Amendment) Act 2020 of Jamaica [27];
- The Land Adjudication Regulation 2019 of Trinidad and Tobago [28];
- The Land Adjudication of Rights CAP228A of Barbados [29]; and
- The Saint Lucia Land Adjudication Act No 11 of 1984 [30]

These items of legislation prescribe broadly how the land registration and tenure arrangements are to be implemented and these can be compared with FFPLA implementation guidelines.

Documents that demonstrated outcomes related to the land registration and titling of the countries in the Caribbean that were assessed for compatibility with the outcomes posited by the FFPLA include:

- Assessing the formal land market and deformalization of property in St. Lucia [1];
- The Impact of Land Titling on Land Transaction Activity and Registration System Sustainability: A Case Study of St. Lucia [2];
- Key challenges and outcomes of piloting the STDM in the Caribbean [12];
- Building the Cadastral Framework: Achievements and Challenges in the English-Speaking Caribbean [14];
- Doing Business 2020 [31]; and
- The National Land Agency of Jamaica's website; [32]

While documentation on the outcomes in a quantitative assessment is scarce, some qualitative information can be indirectly gleaned from current documents on land issues.

4. Results

The FFPLA is a relatively recent statement and application for achieving quick results in land administration reform, primarily in state led systematic adjudication, land titling and registration processes, so that it can better support development. This section presents the results of comparison between existing land registration activities in the Caribbean and the fundamental tenets of the FFPLA.

4.1. Defined Objectives

A significant tenet of the FFPLA focuses on defining an objective of the process and this being not only fit for purpose but having fitness of purpose. This is described as being the 'what' of the process [8]. Countries in the Caribbean are conflicted about specifying a priority purpose for establishing their land administration and, especially with resource constraints, wish the system to achieve all beneficial outcomes at once: economic, social, and environmental. The FFPLA indicates that focus should be on the purpose, which is always achieving security of tenure for all [8]. However, mention should also be made that security of tenure leads to further beneficial development, i.e., social, environmental, and economic outcomes depending on how and where the project is instituted. The FFPLA can therefore be applied 'as little as possible, as much as necessary' in different regions of the country to achieve the purpose of comprehensive security of tenure. Success at achieving security of tenure may or may not lead to successful further achievement, if that is the aim.

Many of the land administration projects and pilot projects are fully or partially funded by international loans or grants with no guarantee of the ability to continue or maintain the project. The project must therefore attain all possible goals at once. Without a clear focus on the priority need, the project will flounder and success is elusive as energies and resources are dissipated in different directions. Very many projects end after the pilot projects, with no state resources to upscale the project into a sustainable programme.

A country's priority in its land governance should be stated in its documented land policy so that specific goals and intermediate success indicators can be derived. The fitness of the purpose for land administration improvement should derive from the research and examination that goes into the development of the land policy document. Conversely, much of the data that is required for the development of the land policy should be gleaned from querying the land administration data, which is usually lacking and therefore a target of improvement. Many Caribbean states do not have a documented land policy nor one that indicates their goal specifically. Saint Lucia is one of the few states in the Caribbean with a documented land policy, albeit one dated 2007 [15]. A revision to this is still in discussion and comes out of an in-depth examination of land issues in the country [16] A major initiative supported by the Australian Agency for International Development (AusAID) from 2012 to 2014 and managed by the Organisation of Eastern Caribbean States (OECS), with technical input from the Global Land Tool Network (GLTN), the University of the West Indies (UWI) and stakeholders within the countries, focused on building the states capacities to develop land policy that could result in achievements in economic, social and environmental development and also in poverty reduction [17]. As a result of this project, the OECS countries produced background papers as a forerunner to developing land policies [16,18]. However, only draft land policies were possible during the short term of the project and the countries were responsible for concluding and formalising the policies in a longer-term participatory process. The larger countries outside of the OECS were also not party to the project.

Jamaica's national land policy is dated 1997 [19]. Land policies should adapt to changes in the economic and social environment and be informed by completed successes so should be periodically revised to maintain currency of focus on new goals. Jamaica's 1997 land policy clearly prioritizes the establishment of a comprehensive land information system to support the data-driven management and planning of land [19]. Overall land use sustainability and land resource allocation are stated to be high on the agenda, so if there has not been a change in policy direction since the time of this published land policy, then boundary location precision is low on the list of specifications for the land administration. The fitness of this purpose for the country can be justified by the need for the country to preserve the environment, as it is dependent on tourism for livelihoods and income. Identified parcels and land uses are the priority data capture requirements here.

Saint Lucia's 2007 land policy states in its vision that the priority is to support economic development in various spheres of activity including agriculture, tourism, and trade [15]. Saint Lucia, however, has already completed its comprehensive land administration database in the 1980s by undergoing systematic adjudication and titling.

Barbados does not have a documented land policy and neither does Trinidad and Tobago nor any of the other Caribbean countries still struggling with establishing their land administration. However, looking at other evidence of actions implemented by the state can allow a discernment of land policy without a specifically constructed document. Barbados can be discerned to have a land policy focused on economic outcomes [20,21]. Like Jamaica, the country is dependent on tourism for economic sustenance. Barbados does have some informal occupation but it is not at the level experienced by some of the other Caribbean countries as it is deemed to be affecting less than 1% of the area of the country [14]. The focus of the registration programme may therefore be centred on areas of marketable and touristic properties. FFP mechanisms should revolve around capturing sufficient identification of parcels to allow quick transactions to occur. Success in this goal was seen early on in the land registration as indicated by Maynard [33]. The drastic reduction in speed and cost of transactions as attained by the introduction of the land registration is remarked on. This early success can be expanded on by using an FFPLA approach to the land registration in progress, with a focus on the areas where transactions are concentrated. Building awareness of this impact through capacity building of the professionals and technocrats charged with developing land policy direction will therefore have positive economic benefits.

Legislation derives from land policy and, while there is not a current land policy document for Trinidad and Tobago, the presence of pro-poor legislation to prevent eviction of squatters on state land [22,23] and the current absence of property tax would support an assumption that social outcomes are the priority in land action. The fitness of a social tenure purpose is unquestionable as the population of informal occupants on state land is high and deemed to be 240,000 in number out of a national population of 1,300,000 [34,35]. A previous document called the Government's Land Policy is subtitled A New Administration and Distribution Policy for Land and focuses on policies surrounding the administrative processes within the state institutions for allocation of state lands to private individuals by lease for agricultural purposes [36].

Table 2 synopsises the status of the selected Caribbean countries with respect to their land policy. Where there is a documented land policy, the objectives of the policy are listed in the table. Where there is no land policy, the land related goals are derived from other actions and documents. It would be helpful to the achievement of successful land administration development if Caribbean countries were to construct clear land policy that directs the land management institutions to identify specific land administration content in order to support implementation of the policy.

Table 2. Presence of a Land Policy.

Country	Presence of Land Policy	Defined Objectives	Fitness of Objectives
Barbados	No	Economic	Livelihoods, economy
Jamaica	Yes	Environmental	Tourism
Trinidad and Tobago	No	Social	Informality
Saint Lucia	Yes	Economic	Tourism

4.2. Legislation

After the 'what' is decided, the 'how' can be addressed to achieve the goals in an efficient way. The structure for the process of constructing or upgrading land administration by adjudication, titling, and registration is supported by legislation. The FFPLA proposes flexibility in the process which requires registration and titling legislation that prescribes relaxed precisions for defining boundaries, and possibility for tenure registration along the continuum. Legislative development follows the development of a land policy. If the land policy is unfocused or misdirected toward a purpose that does not have fitness for the specific jurisdiction within the social, economic, and cultural context, then the legislation may also be fit for purpose but have no fitness of purpose.

The legislation to support the adjudication and titling programme in Trinidad and Tobago was first passed in 2000 with a typical package of a Registration of Title to Land Act [23], a Land Adjudication Act [24] and a Land Tribunal Act [25]. The programme has not yet begun although 20 years have elapsed since the passage of the legislation. During the intervening period amendments were made to the Acts to remove the ability of an occupant to acquire absolute title to state land even after 30 years' occupation, which had been prescribed in the previous version of the Act. This change to the process, when implemented, will have the effect of rendering an estimated 60,000 parcels of state lands, that are currently occupied by informal households, insecure in their documentary tenure even after the programme is completed. The occupant must apply to the Land Tribunal to decide the process to a title. This variation will also increase the resource requirements for determining each instance and slow down the process of regularizing the informal tenure that is prevalent in the country.

The reluctance to part with large volumes of encumbered state land that cannot now be utilised for any other purpose demonstrates the state's commitment to retaining ownership of large quantities of state lands despite the negative perceptions of international agencies toward this practice. Deininger [37] advocates for the devolution of state lands through outright grants, auctions, and sales, or, in default of this, through very long-term leases. Figure 2 shows one of the many informal communities on state land that are of long standing and which will not immediately benefit from the titling as a result of this segment of the legislation. On the positive side, the Land Adjudication Act now, after amendment in 2018 [26], affords persons in occupation of non-state land the ability to acquire absolute title if they have been in occupation for more than 16 years. The Act also allows provisional title for occupation and documentary evidence that does not meet the criteria for absolute title. Systematic titling legislation should seek to title or otherwise facilitate the tenure of informal occupants including family land occupants and squatters on both state and private land. This would involve the flexibility and inclusive characteristics required of the FFPLA.

Figure 2. One of the long-standing informal communities, called Bangladesh, established on state land.

Jamaica, since 2000 has been attempting to perform comprehensive registration using sporadic, voluntary methods under their existing Torrens based title legislation, but since the process did not meet the inclusiveness characteristic of the FFPLA, the attempt has not yet been concluded. Under the process, applicants are required to pay for their processing, which does not serve those who are unable to afford the cost. In 2011 systematic adjudication and titling was attempted in specific areas but, again, since this process required payments by individual land holders and also required precise surveys, it did not sufficiently advance the process. Jamaica in 2018 has, however, decided that the previously voluntary sporadic programme called the Land Administration and Management Programme (LAMP) would now become systematic adjudication and titling managed by the National Land Agency [38,39]. The Registration of Titles Cadastral Mapping and Tenure Clarification (Special Provisions) (Amendment) Act 2020 now allows for absolute titling to those occupants found in possession for more than 12 years [27]. This meets the flexibility requirement advised by the FFPLA.

To address whether the legislation supports flexibility in spatial location, the prescriptions for mapping may be examined. Jamaica's Registration of Titles Cadastral Mapping and Tenure Clarification (Special Provisions) (Amendment) Act 2020 [27] requires that the surveyor 'prepare or cause to be prepared a cadastral map of parcels of land in the systematic adjudication area'. The Director of Surveys, however, is authorized to determine the precisions of the mapping. The Land Adjudication Regulation 2019 [28] of Trinidad and Tobago allows for the preparation of a demarcation map and identification of parcels via a unique ID and also allows for the use of general boundaries of physical features such as a 'fence, hedge, wall, ditch or other physical feature, whether natural or artificial'. The Barbados Land (Adjudication of Rights and Interests) CAP 228A [29] also authorizes the Chief Surveyor to prepare the demarcation map for a declared district. Bestowing the authority for determining the precision of the mapping supports flexibility that can lead to the use of aerial photography or satellite imagery. See Table 3 for the relevant legislation in the selected Caribbean countries.

Table 3. Legislative Support for Titling and Registration.

Country	Legislative Support
Barbados	Systematic Adjudication, Titling, and Registration—CAP 228A
Jamaica	Torrens. Systematic Adjudication, Titling and Registration laws from 2018
Trinidad and Tobago	Torrens. Systematic Adjudication and Titling laws from 2000 not yet implemented to date
Saint Lucia	Systematic Adjudication, Titling and Registration

4.3. Application of FFP Principles

The FFPLA guiding principles [8] indicate that the key characteristics are purpose, flexibility and incrementality, having a clear goal, being flexible in the tenure forms accepted and understanding that there can be gradual improvement with advances in technology and changes in fortunes. The FFPLA goes on to indicate that the main goal is tenure security for all. However, a country's main purpose for developing its land administration may be an economic one that may not be centred on tenure security for all but that, nevertheless, requires a comprehensive parcel information database as a support for intensified land transactions of sale and mortgage, as well as land and property taxation. While the immediate purpose may be economic, the eventual intention may be to benefit society by reduction of poverty. Similarly, where the purpose may be environmental and related to land resource governance, the eventual beneficiaries may be society. Low levels of conflict and high levels of perception of security may reduce the need for documentary security in the form of a land title but in these instances a complete land information system is still necessary to encourage economic development. Table 4 indicates how the country is applying its land titling and registration process, and also indicates where the FFP principles can be seen to be a positive theoretical characteristic of the programme. While the NLA of Jamaica can be demonstrated to be transparent and accessible, allowing information on ownership and value to be searched and viewed online, the other countries do not have the same accessibility. The detailed application and interpretation by the land related responsible national institutions may still not encourage the flexibility which is possible.

Table 4. Application of FFPLA principles.

Country	Process	FFP Principles
Barbados	Formal Adjudication, precise surveying,	
Jamaica	Formal Adjudication, precise surveying	Integrated institutional framework, Transparent land information
Trinidad and Tobago	Formal Adjudication, precise surveying	
Saint Lucia	Formal Adjudication, general demarcation	Physical boundaries, accuracy related to purpose

Jamaica, at 11,000 square kilometres in area is the third largest island in the Caribbean after Cuba and Hispaniola and is therefore the largest English-speaking island in the Caribbean. It is estimated to contain approximately 800,000 parcels but despite having established its land titling process since 2000, it has not significantly advanced the number of new parcels titled since then. The cadastre is primarily more fiscal than juridical so that parcels are considered completely registered when they have gone through the titling and registration process and are on the valuation roll with a tax ID. The tax ID of the individual or entity liable for paying the property tax is also the person's unique ID indicated on the title documents. Table 5 presents the rate of registration in the different parishes in Jamaica over the six years from January 2015 to January 2021. These data are derived from accessible search forms available on the National Land Agency's website at https://www.nla.gov.jm/ accessed on 31 March 2021.

Table 5. Registration progress in Jamaica.

Parish	Registered Parcels at January 2015	Registered Parcels at January 2021	Total Parcels	Percentage Registered	Percentage Change
Kingston	12,751	12,651	13,824	91.51%	0.62%
St. Andrew	82,962	84,388	98,279	85.87%	0.61%
St. Thomas	16,343	17,796	38,347	46.41%	2.88%
Portland	12,623	14,083	33,906	41.54%	0.94%
St. Mary	20,877	24,060	50,044	48.08%	1.29%
St. Ann	31,487	37,190	67,442	55.14%	1.18%
Trelawny	15,874	17,948	32,782	54.75%	0.94%
St. James	43,759	48,503	59,564	81.43%	2.52%
Hanover	10,677	12,475	23,702	52.63%	5.23%
Westmoreland	23,186	25,515	43,642	58.46%	3.26%
St. Elizabeth	23,784	28,888	69,514	41.56%	1.58%
Manchester	30,447	35,120	69,501	50.53%	4.45%
Clarendon	41,112	45,884	95,597	48.00%	3.64%
St. Catherine	108,153	115,238	156,778	73.50%	1.74%
Total	474,035	519,739	852,922	60.94%	3.03%

As Table 5 indicates, the number of new parcels registered in the five years from 2015 to 2021 has increased by only 3% overall. Placed in context, the smaller countries, such as Saint Lucia, have previously performed the entire programme of systematic adjudication and title registration in four years. In that instance, over 33,000 parcels were registered over the four years [1]. It is noteworthy that the capital parish, Kingston, and its close neighbour, St. Andrew, have the highest percentage of registration and this can contribute to economic development in these urban areas. The more rural parishes of Hanover and Manchester have the highest rate of registration in keeping with the land policy focus on environmental management.

Barbados is 430 square kilometres in area and, like Jamaica, can be said to be supported by a fiscal cadastre rather than a juridical cadastre as its valuation roll contains 99 percent of the parcels and the rate of informality is relatively low in comparison with other Caribbean countries. Barbados introduced the land registration application in 1988. The World Bank in its Ease of Doing Business assessment for 2020 determines that the quality of the land administration index in Barbados ranks at 10.5 on a scale of 0 to 30 where Latin America and the Caribbean score at 12.0 and the high-income OECD countries score at 23.2. After years of implementation, only 10% of the parcels were determined to have been titled by 2003 and currently only 15% are estimated to have been titled [14,20]. However, it should be noted that while the fiscal cadastre does not contain precise surveys, it is a sufficient index to support the land administration.

Trinidad and Tobago, with an area of 5100 square kilometres and an estimated number of parcels of 500,000 should focus on achieving security of tenure, as the large number of persons in insecure tenure requires. The application is proposed to be, in the initial pilot project, focused on the island of Tobago, which is noted for having a large number of informal 'family land' parcels. This application is therefore in keeping with the fitness of purpose ideal of the FFP. For the application to be fit for this purpose, participation is key.

Table 6 gives the funding and estimated expenditure on land registration and titling for the selected Caribbean states. The costs identified for the projects exceed the immediate funding available. To this must be added the anticipated inflation over the anticipated time required for adjudication and titling using conventional surveying, as well as opportunity costs for allocating the funds toward this process as opposed to other developmental purposes. A reduction in the project cost is therefore immediately economically beneficial. Land market benefits were modest at the end of Saint Lucia's adjudication and titling process [1,2] and may also be underwhelming for these countries even at completion of these projects.

Table 6. Expenditure on Registration and Titling.

Country	Funding Source	Cost/Parcel in $US
Barbados	Recurrent state expenditure,	No data
Jamaica	National Housing Trust (indirectly State)	934 (only $13 million identified for this) [39,40]
Trinidad and Tobago	Inter-American Development Bank	1100 (only $25 million identified for this) [41] 250/parcel,
Saint Lucia	USAID and Govt. Saint Lucia	total 8 million

The evaluation of the general procedures being implemented by the Caribbean countries for upgrading their land administration through the tool of adjudication and titling indicates that FFPLA approaches are possible but have not generally been adopted. Seven countries have used FFP like approaches between 1967 and 1987 to complete systematic adjudication and titling. Subsequent programmes have largely persisted with sporadic Torrens type titling in slow, tedious and expensive procedures with mandatory boundary surveys that are precise in spatial location and rigid adjudication for rights determination. Some have begun true English type systematic adjudication and titling with the use of demarcation maps and unique parcel IDs, but this implementation has still been very slow and has not emulated the FFPLA.

Even though much of the legislation may be interpreted to allow the application of more flexible tenure acceptance, this is only in terms of informal occupation on private land through typical adverse possession rules for different duration of occupation. More flexible tenure rules have not been incorporated into the process such as titling of 'family land' or automatic conversion of occupation on state land to absolute titles or leaseholds. Flexible tenure rules for flexible recordation as a key principle of the FFP approach should also extend to incorporating existing deeds registries as these are considered to be reliable for the current majority of transactions and, in the cases of Barbados and Trinidad and Tobago, cover the majority of parcels currently registered. Adding a unique identifier to parcels held by deed registration can immediately upgrade the registration without the cost and time required for converting to title registration.

Flexibility of spatial location, while supported by the revised legislation, has not been implemented yet to reduce the lengthy processes. Most countries have relatively recent aerial imagery that can be used to create demarcation maps rather than insisting on precise surveys using professional cadastral surveyors. In most instances deeds are also available that can support in large measure a more relaxed and participatory adjudication, or be acceptable for direct entry into the information system.

The recommendations for the countries, as derived from the key principles of the FFP approach [8,9] are indicated in Tables 7–9.

Table 7. Recommendations for advancing the spatial framework. Adapted from FFPLA [8,9].

Spatial Framework Item	Recommendation
Visible boundaries rather than fixed boundaries	Use walls, fences, to update cadastral index as these exist.
Aerial/satellite imagery rather than field surveys	The countries have recent imagery that can be used
Accuracy relates to purpose	Existing legislation authorises Chief Surveyors to decide on accuracy of the demarcation map. These should be based on existing cadastral index accuracies.
Demands for updating and opportunity for upgrading	Upgrading may be a longer-term requirement.

Table 8. Recommendations for advancing the legal framework. Adapted from FFPLA [8,9].

Legal Framework	Recommendation
Flexible framework	Acknowledge the critical mass of parcels in informal occupation (T&T) and registered in deed registries, and allow for mass migration into one land administration
Continuum of tenure	Integrate family land, leaseholds, and other tenure
Flexible recordation	Integrate informal occupation housed at Land Settlement Agency (T&T) or fiscal cadastre (Barbados) by linking the existing data together
Ensuring gender equity	Begin recording gender as an attribute on transactions, as this is not generally done.

Table 9. Recommendations for advancing the Institutional framework. Adapted from FFPLA [8,9].

Institutional Framework	Recommendation
Good land governance	Review recommendations for treating with large quantities of state land
Integrated institutional framework	Support use of parcel identifier in all agencies as a low-cost step of integrating data
Flexible ICT approach	
Transparent land information	Emulate Jamaica's example of freely accessible land information

In all the countries, there are sufficient walls, fences and other visible boundaries that can be demarcated on the most recent imagery to fill in the existing cadastral index. The aim is to achieve comprehensiveness of the cadastral index. Since the authority is given by the legislation to the Chief Surveyor or Director of Surveys to determine demarcation map precisions, it is up to these professionals to acknowledge that lack of resources within the public institutions will obviate a speedy conclusion to the demarcation process. Upgrading can occur subsequently to the completion of the project as individuals who wish to access credit or otherwise transact in the newly registered land will be required to pay the cost of surveying to normal cadastral survey rules, including precision.

Trinidad and Tobago should acknowledge the large critical mass of parcels in informal occupation particularly on state lands as these lands are already lost to the state without costly and traumatic relocation of communities. Much data exists in separate locations of the land administration institutions that can be taken to support rights that have existed over a period of time and can be acknowledged.

Land management on the large quantities of state lands held by Trinidad and Tobago, in particular, has become ponderous and difficult. Close scrutiny has to be made of the fundamental governance and land policy reasons for retaining state lands before determining how this should be addressed.

5. Conclusions

The assessment of the processes being undertaken by the selected countries of the Caribbean to comprehensively register all land parcels, in keeping with the tenets of the FFPLA, found that a lot of time, money, and effort had already been expended without a predicted time of completion. While some of the smaller countries had accomplished the process successfully prior to the development of the FFPLA application, subsequently countries had moved away from the process. With the current existence of technical tools such as imagery and mobile GPS, it is possible for the countries to improve the rate of roll out of registration by the use of FFPLA principles of reduced accuracies and acknowledgement and integration of separate recording and registration data. However, an important aspect of achieving economically beneficial results is for the country to first identify and publicise a clear objective to be achieved in land that requires the land administration to be effected. Economically beneficial outcomes can be derived from both achieving a low-cost solution to the completion of the comprehensive land registration as well as from the benefits of the use of the completed cadastre. These economically beneficial outcomes, however, appear to be weighted in favour of the savings to be derived from the low-cost solutions rather than from income from land market transaction increases that may be predicted to occur.

Funding: Funding of the publication costs for this article has kindly been provided by the School of Land Administration Studies, University of Twente, in combination with Kadaster International, The Netherlands.

Institutional Review Board Statement: Not applicable.

Informed Consent Statement: Not applicable.

Data Availability Statement: Not applicable.

Conflicts of Interest: The author declares no conflict of interest. The funders had no role in the design of the study; in the collection, analyses, or interpretation of data; in the writing of the manuscript, or in the decision to publish the results.

References

1. Barnes, G.; Griffith-Charles, C. Assessing the formal land market and deformalization of property in St. Lucia. *Land Use Policy* **2007**, *24*, 494–501. [CrossRef]
2. Griffith-Charles, C. The Impact of Land Titling on Land Transaction Activity and Registration System Sustainability: A Case Study of St. Lucia. Ph.D. Thesis, Geomatics Program, University of Florida, Gainesville, FL, USA, 2004.
3. Feder, G.; Onchan, T.; Chalamwong, Y.; Hongladaron, C. *Land Policies and Farm Productivity in Thailand*; The Johns Hopkins University Press: Baltimore, MD, USA, 1988.
4. Feder, G.; Noronha, R. Land Rights Systems and Agricultural Development in Sub-Saharan Africa. *World Bank Res. Obs.* **1987**, *2*, 143–169. [CrossRef]
5. Bizoza, A.; Opio-Omoding, J. Assessing the impacts of land tenure regularization: Evidence from Rwanda and Ethiopia. *Land Use Policy* **2020**, *100*. [CrossRef]
6. Bennett, R.M.; Alemie, B.K. Fit-for-purpose land administration: Lessons from urban and rural Ethiopia. *Survey Rev.* **2016**, *48*, 11–20. [CrossRef]
7. Ho, S.; Choudhury, P.R.; Haran, N.; Leshinsky, R. Decentralization as a Strategy to Scale Fit-For-Purpose Land Administration: An Indian Perspective on Institutional Challenges. *Land* **2021**, *10*, 199. [CrossRef]
8. FIG. *Fit-for-Purpose Land Administration*; FIG Publication 60; FIG Publication: Copenhagen, Denmark, 2014.
9. Enemark, S.; McLaren, R.; Lemmen, C. *Fit-for-Purpose Land Administration: Guiding Principles for Country Implementation*; Report 2/2016; UN Habitat: Nairobi, Kenya, 2016.
10. Zevenbergen, J.; Augustinus, C. Designing a Pro Poor Land Recordation System. In Proceedings of the FIG Working Week, Marrakech, Morocco, 18–22 May 2011.
11. Zevenbergen, J.; Augustinusb, C.; Antoniob, D.; Bennett, R. Pro-poor land administration: Principles for recording the land rights of the underrepresented. *Land Use Policy* **2013**, *31*, 595–604. [CrossRef]
12. Griffith-Charles, C.; Mohammed, A.; Lalloo, S.; Browne, J. Key challenges and outcomes of piloting the STDM in the Caribbean. *Land Use Policy* **2015**, *49*, 577–586. [CrossRef]
13. Williamson, I.; Enemark, S.; Wallace, J.; Rajabifard, A. *Land Administration for Sustainable Development*; ESRI Press Academic: Redlands, CA, USA, 2010.
14. Johnson, S. Building the Cadastral Framework: Achievements and Challenges in the English-Speaking Caribbean. Proceedings of GSDI 10: The Tenth Annual Conference for Geospatial Data Infrastructure GSDI10. 11 Pages. 2008. Available online: https://citeseerx.ist.psu.edu/viewdoc/download?doi=10.1.1.504.4345&rep=rep1&type=pdf (accessed on 25 February 2008).
15. Government of Saint Lucia. National Land Policy. 2007. Available online: https://archive.stlucia.gov.lc/docs/NationalLandPolicy.pdf (accessed on 12 March 2021).
16. Alison, K.; Lendor-Gabriel, G. Saint Lucia Land Policy Issues Paper. 2013. Available online: https://www.oecs.org/our-work/knowledge/library/social-development/slu-land-issues (accessed on 12 March 2021).
17. Global Land Tool Network (GLTN). Land Policy and Management in the Caribbean. 2012. Available online: https://gltn.net/2013/03/21/land-policy-and-management-in-the-caribbean/ (accessed on 12 March 2021).
18. Williams, P. St. Kitts/Nevis Land Policy Issues Paper. 2013. Available online: https://chm.cbd.int/api/v2013/documents/5B5B29D5-A58A-BDE5-963C-08398C894F03/attachments/Land%20Policy%20Issues%20Paper%202013.pdf (accessed on 12 March 2021).
19. Government of Jamaica. National Land Policy of Jamaica. 1997. Available online: https://observatorioplanificacion.cepal.org/sites/default/files/instrument/files/1997.%20NAtional%20Land%20Policy.pdf (accessed on 12 March 2021).

20. Sanjak, J.; Donovan, M. A Methodological Framework for Comparative Land Governance Research in Latin America and the Caribbean. IDB. Technical Note 1026. 2016. Available online: https://publications.iadb.org/publications/english/document/A-Methodological-Framework-for-Comparative-Land-Governance-Research-in-Latin-America-and-the-Caribbean.pdf (accessed on 12 March 2021).
21. Ministry of Finance and Economic Affairs. Barbados Growth and Development Strategy 2013–2020. 2013. Available online: https://observatorioplanificacion.cepal.org/sites/default/files/plan/files/BarbadosSBGDS20132020.pdf (accessed on 12 March 2021).
22. Government of Trinidad and Tobago. State Land Regularisation of Tenure. Act 25 of 1998. Available online: https://rgd.legalaffairs.gov.tt/laws2/Alphabetical_List/lawspdfs/57.05.pdf (accessed on 12 March 2021).
23. Government of Trinidad and Tobago. Registration of Title to Land Act 2000. Available online: http://www.ttparliament.org/legislations/a2000-16.pdf (accessed on 12 March 2021).
24. Government of Trinidad and Tobago. Land Adjudication Act 2000. Available online: http://www.ttparliament.org/legislations/a2000-14.pdf (accessed on 12 March 2021).
25. Government of Trinidad and Tobago. Land Tribunal Act 2000. Available online: http://www.ttparliament.org/legislations/a2000-15.pdf (accessed on 12 March 2021).
26. Government of Trinidad and Tobago. Land Adjudication Amendment Act No. 10 of 2018. Available online: http://www.ttparliament.org/legislations/a2018-10g.pdf (accessed on 12 March 2021).
27. Government of Jamaica. The Registration of Titles Cadastral Mapping and Tenure Clarification (Special Provisions) (Amendment) Act 2020. Available online: https://japarliament.gov.jm/attachments/article/339/The%20Registration%20of%20Titles%20Cadastral%20Mapping%20and%20Tenure%20Clarification%20(Special%20Provisions)%20(Amendment)%20Act,%202020.pdf (accessed on 12 March 2021).
28. Government of Trinidad and Tobago. Land Adjudication Regulation. 2019.
29. Government of Barbados. Land Adjudication of Rights CAP228A. Available online: http://104.238.85.55/en/showdoc/cr/1988_73 (accessed on 12 March 2021).
30. Government of Saint Lucia. The Land Adjudication Act No 11 of 1984. Available online: http://extwprlegs1.fao.org/docs/pdf/stl4302.pdf (accessed on 12 March 2021).
31. World Bank. Doing Business 2020: Economy Profile Barbados. 2020. Available online: https://www.doingbusiness.org/content/dam/doingBusiness/country/b/barbados/BRB.pdf (accessed on 12 March 2021).
32. National Land Agency of Jamaica. Available online: https://www.nla.gov.jm/ (accessed on 29 April 2021).
33. Maynard, T. BARBADOS: Land Reform and the Search for Security of Tenure. In *Land in the Caribbean*; Williams, A., Ed.; Land Tenure Center: Madison, WI, USA, 2003.
34. Griffith-Charles, C.; Rajack, R. The Challenges of Land Governance Assessment with Limited Land Information. In Proceedings of the Annual World Bank Land and Poverty, Washington, DC, USA, 20–24 March 2017.
35. Joint Select Committee on Local Authorities, Service Commissions, and Statutory Authorities. First Report on Inquiry into the Land Settlement Agency in Relation to Squatter Regularisation. Report. 2016. Available online: http://www.ttparliament.org/committee_business.php?mid=19&id=242&pid=28 (accessed on 12 March 2021).
36. Government of Trinidad and Tobago. Government's Land Policy Is Subtitled a New Administration and Distribution Policy for Land. 1992. Available online: https://afraraymond.files.wordpress.com/2015/04/state-1992-land-policy-1.pdf (accessed on 12 March 2021).
37. Deininger, K.; Selod, H.; Burns, A. The Land Governance Assessment Framework: Identifying and Monitoring Good Practice in the Land Sector. Agriculture and Rural Development. World Bank© World Bank. 2012. License: CC BY 3.0 IGO. Available online: https://openknowledge.worldbank.org/handle/10986/2376 (accessed on 12 March 2021).
38. Linton, L. LAMP and NLA to Be Merged. Jamaica Information Service. Available online: https://jis.gov.jm/lamp-and-nla-to-be-merged/ (accessed on 21 March 2018).
39. Henry, B. PM Promises Titling Programme for 20,000 Parcels of Land. *Jamaica Observer* 2018. Available online: https://www.jamaicaobserver.com/front-page/-8216-action-time-8217-pm-promises-titling-programme-for-20-000-parcels-of-land_128527?profile=&template=PrinterVersion (accessed on 12 March 2021).
40. Editor Jamaica Gleaner. NLA to Introduce Systematic Land Registration to Increase Land Titling in Jamaica *The Gleaner*. 2019. Available online: https://jamaica-gleaner.com/article/news/20190414/nla-introduce-systematic-land-registration-increase-land-titling-jamaica (accessed on 21 April 2021).
41. McEachnie, C. Government Moves to Give Land Titles. *Trinidad and Tobago Guardian* 2020. Available online: https://www.guardian.co.tt/news/govt-moves-to-give-land-titles-6.2.1254869.9a0fba3070 (accessed on 12 March 2021).

Article

Fit-For-Purpose Applications in Colombia: Defining Land Boundary Conflicts between Indigenous Sikuani and Neighbouring Settler Farmers

Laura Becerra [1,2], Mathilde Molendijk [2], Nicolas Porras [2,*], Piet Spijkers [2], Bastiaan Reydon [2] and Javier Morales [3]

1 Maestría en Ciencias de la Información y las Comunicaciones, Universidad Distrital Francisco José de Caldas, Bogotá 11021-110231588, Colombia; ljbecerrag@correo.udistrital.edu.co
2 Kadaster International, Land Registry and Mapping Agency of the Netherlands, 7311 KZ Apeldoorn, The Netherlands; mathilde.molendijk@kadaster.nl (M.M.); piet.spijkers@landinpeace.com (P.S.); bastiaan.reydon@kadaster.nl (B.R.)
3 ITC Faculty, University of Twente, 7500 AE Enschede, The Netherlands; j.morales@utwente.nl
* Correspondence: nicolas.porras@landinpeace.com; Tel.: +57-311-5634509

Abstract: One of the most difficult types of land-related conflict is that between Indigenous peoples and third parties, such as settler farmers or companies looking for new opportunities who are encroaching on Indigenous communal lands. Nearly 30% of Colombia's territory is legally owned by Indigenous peoples. This article focuses on boundary conflicts between Indigenous peoples and neighbouring settler farmers in the Cumaribo municipality in Colombia. Boundary conflicts here raise fierce tensions: discrimination of the others and perceived unlawful occupation of land. At the request of Colombia's rural cadastre (Instituto Geográfico Agustín Codazzi (IGAC)), the Dutch cadastre (Kadaster) applied the fit-for-purpose (FFP) land administration approach in three Indigenous Sikuani reserves in Cumaribo to analyse how participatory mapping can provide a trustworthy basis for conflict resolution. The participatory FFP approach was used to map land conflicts between the reserves and the neighbouring settler farmers and to discuss possible solutions of overlapping claims with all parties involved. Both Indigenous leaders and neighbouring settler farmers measured their perceived claims in the field, after a thorough socialisation process and a social cartography session. In a public inspection, field measurements were shown, with the presence of the cadastral authority IGAC. Showing and discussing the results with all stakeholders helped to clarify the conflicts, to reduce the conflict to specific, relatively small, geographical areas, and to define concrete steps towards solutions.

Keywords: fit-for-purpose land administration; indigenous land conflict; Cumaribo; Colombia

Citation: Becerra, L.; Molendijk, M.; Porras, N.; Spijkers, P.; Reydon, B.; Morales, J. Fit-For-Purpose Applications in Colombia: Defining Land Boundary Conflicts between Indigenous Sikuani and Neighbouring Settler Farmers. *Land* **2021**, *10*, 382. https://doi.org/10.3390/land10040382

Academic Editors: Stig Enemark, Robin McLaren and Christiaan Lemmen

Received: 1 March 2021
Accepted: 3 April 2021
Published: 7 April 2021

1. Introduction

1.1. Indigenous Land Boundary Conflicts

One of the most difficult types of land-related conflict is that between Indigenous peoples and third parties, such as settler farmers or companies looking for new opportunities who are encroaching on Indigenous communal lands [1]. Although the number of people with communal rights is relatively small, the areas covered by communal land rights are vast and are often under pressure from the rising global demand for natural resources, raising conflict, and debate [2,3]. This causes conflicts and land boundary disputes between the Indigenous peoples, settler farmers, miners, companies, and other spatially expanding actors [4,5].

This article describes boundary conflicts between Colombian Indigenous reserves (resguardos indígenas) and their neighbours in the Cumaribo municipality in the Vichada department and analyses possible solutions using fit-for-purpose land administration methodology of land conflicts. These resguardos indígenas, or Indigenous reserves, are territorial divisions that, by means of a property title, guarantee a certain Indigenous community the ownership of a territory that is communally owned and traditionally inhabited by them. Boundary conflicts here raise fierce tensions: discrimination of the others, perceived unlawful occupation of land on both sides, stealing of cattle, and removing of fences.

Land in Peace is a project cooperation between Kadaster, Twente University, the Dutch Embassy, and Colombia's Universidad Distrital to develop and propose an efficient land administration system for securing land rights in rural Colombia. The project works together with Colombian land institutions, such the Instituto Geográfico Agustín Codazzi (IGAC), Agencia Nacional de Tierras (ANT), and Superintendencia de Notariado y Registro (SNR). The fit-for-purpose method is applied to jointly measure land with the land users, using a Global Network Satellite Systems (GNSS) antenna and a mobile collector app that was developed, and especially for the purpose improved, by Esri. Kadaster's Land in Peace project analysed the situation in three Indigenous reserves of Cumaribo (2019–2020) in cooperation with IGAC, Colombia's National Geographical Cadastral Institute Agustín Codazzi, who started updating the complete cadastre of Cumaribo. The three Indigenous reserves have different land boundary disputes with their neighbours. The first case describes a situation where Indigenous land overlaps with neighbouring settler farmers' lands. In the second case, Indigenous communities are located outside of the official reserve's boundaries. In the third case, the Indigenous community claims additional land based on local traditional knowledge that has never been recognized by the Colombian government.

The fit-for-purpose (FFP) methodology was applied to map and define the overlapping land claims between the Indigenous peoples and the neighbouring settler farmers (colonos). The land boundaries were measured by each party themselves, and the resulting concrete contested areas were afterwards discussed with all of the parties involved. In several cases, both parties held official ownership documents describing the boundaries in text or with unclear maps, leaving room for different interpretations. The official map of the reserve, for instance, was drawn on a small scale, which means that this limit cannot be related to a precise location in the field. The FFP methodology aims to define, in a participatory way, a clear and transparent definition of the boundaries together with the community, whereby potential huge conflicts can be downsized to a specific area, providing a solid basis for conflict resolution.

This article starts with a description of the Indigenous land situation in Colombia and describes, more specifically, the situation in the pilot region, the municipality of Cumaribo. This is followed by a description of the participatory FFP methodology that was used to measure land boundaries. Finally, the results of the measurements of the Indigenous reserve boundaries in Cumaribo are discussed to show different Indigenous land-related issues that were found based on the three cases before mentioned.

1.2. Indigenous Land in Colombia

The proportion of Indigenous peoples within the Colombian population is low. According to the last population census of 2018, there were 1,905,617 Indigenous peoples in the country, which corresponds with 4.4% of its total inhabitants [6]. This percentage grew by almost 37% since the previous census of 2005, when it was 3.4%, a difference that could be explained by a growing self-consciousness playing a role, since the census question was formulated as asking to which ethnic groups one considers him or herself to belong to. The census established that there are 64 Indigenous ethnic groups in the country with great differences in demographic size. For example, the Wayuu number is 380,000 and the Sikuani is 52,000, while the Bari number is 208 and the Yukuna is 396 [7].

In rural areas, there are 710 Indigenous reserves with legal land titles, which occupy an area of approximately 34 million hectares, which is 29.8% of the national territory [6]. This great discrepancy between the demographic Indigenous peoples' proportion and the proportion of Indigenous land property is explained by the fact that huge parts of these reserves are forest or savanna areas. It is also somehow mitigated by the fact that the inhabitants of the Indigenous reserves also include non-Indigenous people, some by invasion, and others by consent of the legal owners. The limited number of people who declare themselves Indigenous in the census may be another explanation for this discrepancy.

The 1991 Colombian Constitution gave Indigenous peoples the right (the Constitution speaks of Indigenous peoples) to define who can live in their reservations: "The Indigenous territories will be governed by councils formed and regulated according to the uses and customs of their communities" [8]. However, in many cases, the Indigenous peoples do not have the territorial control that legally corresponds to them. For example, in the Indigenous reserve assigned to the Nukak Makú in Guaviare of one million hectares, the Indigenous peoples do not have access. The guerrillas and coca producers in that reserve do not let them enter their own land property.

The institution of Indigenous reserves, a legacy from colonial times, reflects, for the Indigenous, the painful process of the slow occupation of their territories by invading immigrants and marks the historical result of this occupation. However, while the nineteenth century saw a slow degradation of Indigenous communal land property, later constitutional reforms re-established the idea of legal recognition of communal Indigenous lands [8].

Today, there are still occupations, sometimes peaceful, sometimes conflictive, by settlers in search of land in Indigenous reserves. Many Indigenous groups are trying to increase the lands of their reserves, arguing that they should also have control and ownership over the lands surrounding the Indigenous reserves that they consider ancestral lands. From the Indigenous perspective, their position on these vacant, state-owned lands (tierras baldías) is understandable: for them, this category does not exist, since they are the original inhabitants. This problem—the distribution, control, and allocation of territorial property—remains at the heart of the country's history and was one of the principal issues of the recent Peace Accords with the guerrillas [9].

1.3. Cumaribo: Overlap of Indigenous Lands with Neighbouring Settler Lands

The immense forests and savannah lands of Cumaribo are subdivided into one half with 34 Indigenous reserves, while the other half is vacant territory that is property of the Colombian State, with only a few formal assignments of land to private settlers. Cumaribo is situated on the borderline between two large Colombian ecosystems: its northern part belongs to the Orinoco Savannah plains (los llanos) and its southern half is part of the Amazon rainforest basin. With its area of 6.6 million hectares, Cumaribo is the biggest municipality of the country.

Until the middle of the twentieth century, the savannahs of Vichada were mainly populated by semi-nomad Sikuani Indigenous tribes and other minor ethnicities, such as the Piapoco, Cuiba, Saliva, and other Indian tribes. From the 1950s and 1960s onwards, a slow immigration by people from other departments started and Colombia´s Land Reform Institute INCORA (Colombian Institute of Agrarian Reform), in charge of titling state land, granted formal ownership of big parcels of land to these settler farmers. This initiated an increasingly problematic process of land claims between the original Indigenous peoples and the colonists.

According to the anthropologists Metzger and Morey, among the Sikuani there was no notion of private property or communal land tenancy or collective possession [10]. A Sikuani leader expressed that the Sikuani consider land as a collective heritage and not as a tradable good susceptible to being private property. An Indigenous leader of the reserve, Santa Teresita del Tuparro, explained to the IGAC–Kadaster mission the Indigenous position on the so-called vacant lands (not privately owned and thus, consequently, state-owned land) as follows: "State land does not exist" and "that the state land is owned by nobody is not true," meaning that the land was always theirs, and that they were there before anybody else.

During the last half of the twentieth century, many traumatic changes occurred in the Sikuani society. Apart from the settlers who started huge cattle farms and introduced more and more barbed wire fences, towards the end of the century, wide-scale coca leaf production was introduced, and the south of the municipality became an important route for the trafficking of cocaine to neighbouring Brazil and Venezuela. Both guerrilla and paramilitaries made their entrance after the drug mafias, with whom the guerrillas became more and more mixed up [11].

Additionally, the tropical plains soils, until recently regarded as practically useless apart from extensive cattle ranching, came in the purview of national and international agricultural investors. The savannah region of Colombia has recently and increasingly been compared to the Cerrado region of Brazil, where, under the same soil and climate conditions, an impressive agricultural production came into being and which catapulted Brazil into one of the principal food exportation nations of the globe [12]. The Colombian savannah region has been named the last agricultural frontier of Colombia, while the country has been included, by the Food and Agriculture Organization of the United Nations (FAO), among the world's five global potential food producers. During the last decade, maize, soya, and oil palm plantations have been established in Cumaribo. Moreover, Colombian development economists argue for the need to construct a road connecting Bogotá with the Orinoco river a thousand kilometres further afield. Nevertheless, the current situation in Vichada and Cumaribo is still one of economic stagnation and poverty, with a lack of basic services and infrastructure. On the contrary, an agricultural boom in the near future, such as in Brazil, is a real possibility and an opportunity for an increasing number of agricultural investors in the area [13].

1.4. Fit-for-Purpose Participatory Land Administration

Different studies have addressed participatory mapping with Indigenous peoples [14–18]. These studies indicate that to understand the situation and tradition of the Indigenous peoples, their communal knowledge should be used to map land boundaries and important places by mapping together with the Indigenous peoples. They must be present in the process to extract and map local knowledge. Álvarez and McCall [14], for example, stated that this approach helps local Indigenous groups to be incorporated as active subjects in the registration of their cultural heritage, as well in the defence and management of it.

Additionally, Sletto [19] mentioned how the development of participatory mapping increases the possibilities for Indigenous peoples to use cartography to better represent their conceptions of space and place. Nowadays, Indigenous land rights are included more, after years where state maps often did not include Indigenous toponyms, where Indigenous lands were labelled as "empty", and where contiguous Indigenous land use zones had been fractured in isolated reserves or donated to other land users, such as settler farmers, as described by Bryan [20]. From Braceras' [21] study, where different participatory mapping methods to capture land perceptions of Indigenous peoples were compared, it can be concluded that a social cartography session is a useful tool for conflict resolution and for increasing Indigenous empowerment. However, this will only be reached with involvement of all stakeholders, such as neighbours and the State. The State's reality in official registries often differs from the Indigenous people's reality, and makes decisions based on their information; therefore, involvement of the State is essential to establish and guarantee Indigenous land rights. Other researchers [22] stress that the government has to protect Indigenous land rights to prevent market forces, such as settler farmers and companies, from grabbing more Indigenous lands and thereby undermining communal land rights.

An approach with this participative, community-based focus is the FFP land administration approach. It is used to reach agreements between neighbours on their mutual boundaries and easements [23–25]. Following Enemarks et al.'s guidelines [25], FFP focuses on flexible, efficient, affordable, and transparent land recording methods, including all types of land occupation, to be able to improve land tenure security. This FFP methodology contributes to some of the Sustainable Development Goals' (SDGs) objectives, as Enemark highlighted. According to Enemark [25], land administration systems should be designed for the needs of the people when identifying the occupation and use of land, whether these land rights are legal or locally legitimate. To meet the actual needs in society today, the systems need to be flexible in terms of the legal regulations, as well as the institutional arrangements. Enemark [25] stated that good land administration is essential to be able to reach the accorded Sustainable Development Goals, because it provides a country with an infrastructure for implementing land management strategies in support of sustainable development.

The FFP approach was also applied in the Cumaribo project. Since conventional Colombian land data collection methods are expensive and time-consuming, an efficient and user-friendly data collection scheme was developed for the current Colombian situation, using advanced but easy-to-use technological tools [26,27]. Parcels are always measured together with the land users and community members, who are trained to carry out the work with professional supervision. After the data collection, a public inspection is held to show the results to the community and to collect signatures of approval on the measured limits.

The literature on the FFP concept for Indigenous boundary disputes is scarce and does not go far beyond the remark that Indigenous or customary land rights should be included in the land administration [24]. An FFP approach might be beneficial to safeguard the rights of customary groups, since data are collected of all claims in a certain area, with the stakeholders participating. This can be used as a basis for conflict resolution. Contrary to more traditional approaches, the collaborative FFP methodology is very much in line with the intent of the UN's Free, Prior and Informed Consent guidelines (FPIC) [28]. The FPIC guidelines provide a framework that ensures that the rights of Indigenous peoples are guaranteed in any decision that may affect their territories. The framework comprises four components: (a) free, without coercion or bribery; (b) prior consultation with the community-found place before starting; (c) informed information to the community is provided in an understandable way; and (d) consent—the community has the right to give or withhold their consent to any decision [28]. The FFP approach operationalises these guidelines, because the Indigenous peoples themselves are the prime actors in the process, including their responsibility for the field data collection, and the community is always informed of the whole process before starting. Moreover, the FAO's Voluntary Guidelines on the Responsible Governance of Tenure (VGGT) emphasize the importance of Indigenous land tenure [29]. Section 9, and more specifically 9.4, of the VGGT focuses on the customary land rights of Indigenous peoples and points out the responsibility of the State to protect the land used by the Indigenous peoples.

The work in Cumaribo is based on experiences of FFP applications in rural areas in Colombia and elsewhere [27,30–33]. These have been adapted to map traditional land of Indigenous reserves. The next section (Section 2) describes the steps of the FFP approach applied in the Colombian project with Indigenous peoples.

2. Materials and Methods

Being a participative land administration approach, the community is always involved in each step of the FFP application in Colombia, as can be seen in Figure 1. The point of departure in the application of FFP in Colombia is a socialisation process with all the stakeholders present, followed by a social cartography session. Here, the land users first indicate their parcel boundaries on a large, printed map on which the official parcel information on top of a satellite image is visible.

Most parcel boundaries are defined by natural boundaries, such as rivers or streams, but depending on the available imagery of the specific region, these boundaries cannot always be identified in the satellite image. In case there is not much vegetation in the area and high-resolution imagery is available, land boundaries can be identified more precisely during the social cartography session.

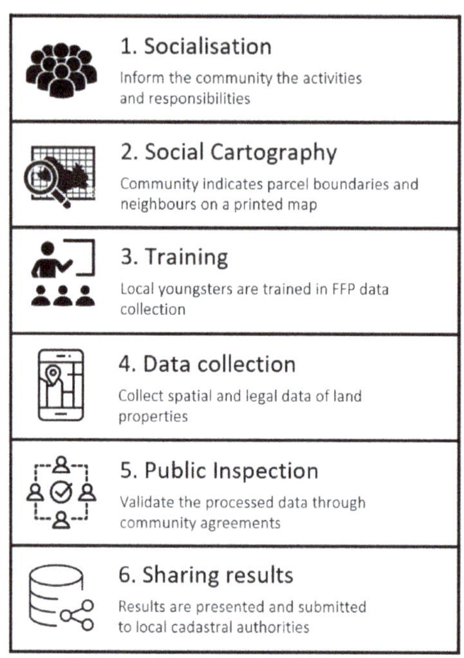

Figure 1. The fit-for-purpose (FFP) methodology steps as applied in the Colombian Land in Peace Project. Elaborated by the authors. Source: Land in Peace Project, Kadaster.

When more precise boundary definitions are necessary, the community is invited to measure the boundaries in the field. During the field measurements, the vertices of the parcel boundaries are indicated by the land user and the coordinates of each vertex are recorded with a Global Network Satellite Systems (GNSS) antenna. These points are recorded with the mobile application Esri Collector on a standard smartphone (Android or iOS). This application is connected by Bluetooth to the GNSS antenna to capture the coordinates (latitude and longitude), as well as the horizontal and vertical accuracy and the number of found satellites, among other things. The antenna is always handled by the land user themselves—in this case, the Indigenous leader or settler farmer—with the objective that they can measure the boundaries of their property according to their local knowledge and perceptions. The Indigenous leader decides what information is collected and communicates this with his community. This generates confidence in the processes since the perception of the Indigenous peoples is deciding. The Esri Collector app is operated by a grassroots surveyor, a local youngster who has received a short FFP data collection training session. The grassroots surveyor is supervised by a professional surveyor for quality control, technical assistance, and coordination.

Both spatial and legal information are gathered in one run per land user in the Esri Collector app, which assures integrated data collection. The measurements of the parcel are visualised in the app as polygons, which makes it possible to directly show the shape and size of the property to the land user. The legal collected data are names, ID number, and date of birth of the right holder(s) with a photograph of the person(s) and their ID card(s); a picture of the parcel; a description of the right holder(s) relationship with the parcel; and photographs of the documents that provide proof of some kind of land right or long-term land use. After data collection, all data are uploaded to a PostgreSQL database. To protect Indigenous Data Sovereignty, the CARE principles (Collective benefit, Authority to control, Responsibility and Ethics) are taken into account [34]. The Indigenous peoples decide what is captured in the field. This includes the boundaries of their reserve, based on local and historical knowledge, but also traditional and sacred places for the Indigenous peoples. Collected data are afterwards checked with the Indigenous leaders, maps of all measured places are shown, and they have the opportunity to modify and add information that is important to their community. Colombian governmental institutions can only use data that are officially approved by the Indigenous peoples.

The technology that is used in the FFP application in Colombia was developed to meet the purpose, to develop a participative approach of capturing relationships between land and its users, for the Colombian situation. The hybrid technological basis between adjusted proprietary (Esri) and open source (PostgreSQL) software has been found to be most effective [35] and has made it possible to create different applications around it for data collection, processing, and validation, and for their integration into the data systems of State agencies. The developed applications for the FFP approach in Colombia are visually oriented. The applications show maps of the parcel boundaries and include the names and pictures of the neighbouring settler farms, pictures that were taken in the field and a satellite image that shows the natural elements that can be recognized by the Indigenous peoples, such as rivers and vegetational corridors. This makes it easier to communicate field results with the Indigenous peoples.

Using these innovative technologies and adjusting the mobile application to this purpose has led to several advantages, which can be summarized as follows: (i) it makes it possible to capture physical and legal administrative data in an integrated way; (ii) it is user-friendly; (iii) it is robust enough to guarantee the capture of the spatial, semantic, and topological relationships of the data without the need for direct intervention of the data collector; (iv) it can work in a disconnected way in areas without internet; (v) it makes it possible to centralize information from a database in the cloud, without intermediate files; and (vi) the collected data comply with the profile of the Land Administration Domain Model (LADM) standards for Colombia.

The used GNSS antenna, Trimble R2, is equipped with a FieldPoint real-time extended (RXT) correction service. It communicates with an operations centre through a satellite link or through the internet and provides the antenna the real-time position with a submeter accuracy [35]. The accuracy of the vertices that are measured in the field meet the official standards in Colombia, which have been adjusted recently to comply with the realities in the field. No geodetic network needs to be established before starting the field measurement, which facilitates the work in remote, Colombian rural areas. Moreover, the fact that the antenna is easily operable by land users themselves enhances the participative process. This increases trust of the community in the collected field data.

After data collection, the results of the field surveys per community are shown and publicly discussed in a so-called public inspection: all land users can judge the results concerning their own parcels and other parcels. Conflicts or disputes can be settled during the public inspection, where Colombian cadastral authorities are also present. The status of the moment is captured, both agreements and disagreements. This information is then handed over to the local Colombian land institutions, who are responsible for resolving the land conflicts.

3. Results

Three Indigenous reserves in Cumaribo were visited in cooperation with IGAC: Santa Teresita del Tuparro, Muco Mayoragua, and San Luis del Tomo (see Figure 2). There was an unclear boundary situation in each reserve that was analysed, whereafter the FFP approach was applied to clarify the boundary situation and to find a possible solution. The following paragraphs describe the situation and applied methodology in each reserve.

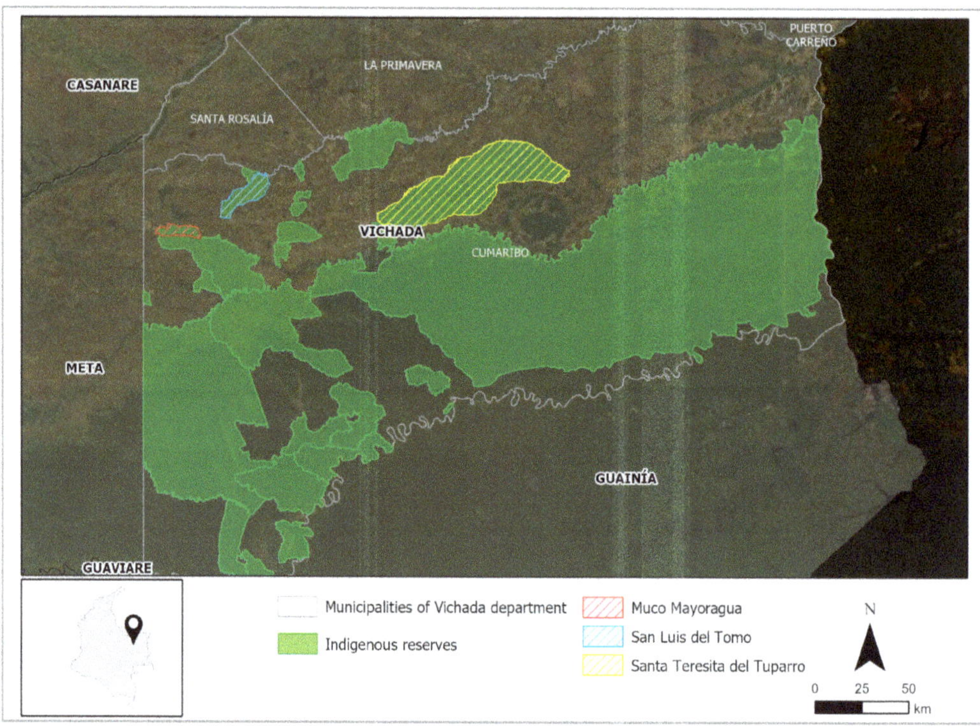

Figure 2. The three Indigenous reserves that were visited in Cumaribo: Muco Mayoragua, San Luis del Tomo, and Santa Teresita del Tuparro in the municipality of Cumaribo, 2020. Source: Land in Peace Project, Kadaster.

3.1. *Santa Teresita del Tuparro*

The Indigenous Santa Teresita del Tuparro reserve consists of approximately 200,000 ha with 86 small communities. Most of the reserve's borders are defined by rivers and creeks; on the remaining part, a land stretch of approximately 20 km, conflicts with settler farmers have arisen. Although this only represents 9% of the total border of the Indigenous reserve, this stretch of unclear boundaries fuels serious land conflicts.

Santa Teresita was established as an Indigenous reserve in 1978, and in 1983, it was given legality. The official documents of the Indigenous reserve state that the road, south of the reserve, represents the boundary. However, in the savanna reality of Cumaribo, a road is a moving boundary, without exact limits (Figure 3). This leaves room for different interpretations. It is impossible to know the location of the road in the year 1978 to which the land title refers.

Additionally, there are inaccuracies in the existing official State information that lead to confusion between the land users: the Santa Teresita del Tuparro reserve of 1978, for instance, had an official size of 180,000 ha, while the later document of 1983 reveals 220,000 ha for the same reserve.

The case of Santa Teresita del Tuparro illustrates the further encroachment of settler farmers into the reserve, which is a result of unclear boundaries and the need for the government to align and update the existing official government data (see Video S1).

Figure 3. Network of roads that form the southwest limit of the Indigenous Santa Teresita del Tuparro reserve, Cumaribo, 2020. There is not one road that can be indicated as the main road. Source: Land in Peace Project, Kadaster.

The FFP approach was used to clarify the land situation by defining the boundaries according to the perception of each party in the field itself. First, a socialisation with the Indigenous peoples was organized, with the presence of the traditional leaders (caciques) and authorities of Indigenous communities close to the borders with the settler farms. The next day, a similar meeting was organized with the settler farmers. During these socialisation sessions, the Indigenous leaders expressed that the settler farmers are invading their territory around the southern border of the reserve. The settler farmers claimed that they have rights to this land too: four have official property titles and others have been settled there for many years. To clarify the situation, the FFP approach was proposed and explained to both the Indigenous peoples in the local Sikuani language and to the settler farmers. They agreed to measure the limits of the reserve and farms, based on their perceptions, in the field. The Indigenous peoples also gave permission to the neighbouring settler farmers to enter the reserve to be able to measure the farms. Subsequently, several measurement crews went out to measure the limit of the Indigenous reserve and to measure the limits of the eight neighbouring parcels.

Traditional leader Horacio Bonilla, who was present at the establishment of the reserve in 1963, led the Indigenous representatives along the borders and indicated each point with its local Sikuani name. With the captured coordinates, the perceived boundary according to the Indigenous peoples was constructed. The fieldwork was also an opportunity for Indigenous youngsters to learn about the history of their land and about the FFP data collection, which was carried out with Trimble R2 antenna and the Esri Collector app (see Figure 4). Likewise, the neighbouring settler farmers measured the limits of their parcels using the antenna and Esri Collector app. Due to the complexity of the terrain, these measurements were made by car or, in some cases, by motorcycle.

The field data collected by the Indigenous representatives of Santa Teresita del Tuparro and the neighbouring settler farmers themselves clearly show the overlapping areas (see Figure 5). There are large parts of land of the Indigenous reserve that are claimed as well by the neighbouring settler farmers. For example, the owners of an informal parcel sold out small plots of their land to others, despite the fact that these lots are partly located on the land that is also claimed by the Indigenous peoples and, above all, are located on top of a moving road (see Figure 6). In total, 1621 ha of the neighbouring parcels overlap with the Indigenous reserve.

Figure 4. Measuring the boundary of the Santa Teresita del Tuparro reserve with Indigenous leader Horacio Bonilla using a Global Network Satellite Systems (GNSS) antenna. Cumaribo, 2019. Source: Land in Peace Project, Kadaster.

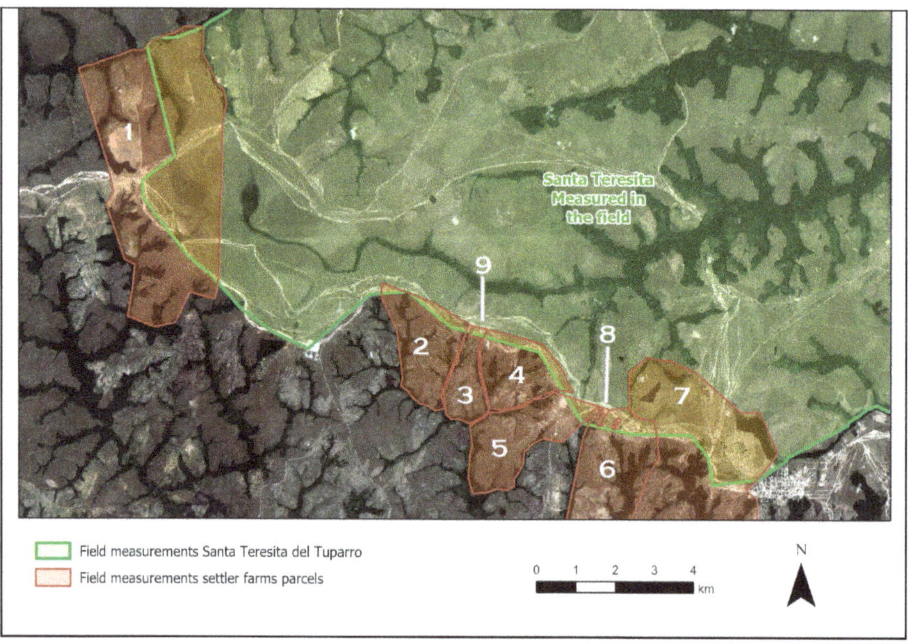

Figure 5. The Indigenous Santa Teresita del Tuparro reserve, as measured in the field by the Indigenous leaders compared to the parcels of the neighbouring settler farmers, as measured in the field: 1, La Guajira; 2, Buenavista; 3, Cuatro Vientos; 4, La Estancia; 5, Los Moriches; 6, Patio Bonito; 7, military base; 8, sold lots; 9, El Basurero (the municipal garbage belt). Cumaribo, 2020. Source: Land in Peace Project, Kadaster.

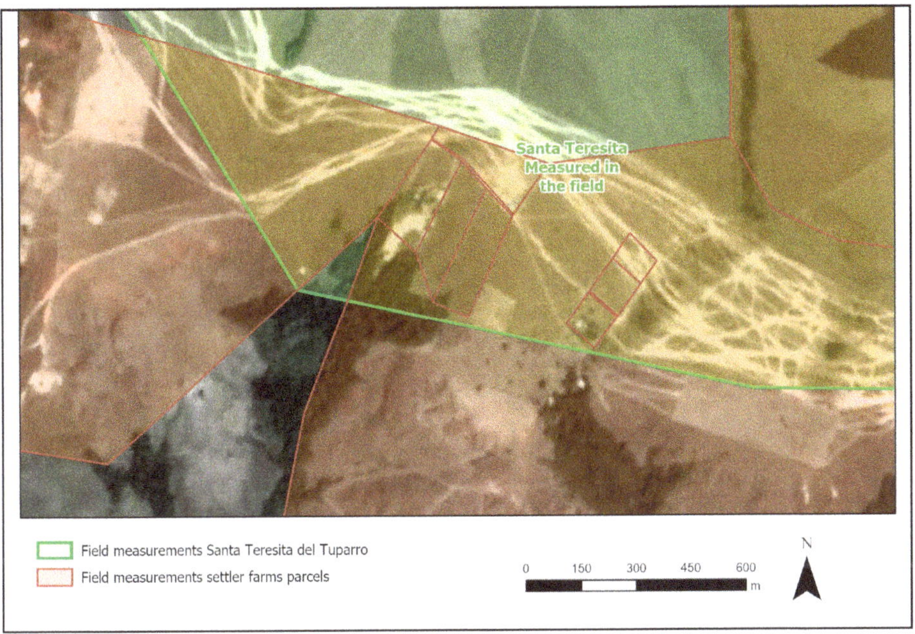

Figure 6. Sold lots on the land that overlaps with the Indigenous Santa Teresita del Tuparro reserve. The lots overlap with the network of roads that is officially the limit of the reserve. Cumaribo, 2019. Source: Land in Peace Project, Kadaster.

Table 1 shows the areas of each parcel according to the cadastral institutions in Colombia compared to the areas that resulted from the field measurements of 2019. It also shows the area of the overlaps between the corresponding parcel and the Indigenous reserve. What can be seen is a significant difference in the areas known by the different governmental land institutions, such as IGAC and, for example, INCORA, the former National Land Agency. Furthermore, there are many differences between the areas in the official State registries and the areas indicated by the land users themselves in the field. For example, according to the documents of the parcel La Guajira, the parcel measures 1324 ha. However, the FFP field measurements show that the parcel has an area of 1900 ha.

Table 1. The size of the measured parcels in hectares and the number of hectares overlapping with the Indigenous Santa Teresita del Tuparro reserve.

	La Guajira Farm		Military Base		Municipal Garbage Dump		Patio Bonito Farm		Los Moriches Farm	
Source	**Year**	**# Hectares**	**Year**	**# Hectares**	**Year**	**# Hectares**	**Year**	**# Hectares**	**Year**	**# Hectares**
IGAC	2012	200 (182 + 18)	2012	1214	2012	0	-	-	-	-
INCORA	1998	1324	1964	1596	2007	10	1988	1300 [1]	2012	311 [1]
Parcel measured by owner	2019	1900	2019	1392	2019	9.6	2019	1955	2019	412
Overlap title parcel with title Indigenous reserve	1998	1008	1964	430	2007	0	-	-	-	-
Overlap parcel measured by owner with title reserve	2019	850	2019	614	2019	6.5	2019	66	2019	23

[1] The deed is not registered; the area is extracted from informal transaction documents. IGAC, Instituto Geográfico Agustín Codazzi. INCORA, Instituto Colombiano de Reforma Agraria.

At the following public inspection, three mapping realities were shown: the borders as defined by the government, the borders as defined by the Indigenous peoples, and those as defined by the settler farmers. Seeing the overlaps, discussions between the Indigenous peoples and settler farmers started in a harmonious way, based on the different mapping realities.

Showing maps of the measured limits made the dispute focus on the real area of conflict. Overlaps with informal neighbouring farms could be solved by the government when these farmers enter the formalisation process. The government can, by law, not adjudicate land to farmers that overlap with the reserve. For the formal landowners, the problem is more complicated. It became clear that the land dispute is part of a bigger problem: the inconsistent data of different governmental institutions at different times. The people of the Indigenous reserve have valid claims, as do the formal owners, as both parties have official documents in which the land claims overlap. The state—in this case, INCORA—gave out the same piece of land to the Indigenous peoples, to the military base, and to the settler farmers.

It can be concluded that there is an urgency to align official governmental data regarding the same parcel collected by different governmental agencies at different points in time. Therefore, it is necessary for the government to update the existing official State data, not only regarding the location of this moving road, but also for solving the discrepancy between the official documents for the same Indigenous reserve.

3.2. *Muco Mayoragua*

Muco Mayoragua is an Indigenous reserve in West Cumaribo of approximately 10,000 ha, located between the Muco river and the Mayoragua stream. The reserve has officially been recognized as an Indigenous reserve since 1997 by INCORA.

As part of the cadastral updating of the Indigenous reserve with IGAC, a socialisation meeting was planned in the community of San Miguel within the Indigenous reserve. Upon arrival at the location of the community, it turned out that the community is located outside of the official boundaries of the Muco Mayoragua reserve (Figure 7).

In the socialisation with the Indigenous leaders of Muco Mayoragua, they expressed that all communities are within the limits of the reserve, including San Miguel, and that the official limits of the State do not represent the reality. After the social cartography session, an agreement was reached to measure the boundary of the reserve in order to establish the actual limit according to the Indigenous peoples. The three communities that seemed to be located outside the reserve are located on what appears to be a private parcel of a neighbouring settler farmer, on the northwest of the Indigenous reserve. For that reason, the caretaker of the farm, called El Palmarito, was invited to the social cartography session. The caretaker expressed agreement with the Indigenous peoples and indicated that the State map of the reserve is outdated.

According to the FFP methodology, two field surveys of the north western limit were realised, one with the Indigenous leaders and one with the person in charge of the El Palmarito farm. The two measurements were carried out with the Trimble R2 antenna, whereafter legal documents of the reserve and the neighbouring parcel were collected.

While carrying out the field measurements, it was observed that the limit is a fence of concrete poles placed by the neighbour of the El Palmarito farm. Both the Indigenous leaders, as did the farmer, indicated the fence as the real limit of the Muco Mayoragua reserve (see Figure 8).

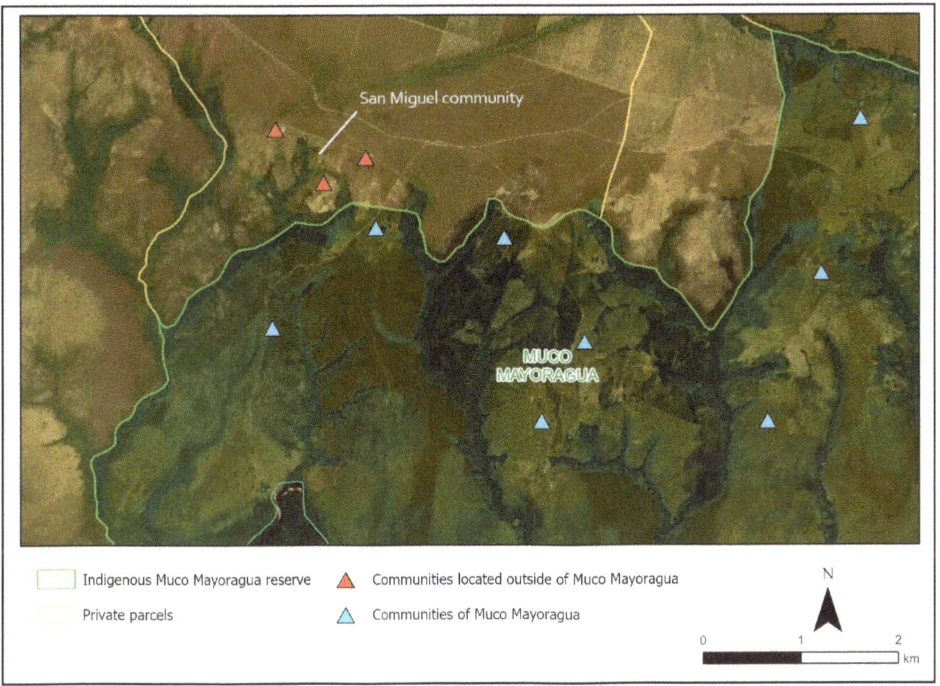

Figure 7. The Indigenous Muco Mayoragua reserve. Three communities (in red) are located outside of the official limits of the reserve but claim to be part of the reserve. Cumaribo, 2020. Source: Land in Peace Project, Kadaster.

Figure 8. Measuring the boundary of the Muco Mayoragua reserve with one of the Indigenous leaders. Cumaribo, 2020. Source: Land in Peace Project, Kadaster.

The results (see Figure 9) were shown at the public inspection, where both sides agreed that the three communities are indeed located within the reserve. They expressed that the measured area in the field, an area of 245 ha, belongs to the Indigenous reserve. Only in the outdated official State data are the three communities located outside the reserve. In reality, both parties agree on the boundaries, but the State data do not reflect these realities in the field. Including the three communities in the reserve, as they should be, means adjusting the official map of the reserve, as well as adjusting the official map of the private land of the neighbouring farmer.

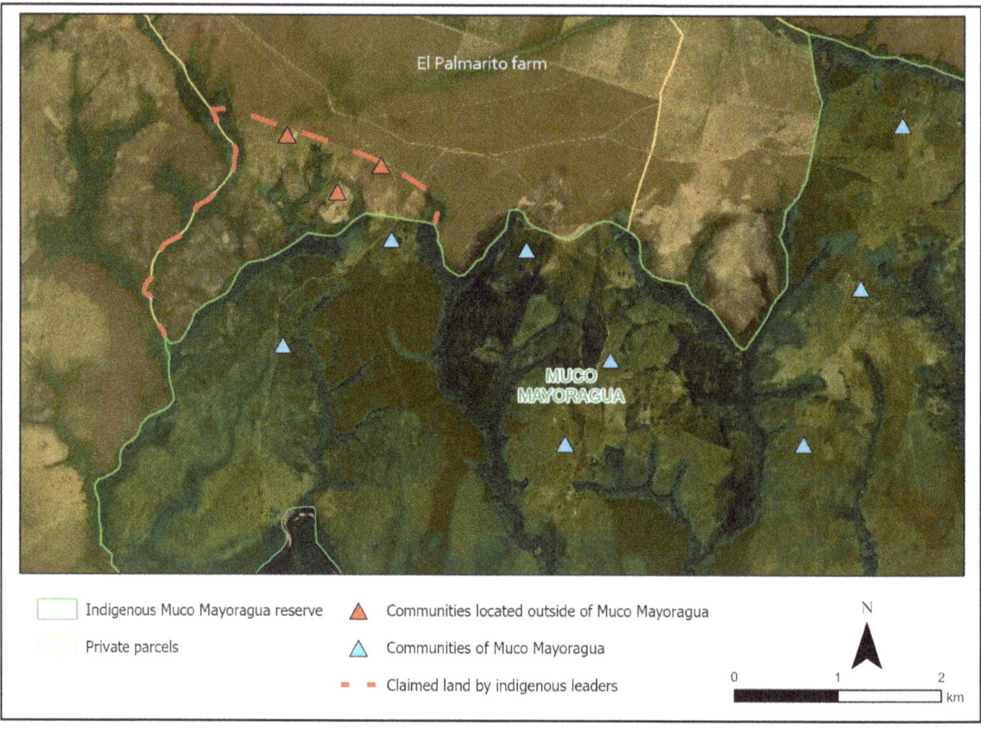

Figure 9. The area that is claimed by the Indigenous Muco Mayoragua reserve (red dotted line). They claim 245 ha of land that is private land (El Palmarito private farm), according to the government. There are three committees of the reserve that are located in these 245 ha. Cumaribo, 2020. Source: Land in Peace Project, Kadaster.

3.3. San Luis del Tomo

San Luis del Tomo is an Indigenous reserve in Cumaribo of approximately 25,000 ha, with its limits being the Tomo river, Samarro river, and Urimica stream, which received legal status by INCORA in 1983. The official deed states that the reserve has natural boundaries on all sides, except for a southern boundary with a length of 5 km, whose limit meets three farms.

A socialisation with the Indigenous leaders of the Indigenous reserve was held in the Llano Lindo community within the reserve. When observing the official map of the reserve, the leaders expressed that the map of 1983 is not correct because INCORA defined the boundary following a tributary of the Samarro river and not the main river, as mentioned in the title deed of the reserve.

The social cartography with the community served to understand the situation. Indigenous leaders, through their ancestral knowledge, drew the main course of the Samarro river (see Figure 10). Additionally, using a satellite image of the area, governmental data of waterbodies, and the indications of the Indigenous leaders, a map was created together with the Indigenous peoples. Figure 11 shows the official limits of the reserve and the limits according to the Indigenous peoples of San Luis. The boundary, as indicated by them, differs from the official limit due to the fact that the community claims that the main cause of the river is not the stream that is currently indicated as the boundary, but is instead located east of the current boundary. The additional area that is claimed with this is approximately 3000 ha. Based on the approximate size of the reserve, the Indigenous peoples claim a 12% expansion of their reserve. Comparing the area with the governmental cadastral data of IGAC shows that there is already one privately owned farm and four other farms without an official property title located on this piece of land.

Based on the produced maps and the resulting discussion, the Indigenous peoples of San Luis del Tomo requested a correction of the title that was granted in 1983 by INCORA, to incorporate the missing 3000 ha into their Indigenous reserve.

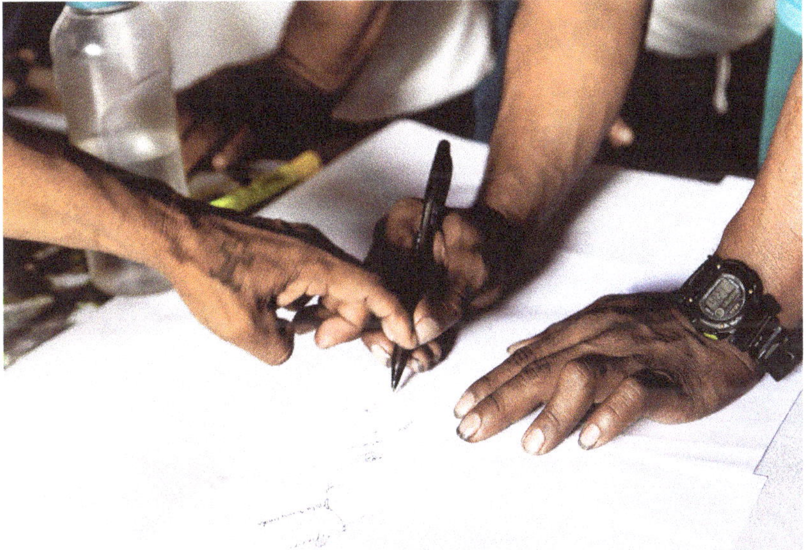

Figure 10. Drawing of the boundary perception (river) by Indigenous peoples. Cumaribo, 2020. Source: Land in Peace Project, Kadaster.

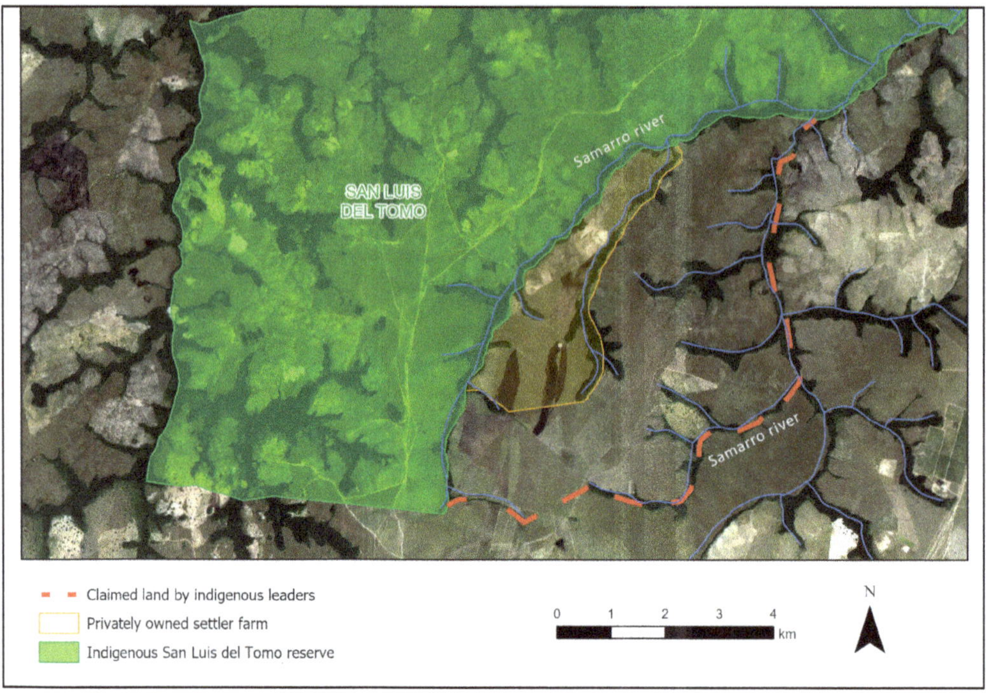

Figure 11. The Indigenous San Luis del Tomo reserve. The red line indicates the limit of the reserve according to the Indigenous peoples. Approximately 3000 ha are claimed by the Indigenous peoples. There is already one private parcel in the claimed area. Cumaribo, 2020. Source: Land in Peace Project, Kadaster.

4. Discussion and Conclusions

The three cases that were analysed with the FFP approach show that conflicts concerning Indigenous land boundaries are created due to inconsistencies between the reality and governmental data.

The limits of Indigenous reserves that are known by the State are often outdated and do not match the reality in the field. The resolutions that provide the reserves their legal status contain unclear maps and vague descriptions of boundaries. These resolutions were handed out in the 1970s, when there were not many technologies available for precise field measurements. This resulted in uncertainty about the real limit, which has created overlapping claims between Indigenous peoples and their settler neighbours.

The Santa Teresita del Tuparro case offers an example of how inconsistencies between State data and reality causes conflicts between neighbours. The State gave out land properties to the Indigenous peoples and to the settler farmers on the same land. The vague description of the moving road and the imprecise definition of the boundaries in these titles make it impossible to define the actual limit.

Other cases, such as the situation in Muco Mayoragua, show that not only precision of the boundary plays a role, but that State data do not represent the real extent of the Indigenous reserve. Indigenous peoples are claiming large parts of land to be part of their reserve. The State has never analysed these claims; therefore, communities are located outside of the official limits of the reserve. From the experience in San Luis del Tomo, where the Indigenous peoples state that the limit is another course of the river, it can also be concluded that the State did not include perceptions of the Indigenous peoples when formalising the reserve.

These inconsistencies cause potentially fierce conflicts between Indigenous peoples and neighbouring settler farmers. The Colombian government is the authority that solves these boundary conflicts. Informal parcels of farmers simply cannot overlap with an Indigenous reserve in future formalisation processes, and databases need to reflect the reality. Therefore, the presence of the Colombian state in the process is important for solving discrepancies between boundaries because the State is the only party that can guarantee an official solution or clarification of the situation.

Based on the work in the border regions between the Indigenous reserves and their neighbouring settler farms, it can be concluded that the FFP approach works well in such difficult and deeply-felt situations of land conflicts. The FFP approach, with the cooperation of the Indigenous peoples and social cartography, can be used to map the physical reality and to understand the existing boundary conflicts, overlaps, and errors that were made at the time of awarding the reserves by the government. Additionally, the involvement of the neighbours at the socialisation of the results at the public inspection, and with this establishing local agreements, is crucial for conflict resolution. What can be concluded is that the participatory approach leads to trust in the collected field data.

Several researchers, such as Braceras [21] and Alvarez and McCall [14], stated that the community has to be involved in the mapping of Indigenous land rights and that participatory mapping is a useful tool for conflict resolution and for increasing Indigenous empowerment at a local scale. Capturing local knowledge is important; however, this is not always sufficient for guaranteeing tenure security for the community for the long term or on a sustainable basis. Braceras [21] stated that it is important to notice that the State is an important stakeholder, a conclusion that can also be drawn from the experiences in Cumaribo. Looking at real land boundary conflicts in the described cases, many conflicts are caused by inconsistencies in official State data. For that reason, these State data need to be updated to conform to the reality. Participatory mapping studies focus mainly on capturing Indigenous traditional knowledge, while the applied FFP approach provides an example of how information is collected after being shared and discussed with all stakeholders, including the Indigenous peoples, the neighbours, and the State, to solve the conflict on different levels.

FFP distinguishes itself from other, often static, participatory mapping applications by being flexible and adaptable to the circumstances. Flexibility and efficiency are important aspects of Enemark et al.'s FFP guidelines [25]. The FFP approach needs to be adapted to the local situation and to the community. The work in Cumaribo showed the need of adapting the data collection approach to the geographic circumstances in the Indigenous reserves. In San Luis del Tomo, for example, the field measurements were replaced with the use of satellite images, due to the fact that a river, the limit according to the Indigenous peoples, is more easily identifiable on images than by field measurements. By organising a social cartography session and inviting the Indigenous peoples to indicate the actual limit along the Samarro river, the time and costs of fieldwork were spared. In the Santa Teresita del Tuparro case, the boundary was not visible on the map and, thus, participatory field measurements with GPS were necessary.

The applied FFP approach also meets the standards of the UN's FPIC guidelines [28]. The Indigenous peoples were informed of the process before starting any data collection was fully involved in each case, without coercion. Both the Indigenous peoples and neighbours were invited to sign for their agreement or disagreement at the public inspection, to guarantee their consent to any information passed on to the government. Moreover, applying the FFP land administration approach, in cooperation with the State, is in line with Section 9.4 of the FAO's Voluntary Guidelines of Governance Tenure [29]. Mapping boundary conflicts together with the community and proposing solutions for securing land tenure of Indigenous land supports the responsibility of the government to provide appropriate recognition and protection of the Indigenous land rights.

The differences between the Indigenous cultures, and the moving agricultural frontier—in this case, cattle ranchers—remain at the heart of the matter. This fundamental clash of values often materialises in boundary conflicts. Although the challenge remains regarding how Indigenous peoples in reserves can develop economically while maintaining their own cultural identity, the experiment with the FFP approach shows that it can greatly help to focus the dispute on specific contested areas, followed by a common discussion on the land dispute at hand and practical handouts for the government to solve the land disputes.

The cadastral actualization project with IGAC in the Indigenous reserves has been finished. All data that are collected are included in IGAC's database. The differences between the official State data and the measurements in the field are included as an informal data layer in their database. This information is also socialized with the Indigenous communities, which means that they understand the situation and are aware of what information is included in the State's registers. The leaders of each Indigenous reserve have received elaborated maps with the information of the situation according to the State's official documents and according to the field measurements that were realized with the community. This means that the collected information is not only in hands of the governmental institute IGAC, but also in hands of the Indigenous community. During the final socializations, it was explained to the Indigenous leaders that the measurements of the real limits in the field are recognized and included by IGAC.

However, the next step is the modification of boundaries and resolution of conflicts at the State level, to guarantee the protection of Indigenous land rights in the long term. When a parcel of a settler farm is sold, the new owner has to be aware of the updated situation and not ignore the agreements that were made. This needs to be done by the Agencia Nacional de Tierras, Colombia's National Land Agency (ANT), which is the only authority that can modify boundaries and grant new parts to existing Indigenous reserves. The collected data by IGAC are shared with ANT, with approval of the Indigenous communities. Therefore, the project continues to work with ANT to analyse how these local agreements are processed at the State level.

Based on the findings of this project, several recommendations for further research arise. First, it is important to continue monitoring the State's participation and its commitment to identify the reality of Indigenous land's borders. This project has been carried out with several Sikuani Indigenous reserves, but Colombia has over 60 different ethnic communities, which have other characteristics than the Sikuani community. Similar studies in other ethnic groups would broaden the knowledge on present land conflicts as a result of ongoing encroachment on historic indigenous land rights in Colombia and other Latin American countries. There are also many Indigenous communities whose land rights have never been officially granted. They live on State-owned lands and their rights on these ancestral lands are not recognized. A follow up program could be the FFP approach applied in a certain region to protect Indigenous lands that are not legalized.

Additionally, a challenge that many Indigenous communities are facing nowadays, is the expansion of the population within the Indigenous reserves. The number of communities within the Indigenous reserves is increasing, and, therefore, land for cultivating crops is becoming scarcer. The lands surrounding these reserves are often already occupied, which makes it difficult to expand the Indigenous territory. To be able to provide for the whole community, further research should be focused on sustainable use of the indigenous lands.

Supplementary Materials: The following is available online at https://www.youtube.com/watch?v=HNTbjPrbLA8: Video S1: Land conflicts in Cumaribo (Colombia) between Indigenous peoples and cattle ranchers.

Author Contributions: Conceptualisation, L.B., M.M., N.P., P.S. and B.R.; methodology, L.B., M.M., N.P., P.S., B.R. and J.M.; software, J.M.; investigation, L.B., N.P. and P.S.; writing—original draft preparation, L.B. and N.P.; writing—review and editing, L.B., N.P., M.M., P.S. and B.R.; supervision, M.M. and P.S.; funding acquisition, M.M. All authors have read and agreed to the published version of the manuscript.

Funding: The funding of the publication costs for this article has been kindly provided by the School of Land Administration Studies, University of Twente, in combination with Kadaster International from the Netherlands.

Institutional Review Board Statement: Not applicable.

Informed Consent Statement: All subjects gave their informed consent to the Land in Peace Project for using their photographs before these have been published.

Data Availability Statement: Not applicable.

Acknowledgments: The authors of this article want to thank IGAC for the cooperation in Cumaribo and to express their gratitude to the Sikuani Indigenous communities that participated in each of the case studies.

Conflicts of Interest: The authors declare no conflict of interest.

Ethics Disclaimer: The authors declare that the Indigenous communities of the three Indigenous reserves have been priorly informed of the process. Data and pictures published in this article are socialized in the field and approved, according to the FPIC principles, for use by the Indigenous leaders.

References

1. United Nations Human Settlement Programme (UN-HABITAT). *Securing Land Rights for Indigenous Peoples in Cities*; United Nations Human Settlement Programme: Nairobi, Kenya, 2011. [CrossRef]
2. Anthias, P. Ambivalent cartographies: Exploring the legacies of Indigenous land titling through participatory mapping. *Crit. Anthropol.* **2019**, *39*, 222–242. [CrossRef]
3. Chirif, A.; García Hierro, P. *Marcando Territorio: Progresos y Limitaciones de la Titulación de Territorios Indígenas en la Amazonía*; IWGIA (International Work Group for Indigenous Affairs): Copenhagen, Denmark, 2007.
4. Barthel, K.; Cespedes, V.; Salazar, B.; Torres, R.; Varón, M. *Land and Rural Development Policy Reforms in Colombia: The Path to Peace*; World Bank Conference Land and Poverty: Washington, DC, USA, 2016.
5. Sletto, B. Indigenous people don't have boundaries': Reborderings, fire management, and productions of authenticities in Indigenous landscapes. *Cult. Geogr.* **2009**, *16*, 253–277. [CrossRef]
6. DANE. *Población Indígena de Colombia: Resultados del Censo Nacional de Población y Vivienda 2018*; Dirección Nacional de Estadística: Bogotá, Colombia, 2019.
7. Rojas, J.M. Ocupación y recuperación de los territorios indígenas en Colombia. *Análisis Político* **2000**, *0*, 69–83.
8. Mora, J.P.; Barrera, C.H.; Correa, J.d.C. Indigenous communities in colombia. For an understanding in constitutional history. Case: 1991 constitution. *Rev. Incl.* **2021**, *8*, 470–487.
9. Braconnier, L. Los Derechos Propios de Los Pueblos Étnicos En El Acuerdo de Paz de agosto de 2016. *Rev. Derecho Estado* **2017**, *40*, 113–126. [CrossRef]
10. Morey, R.; Metzger, D. The Guahibo: People of the Savanna. *Acta Ethnol. Linguist.* **1974**, *31*, 86–88.
11. UNHCR; ACNUR. *La Lucha Del Pueblo Indígena Sikuani En Colombia, Por La Pervivencia y Su Derecho Ancestral Al Territorio*; The UN Refugee Agency, United Nations High Commissioner for Refugees: Geneva, Switzerland, 2017.
12. Fernandes, V.; Reydon, B.P.; Silva, D.; Bueno, A.P. Land governance, land policy and Indigenous people land use and access rights in the Brazilian Amazon and Matopiba after the constitution of 1988. In Proceedings of the Annual World Bank Conference on Land and Poverty, Washington, DC, USA, 20–24 March 2017; The World Bank: Washington, DC, USA, 2017.
13. Reydon, B.; Molendijk, M.; Porras, N.; Siqueira, G. The Amazon Forest Preservation by Clarifying Property Rights and Potential Conflicts: How Experiments Using Fit-for-Purpose Can Help. *Land* **2021**, *10*, 225. [CrossRef]
14. Álvarez Larrain, A.; McCall, M.K. Participatory Mapping and Participatory GIS for Historical and Archaeological Landscape Studies: A Critical Review. *J. Archaeol. Method Theory* **2019**, *26*, 643–678. [CrossRef]
15. Louis, R.P.; Johnson, J.T.; Pramono, A.H. Introduction: Indigenous cartographies and counter-mapping. *Cartographica* **2012**, *47*, 77–79. [CrossRef]
16. Ramirez-Gomez, S.O.; Brown, G.; Fat, A.T.S. Participatory mapping with Indigenous communities for conservation: Challenges and lessons from Suriname. *Electron. J. Inf. Syst. Dev. Ctries.* **2013**, *58*, 1–22. [CrossRef]
17. Reyes-García, V.; Orta-Martínez, M.; Gueze, M.; Luz, A.C.; Paneque-Gálvez, J.; Macía, M.J.; TAPS Bolivian Study Team. Does participatory mapping increase conflicts? A randomized evaluation in the Bolivian Amazon. *Appl. Geogr.* **2012**, *34*, 650–658. [CrossRef]
18. Rye, S.A.; Kurniawan, N.I. Claiming Indigenous rights through participatory mapping and the making of citizenship. *Political Geogr.* **2017**, *61*, 148–159. [CrossRef]
19. Sletto, B. Indigenous rights, insurgent cartographies, and the promise of participatory mapping. *Portal* **2012**, *7*, 13–15.
20. Bryan, J. Walking the line: Participatory mapping, Indigenous rights, and neoliberalism. *Geoforum* **2011**, *42*, 40–50. [CrossRef]
21. Braceras, I. Cartografía Participativa: Herramienta de Empoderamiento y Participación por el Derecho al Territorio. Master's Thesis, Universidad del País Vasco, Bilbao, Spain, 2012.

22. Hak, S.; McAndrew, J.; Neef, A. Impact of Government Policies and Corporate Land Grabs on Indigenous People's Access to Common Lands and Livelihood Resilience in Northeast Cambodia. *Land* **2018**, *7*, 122. [CrossRef]
23. Enemark, S.; Bell, K.C.; Lemmen, C.; McLaren, R. *Fit-for-Purpose Land Administration*; FIG Publications No 60; International Federation of Surveyors: Copenhagen, Danmark, 2014.
24. Enemark, S.; Clifford, K.; Lemmen, C.; McLaren, R. *Fit-for-Purpose Land Administration: Guiding Principles for Country Implementation*; UN-Habitat, Kadaster, GLTN: Nairobi, Kenya, 2016.
25. Enemark, S. *From Cadastre to Land Governance: The Role of Land Professionals and FIG*; World Bank Conference Land Policy and Administration: Washington, DC, USA, 2010.
26. Jones, B.; Lemmen, C.; Molendijk, M. *Low Cost, Post Conflict Cadastre with Modern Technology*; World Bank Conference Land and Poverty: Washington, DC, USA, 2017.
27. Molendijk, M.; Morales, J.; Lemmen, C. Light Mobile Collection Tools for Land Administration—Proof of Concept from Colombia. *GIM Int.* **2015**, *27*, 20–23.
28. United Nations REDD Programme. Guidelines on Free, Prior and Informed Consent. Available online: https://www.unredd.net/documents/un-redd-partner-countries-181/templates-forms-and-guidance-89/un-redd-fpic-guidelines-2648/8717-un-redd-fpic-guidelines-working-final-8717.html (accessed on 28 January 2021).
29. Food and Agriculture Organization of the United Nations. *Voluntary Guidelines on the Responsible Governance of Tenure of Land Fisheries and Forests in the Context of National Food Security*; Food and Agriculture Organization of the United Nations: Rome, Italy, 2012.
30. Ambani, S.; Kalinga, J.; Kiambuthi, M.; Molendijk, M.; Tomberg, M.; Unger, E.M.; Lemmen, C.; Jones, B. Fit-for-Purpose and Fit-for-Future Technology for Cadastral Systems. Example Cases from Kenya and Colombia. In Proceedings of the Fit-for-Purpose in the Global Diversity, FIG Conference, Helsinki, Finland, 29 May–2 June 2017.
31. Balas, M.; Carrilho, J.; Joaquim, S.; Lemmen, C.; Murta, J.; Matlava, L.; Mario Ruy, M. *A Fit for Purpose Land Cadaster in Mozambique*; World Bank Conference Land and Poverty: Washington, DC, USA, 2017.
32. Jones, B.; Lemmen, C.; Molendijk, M.; Gorton, K. *Fit-for-Purpose and Fit-for-Future Technology for Cadastral Systems*; World Bank Conference Land and Poverty: Washington, DC, USA, 2017.
33. Lemmen, C.; Van Oosterom, P.; Kalantari, M.; Unger, E.M.; Teo, C.H.; de Zeeuw, K. Further standardization in land administration. In *Responsible Land Governance–Towards an Evidence-Based Approach*; World Bank Conference Land and Poverty: Washington, DC, USA, 2017.
34. Hudson, M. The CARE Principles for Indigenous Data Governance. *Data Sci. J.* **2020**, *19*, 43. [CrossRef]
35. Morales, J.; Lemmen, C.; de By, R.; Molendijk, M.; Oosterbroek, E.P.; Ortiz, A.E. On the Design of a Modern and Generic Approach to Land Registration: The Colombia Experience. In Proceedings of the 8th International FIG workshop on the Land Administration Domain Model, Kuala Lumpur, Malaysia, 1–3 October 2019.

Article

Assessment of Land Administration in Ecuador Based on the Fit-for-Purpose Approach

Dimo Todorovski [1,*], Rodolfo Salazar [2] and Ginella Jacome [2]

1 Faculty ITC, University of Twente, 7514 AE Enschede, The Netherlands
2 Department of Earth Sciences and Construction, University of the Armed Forces ESPE, Sangolqui 171103, Ecuador; rjsalazar@espe.edu.ec (R.S.); gijacome@espe.edu.ec (G.J.)
* Correspondence: d.todorovski@utwente.nl; Tel.: +31-53-4874-329

Abstract: Land administration is established to manage the people-to-land relationship. However, it is believed that 70% of the land in developing countries is unregistered. In the case of Ecuador, the government has an ambitious strategy to implement a national cadaster on the full territory in a short time period. Therefore, the objective of this study was the assessment of land administration in Ecuador based on the fit-for-purpose approach as an assessment framework. A literature review was performed on the topic of land administration, including guidelines for improvement and assessment frameworks. The basic concept of fit-for-purpose land administration was reviewed with the three frameworks, which are: spatial, legal, and institutional. Interviews and focus group discussions were performed in Ecuador for collecting primary and secondary data about land administration in this country. Results from these activities are presented and discussed using the structure of the basic concept of fit-for-purpose land administration with the three frameworks. It was found that during the field data collection precise land survey of fixed boundaries was performed and around 55–60 attributes per parcel were collected as a part of the field land survey in Ecuador. Based on the findings, discussions were developed, and a score table was created identifying which principles should be addressed if rapid mapping and land registration are desired by the government of Ecuador to be implemented on the whole territory in a short time period. Finally, the paper ends with conclusions and recommendations.

Keywords: fit-for-purpose land administration; Ecuador

Citation: Todorovski, D.; Salazar, R.; Jacome, G. Assessment of Land Administration in Ecuador Based on the Fit-for-Purpose Approach. *Land* **2021**, *10*, 862. https://doi.org/10.3390/land10080862

Academic Editors: Stig Enemark, Robin McLaren and Christiaan Lemmen

Received: 30 March 2021
Accepted: 24 July 2021
Published: 17 August 2021

1. Introduction

The importance of land and its administration is recognized globally, and this is embedded in the United Nations Sustainable Development Goals. It is believed that 11 out of 17 sustainable development goals have a relation with the land component. This gives clear guidelines to United Nations member states that practicing efficient land administration leads to a sustainable future. Social and economic benefits from good land administration are continuously repeated by policymakers of developed countries as an important element for the wellbeing of their citizens. All this is showing that land and its administration are high on the global agenda. But on the other hand, there is a big imbalance or so-called "security of tenure gap" between countries that have in place efficient and effective land administration systems and those that do not. There is an estimation that 70 per cent of the land in developing countries is unregistered [1].

Therefore, an innovative approach for addressing land issues in developing countries is needed. This should be based on affordable, sustainable, and rapid land mapping and registration of land rights using the most appropriate and recent technology relating this to the purpose. An approach that can address all land tenure types, including informal tenure effectively, as well as assist in recovery and reconstruction processes after catastrophes. In this regard "fit-for-purpose land administration" is emerging as an acceptable concept [2].

The fit-for-purpose land administration concept, with its 12 principles, could be used as an assessment framework, as it is in this paper.

Ecuador is a developing country in Latin America with an emerging economy. It is currently utilizing several systems of land administration, using the conventional method by collecting many attributes during the systematic land survey and land registration. Land administration is a function that is the responsibility of municipalities. Municipalities are implementing land administration, mainly focusing on the property tax register, which is helping them with collecting property taxes. In many cases, the property tax represents 80% of the budget of the municipalities. Implementing land administration like this brings more benefit to the municipalities and government rather than to the citizens, who do not fully perceive economic and social benefit from this state function. Examples of those benefits are: registered ownership enables easier access to loans and credits that are leading to economic development or proper land use mapping/registration leads to improved urban/rural planning and better quality of living.

The Ecuadorian government updated its constitution in 2008 with a part about the establishment of a national cadaster with an ambitious strategy to implement this in a short period of 4–5 years. In the last couple of decades, Ecuador has been establishing land administration in the urban areas with approximately 70–75% of the municipalities and 25% coverage of the land administration in the rural areas. Looking at these figures, it can be derived that if it continues with the same pace and speed, it will take many years to finish its full land administration coverage. Performing like this, it is not fully supporting economic growth, food security, natural conservation, reconstruction after disasters, and poverty reduction. Therefore, a fit-for-purpose approach could be considered for land administration in Ecuador focusing on the requirements and benefits for all.

The objective of this study was the assessment of land administration in Ecuador based on the fit-for-purpose approach as an assessment framework. Literature about land administration, guidelines for its improvement, and assessment frameworks for land administration are reviewed in Section 2. In continuation, the fit-for-purpose land administration basic concept with three frameworks, which are spatial, legal, and institutional, are further elaborated; each of the three frameworks contains four principles of fit-for-purpose land administration. The methodology for this research and various activities for collecting data are described in Section 3. In Section 4, empirical data about the status of land administration in Ecuador are presented. Section 5 develops discussions using the principles of the fit-for-purpose approach as an assessment framework. Finally, the paper ends with drawing conclusions and recommendations regarding fit-for-purpose land administration for the case of Ecuador.

2. Literature Review

This section features a literature review performed on the topics of land administration, guidelines for its improvement, and land administration assessment frameworks. In addition, the concept of fit-for-purpose land administration with its principles is presented with justification for the selection of this concept to be used as an assessment framework for our case.

2.1. Land Administration

In the literature about land administration, several definitions of land administration can be found. According to [3], land administration is defined as "the process of determining, recording, and dissemination information about the tenure, value, and use of land when implementing land management policies". Another definition is: "land administration is the process of regulating land and property development and the use and conservation of the land; the gathering of revenues from the land through sales, leasing, and taxation; and resolving of conflicts concerning ownership and use of land" [4]. Later in 2010 [5], provided the following definition: "the processes run by the government using public- or private-sector agencies related to land tenure, land value, land use, and land development". For our case, the definition from [3] is most applicable because Ecuador is still in the phase of creating a national cadaster on the whole territory (as elaborated in Section 4).

In order to increase understanding of land administration worldwide a "Continuum of land rights" was designed [6]. This continuum provides a range of forms of land rights and has a varying set of rights, degrees of security. and enforcement. On the left side of the ladder, we can see the informal land rights; although they have social legitimacy, informal land rights are not recorded officially in many countries of Asia and Africa. Some countries have customary or communal land tenure types that may not be recorded or recognized by the formal systems and legal frameworks. However, they exist, and land transactions of these land tenure types continue informally [6]. "A continuum of land rights offers practical recordation of land rights that allows people to get onto this tenure rights ladder. It provides an incremental approach of upgrading land rights over time in response to available technology and resources" [7].

2.2. Guidelines for Improvement and Assessment Frameworks for Land Administration

In line with the identified importance of having in place good land administration, several international non-governmental organizations, professional associations, and academia have been working on its evaluations and recommendations for improvement. The Food and Agriculture Organization (FAO) of the United Nations, working on the land topic, developed voluntary guidelines that can be used as an evaluation tool as well [8]. The World Bank developed the Land Governance Assessment Framework (LGAF), which is a diagnostic instrument to assess the status of land governance at the country or sub-national level [9]. Recently, the United Nations Committee of Experts on Global Geospatial Information Management (UN-GGIM) has developed the Framework for Effective Land Administration (FELA). It is a framework that can be used as a guideline to develop, reform, modernize, and monitor land administration [10]. These guidelines and assessment frameworks are elaborated here just to name a few that were recently published.

In the last decade, many developments have been achieved in the area of land administration with regard to publications and literature. Several books have been published [5,11]; international guidelines, evaluation, and assessment frameworks have been presented and published. From the initiatives by international organizations' publications and literature, it can be concluded that conventional land administration systems have several limitations when an attempt is made to implement them in developing countries. This is mainly because Western-style land administration systems were implemented for many years with a focus on fixed boundaries, accurate mapping, and surveying. Western-style land administration has complex bureaucratic procedures, involving sophisticated technologies, which does not perform well with the developing countries' needs and services. When implementing conventional land administration in developing countries it usually takes more time and resources than initially planned [11]. Therefore, international organizations are developing alternative methods and approaches for rapid mapping and land registration based on countries' contexts and purposes.

2.3. The Concept of Fit-for-Purpose Land Administration

Looking at the land administration in less developed countries, only around 30% of the land is included in the formal land administration systems [1], and this is not supporting the appropriate economic and social wellbeing of their citizens. With this, an urgent need is identified for building rapid and simple systems that would capture the optimal elements about the people-to-land relation. "When considering the resources and capacities required for building such systems, the more advanced concepts as predominantly used in developed countries may well be seen as the end target but not as the point of entry. When assessing technology and investment choices, the focus should be on a "fit-for-purpose approach" that will meet the needs of society today and can be incrementally improved over time" [2]. Article [2] has references to the voluntary guidelines from the

FAO [8] and LGAF from the World Bank [9], and it can be derived that this publication is a continuation of the work on evaluations and recommendations for improvement in the area of land administration. The suggested approach meets the requirements like affordable, fast, and sustainable methods of land mapping, land registration, and titling, which can address all land tenure types, including informal tenure, effectively as well as aid provision in the recovery and reconstruction processes after catastrophes. This approach is designed to fit the purpose of the society over being in line with the existing rules and methods of conventional land administration. Elements of the fit-for-purpose approach are flexible, inclusive, participatory, affordable, reliable, attainable, and upgradable [2]. The concept of fit-for-purpose land administration contains three interrelated core frameworks that work together to deliver the fit-for-purpose approach: the spatial, the legal, and the institutional frameworks. The fit-for-purpose land administration approach includes four core principles for each of the three frameworks. See Table 1 below, which shows an overview of the "key principles of the fit-for-purpose land administration approach":

Table 1. The key principles of the fit-for-purpose approach (taken from [2]).

Key Principles		
Special Framework	**Legal Framework**	**Institutional Framework**
Visible (physical) boundaries rather than fixed boundaries	A flexible framework designed along administrative rather than judicial lines	Good land governance rather than bureaucratic barriers
Arial/satellite imagery rarher than field surveys	A continuum of tenure rather than just individual ownership	Integrated institutional framework rather than sectorial silos
Accuracy relates to the purpose rather than technical standards	Flexible recordation rather than only one register	Flexible ICT approach rather than high-end technology solutions
Demands for updating and opportunities for upgrading and ongoing improvement	Ensuring gender equity for land and property rights	Transparent land information with easy and affordable access for all

Depending on the purpose and the end users' requirements, it is important to choose the appropriate fit-for-purpose strategy for fieldwork data collection. If many attributes are collected during the fieldwork, then many attributes have to be maintained; this means that there should be awareness for this "multiplier" effect [12].

Implementation of the fit-for-purpose land administration approach is evident in many country cases. Rwanda, for example, has covered the whole country using a fit-for-purpose land administration approach within five years and for a cost of around 6 USD per parcel/spatial unit [1]. In [13], lessons learned from implementing fit-for-purpose land administration in three developing countries, Indonesia, Nepal, and Uganda, are elaborated; in Indonesia, the introduction of fit-for-purpose land administration is supported by the president. A recent study has critically analyzed three land formalization initiatives in India that have employed flexible recording approaches and where decentralization was used to scale implementation [14]. In the territory of Latin America, several pilots, proof of concept, and implementation projects have started in Colombia [15], and the fit-for-purpose land administration approach has been introduced in Ecuador [16].

2.4. Land Administration System in Ecuador

Ecuador is executing its land administration function using a conventional method, precisely measuring the fixed boundaries during the land survey and collecting many attributes within the first land registration. Organizations that support the first land registration and survey are the Ministry for Urban Development and Housing (for urban areas) and the Ministry of Agriculture (for rural areas). Consequently, after these activities, the land administration function is transferred as a responsibility of municipalities. Municipal offices are focusing on the property tax register, which is helping them with collecting property taxes. Municipal offices are also responsible for the maintenance, update, and delivery of products and services regarding cadastral maps and land registration. Until the moment of this research, Ecuador has been establishing land administration in the urban areas with approximately 70–75% of the municipalities and 25% coverage of the land administration in the rural areas. Because there is no national land administration system in Ecuador, there are no precise numbers of registered/unregistered parcels. The estimation by the authors, based on presentations provided by the MIDUVI and MAG, is that there are in total 8.5 million parcels, approximately 4 million parcels in rural areas (1.2 million registered) and 4.5 million parcels in urban areas (3.2 million registered). The government of Ecuador updated the constitution in 2008 with a part about the establishment of a national cadaster with an ambitious strategy to implement this in a short period of 4–5 years. Following the update of the constitution, additional laws and bylaws were enforced to support the establishment of a national cadaster, including six tenure types that are legally and socially accepted. Looking at the status of land administration and the figures, it can be derived that if it continues with the same pace and speed it will take many years to finish its full land administration coverage (Section 4 presents in detail the land administration in Ecuador). Therefore, a fit-for-purpose approach could be considered for land administration in Ecuador focusing on the requirements and benefits for all.

The fit-for-purpose land administration concept with its principles is used when countries are rethinking the purpose of their land administration, adopt new implementation strategies with ambitious timelines, and as an assessment framework. Because Ecuador is in a similar situation, the fit-for-purpose land administration approach was used as an assessment framework in this paper.

3. Methodology, Case Study, and Data Collection Methods

The qualitative methodology using a case study approach was most suitable for this research. Ecuador was selected as a case study for this research guided by the fact that although land administration was introduced in the 1940s there is still no full country coverage. In addition, the government of Ecuador has an ambitious strategy to implement a national cadaster on the full territory in a period of 4–5 years. Here follows a map of Ecuador as Figure 1:

Figure 1. Map of Ecuador available online: https://www.worldometers.info/img/maps_c/EC-map.gif (accessed on 16 August 2021).

The qualitative methodology involves observing a phenomenon, which is considered a contemporary practice, where the boundaries between the phenomenon and the environment are not clear [17]. In our research, the boundaries between the phenomenon of fit-for-purpose land administration and the environment are not clear, therefore we are observing fit-for-purpose land administration as a contemporary practice. Article [18] elaborates a case study approach to research, empirically investigating a contemporary phenomenon in the real-life context to understand precipitating such a phenomenon.

In order to apply the qualitative methodology to gain an in-depth understanding of fit-for-purpose land administration, this research is based both on the literature review and on primary and secondary data sources. The latter were collected during one-week data collection activities and write shops at the end of 2017 in Quito, Ecuador. Primary data were collected during the interviews and focused group discussions via expert consultation. A semi-structured interview form was used consisting of three topics, namely the frameworks of the fit-for-purpose land administration concept: spatial, legal and institutional frameworks. In continuation, detailed questions/discussions were developed regarding the twelve principles, as presented in Table 1, which provided a solid base for doing the assessment of the land administration in Ecuador. Secondary data were collected from governmental organizations at ministerial and municipal levels. Using qualitative methodology, we collect data that is more in a descriptive data manner rather than statistical.

During the one-week data collection activities, firstly the fit-for-purpose land administration concept was presented and then interviews were performed using the expert consultation method with 16 land professionals. Focused group discussions were applied during the study visits to the two ministries. Additional stakeholders were invited for interviews, but they were not available during the week of data collection. Table 2 presents an overview of the number of experts consulted during the data collection activities.

Table 2. An overview of consulted number of experts during the research.

Ministry for Urban Development and Housing (MIDUVI)	6
Ministry of Agriculture (MAG)	4
Municipal cadastral office of the capital city of Ecuador—Quito	2
One municipality that was affected by the 2016 earthquake	2
University ESPE	2

Data verification from the interviews and expert consultations was implemented using the write shop approach, where participants from the MIDUVI, MAG, and municipal cadastral office of the capital city Quito drafted 3–4 page documents about their organization's performance. The documents are in the structure of the basic concept of fit-for-purpose land administration with the three frameworks and can be received on request from the authors. Collecting data and storing it in the structure of the fit-for-purpose basic concept was done so with the intention of having a comprehensive set of required data and information about land information in Ecuador for the coming assessment and analyses. Assessment and analyses of the data were conducted via discussions where results were discussed against the fit-for-purpose frameworks using the same structure of the three frameworks and twelve principles for better interpretation (in Section 5 of this paper). A qualitative methodology based on literature review, interviews, and focus group discussions for data collection, data verification via a write shop, and a score table as a tool was successfully implemented for the case of Ecuador and this could be repeated as a method for other cases when performing similar assessment studies.

After the interviews and the write shop for data collection, two workshops were implemented in Ecuador under the School for Land Administration Studies (a joint initiative with the Netherlands Kadaster International and Faculty ITC, University of Twente). The first workshop was implemented in Ecuador on a national level and included 36 participants on the topic of the introduction of the fit-for-purpose land administration approach at the University of

Armed Forces ESPE, Sangolquí, at the end of 2017. In 2018, the second international workshop, with approximately 80 participants from seven Latin American countries and two countries from Europe, dealt with the topic of fit-for-purpose land administration and LADM+ and was performed at the Instituto Geografico Militar in Quito. Further planned activities piloting fit-for-purpose land administration in several locations and presenting results to the authorities in Ecuador have been impacted by the COVID-19 pandemic and postponed.

Findings from this research are presented as results in the next section under the structure of the headings of the three fit-for-purpose frameworks with a link to the source.

4. Results: Land Administration in the Case of Ecuador

This section presents the results of this study based on expert consultation with land professionals during the interviews and focused group discussions. It follows the subsection structure of the three frameworks of fit-for-purpose land administration, and within the subsections, firstly national then urban and finally rural land administration are elaborated. This writing logic was adopted because first land registration is done differently and by different actors for urban and rural areas, focusing on details that can be related to the four principles within each framework.

4.1. Institutional Framework

On the national level, land administration in Ecuador is a state function that is performed by municipalities. Since 1940s, municipalities have been responsible for implementing and maintaining land administration, and the land administration system was designed mainly to support the processes of valuation and property tax collection by municipalities. The majority of municipalities depend on this tax collection, which represents, in many cases, up to 80% of their budget. Within the municipalities there are three land-administration-related data sets: (1) property tax records, (2) cadaster with map representation, and (3) records of owners (persons and legal entities). In half of the municipalities, these three data sets are not properly linked and they perform like isolated islands.

All municipalities in Ecuador maintain and manage the data from their textual and numeric dataset for land administration in their urban areas, commonly known as the urban cadaster. Since 2011, the Ecuadorian Development Bank has financially supported municipalities to establish their cadasters. This support is via financing projects that are unified and established using the same standards and the georeferenced methodology. Following this methodology, appropriate linked spatial and alphanumeric datasets are created. Data collection and maintenance for urban cadaster are elaborated in detail in the National Technical Standards for the Cadaster of Urban-Rural Real Estate and Property Appraisals [19] prepared and published by the MIDUVI.

Land administration in rural areas or the so-called "rural cadaster" was not the focus of attention of the Ecuadorian government and municipalities for a long time; this accumulated many challenges for its administration. Therefore, the project "National System of Information and Management of Rural Lands and Technological Infrastructure" (SIGTIERRAS) was developed and implemented by the MAG (2011–2015). The Ecuadorian Government received a loan of 10 million US dollars from the International Development Bank for completing the project. The aim of SIGTIERRAS was an implementation of an efficient system of rural land administration and management. As a long-term vision, the system was designed to support: "updating the cadaster, providing legal security to property rights, supporting legalization and regularization of land tenure, applying fair and equitable tax policies, and providing information for the planning and territorial organization of the rural area" – based on the focus group discussion with SIGTERRAS personal. In the project period, 47 municipalities were completed with the establishment of a rural cadaster representing 25% of the rural area of the country. In the period before 2010, sporadic parcel registration was possible in municipal offices for the rural areas, and it can be derived that systematic land registration started with the project in 2011.

SIGTIERRAS was created to generate land information in an appropriate information system for land administration and management. In the process of creation, a collaboration was established between many related actors responsible for standardization and metadata, and as beneficiaries of the system and the data collected and processed. Actors dealing with standards and metadata are the Military Geographic Institute, Ecuadorian Space Institute, National Secretariat of Planning, and National Council of Geomatics. Beneficiaries of the information system and the data created are municipalities of local self-government, the MAG, the MIDUVI and the SOT. Organization and management of the project, including fieldwork for data collection, was performed by the specialized Executing Unit of the MAG.

4.2. Legal Framework

Observing the legal framework on the national level, in 2008 a new constitution was enforced in Ecuador [20]. In the new constitution, there is a part acknowledging the need for the establishment of a national cadaster that would support the service and information provision on a municipal level. The need for the establishment of a national cadaster is based on the evidence from the past that municipalities were creating and maintaining cadasters on their own without unified standards for the whole state. In addition to the new constitution, a presidential decree was issued in 2011 [21] giving guidelines about the national cadaster. From the presidential decree, we extract the following: "the objective of the National System of the Integrated Geo Cadaster of Habitat and Housing is to: (...) register systematically, logically, geo-referenced and ordered, in a comprehensive and integrated database, urban and rural cadasters, which serves as a tool for the formulation of urban and rural development policies".

From 2016 onward, the MIDUVI has had a major role in the standardization and construction of the national cadastral system. This was manifested via preparing and providing standards on a national level that support municipalities in their establishment of a unified national cadaster. Following the older legal framework in Ecuador, in 2016 the Organic Law of Territorial Land Use and Management, SAN-2016-1196 [22], came into force. According to this law, land administration in Ecuador includes six land tenure types and these are both legally and socially accepted: (1) state property, (2) private property, (3) associative property, (4) cooperative ownership, (5) mixed-ownership, and 6) community property—this tenure type can be registered and titled in favor of communes, communities, groups of people, and nationalities (as explained in article 85 of [22]). In addition, within the same law, gender equity is elaborated. Within this so-called Organic Law, the MIDUVI was appointed as a governing body for the establishment and maintenance of the national cadaster of Ecuador. The performance of the MIDUVI is supported by one more governmental organization: the Superintendence of Territorial Planning (SOT). The SOT was created in 2016 with the authority to work on capacity development, quality control, and, in addition, has a regulatory role for the cadaster in Ecuador for the performance of all municipal cadasters.

The Organic Law of Territorial Land Use and Management [22] in Article 100 established: "The National Registered Integrated Geographical Cadaster must be updated continuously and permanently and will be administered by the governing body of habitat and housing, which will regulate the conformation and functions of the system and will establish standards, protocols, deadlines, and procedures for the collection of cadastral information and valuation of real estate taking into account land classification, land uses, among others."

For the land administration in the rural areas, land registration was initiated with Ministerial Agreement 160 in 2008 [23], which established the Executing Unit of the MAG for implementation of the project SIGTERRAS. The additional legal framework with regard to rural land registration and administration was addressed in the New Constitution of the Republic of Ecuador in 2008 [20] and the Organic Law [22] under Article 42. Within the law [22] under Article 85 the six land tenure types are addressed and in the same article the achievement of gender equity and registration of vulnerable and indigenous groups with regard to land and property rights are elaborated.

4.3. Spatial Framework

On the national level, two main governmental initiatives were implemented to support the creation of the national cadaster. The first initiative is the project SIGTIERRAS supported and implemented by the MAG, and the second initiative allows the construction of the "urban cadaster" (term used by our local respondents) in 140 municipalities. The second initiative is financially supported by Ecuadorian Development Bank. Both initiatives have their contribution to the generation and increasing the cadastral coverage of the country following the existing regulations.

As for the land administration field data collection, maintenance, and dissemination, Ecuador is employing several methods. Conventional methods for the land survey are followed by collecting many textual attributes during the survey about every single parcel both in urban and rural areas. First registration of land rights and provision of land administration products via public presentation is a lengthy procedure. Rules and regulations for these activities are addressed in several laws and details for their implementation are described in the manuals for the urban [24] and rural [25] cadasters.

For the land administration in urban areas, at the period of the activities for this research, 140 of the 221 municipalities were implementing projects of the georeferenced urban cadaster. Consequently, a cadastral updating is based on the definition of the urban areas of each municipality clearly delineated apart from the rural areas. For these 140 municipalities, the same standards for data collection and mapping procedures are applied following the manual for urban cadastral projects [24]. Regarding the land survey in urban areas, "the guidelines include exhaustive data collection in the field in order to collect data for the first registration of the real estate with a comprehensive level of details needed to form a cadastral map in scale R = 1:1000," as expressed in the focus group discussion at the MIDUVI.

Looking at other municipalities, 33 municipalities (out of 221) are implementing their own local urban cadasters. These municipal cadasters are following their own standards and systems combined with the spatial component—maps in local coordinate systems. Among the rest, 34 out of 221 municipalities do not have georeferenced urban cadaster systems or areas covered by cadastral maps. During the focused group discussion at the MIDUVI, there was an evident awareness that "this way of performance with non-integrated computer solutions brings, consequently, a lot of overlap and duplication of functions within the data collection, map production, analysis, and provision of cadastral data".

Observing the numbers of municipalities (140 + 33), our respondents replied that an approximate coverage of the urban cadaster in Ecuador is 70–75%. It remains unclear what is the actual cadastral coverage within these urban cadastral areas having in mind the fact that this activity started with the implementation of the presidential decree in 2011.

For the land administration in the rural areas, in the period 2011–2014 an aerial survey was completed with coverage of 225,448 km^2, representing 89% of the territory of Ecuador. The result from the aerial survey was orthophotos on a scale of 1:5000. After their production, orthophotos were used for cadastral land surveying. In addition, the aerial survey produced thematic cartographic maps and a digital terrain model of the country. For cadastral purposes, a massive fieldwork activity engaged hundreds of smaller teams for the fieldwork land survey and administrative data collection. This part of the project required complex logistics, outsourcing, and experts in different areas in order to manage and organize these teams, as well as proper data and maps storing, under present rules and regulations. Several participants in this project, during the interviews, mentioned that the project needed more human engagement than initially planned and it would be unlikely to repeat the similar or larger project in a similar manner. In total, one million parcels were surveyed with fixed boundaries using different land survey equipment (tape surveys, total stations, real-time kinematic (RTK) devices, GNSS, and handheld GPS) and textual data were collected. Regarding textual data, a two-page form was used (see Appendix A source MAG SIGTIERRAS). Around 55–60 textual attributes were collected for every single rural parcel. The form, it appears, was created for the purposes of the tax cadaster. It contained the possibility to enter data in the following parts: general info about the parcel/form, 6 attributes; part (1) parcel key information, 12 attributes; part (2) private person, marital status, and spouse info, 11 attributes, or legal entity, 9 attributes; part (3) legal information about the parcel with title, 16 attributes, and if the parcel was without a title then 13 attributes; part (4) land use type, natural coverage, and ecosystem, 8 attributes. After the period for aerial and land surveying and textual data collection, a public verification procedure was implemented as an interactive and participatory event for legalization of the derived results and provided a map extraction to the owners.

The information system for land administration and management developed with the project was named "SINAT". After the first land survey, mapping, land registration, and land records verification in practice were completed, and transfer of knowledge, system, and datasets were handed over to the municipalities. From that moment, it is the responsibility of the municipalities to maintain and keep the rural cadaster up to date. At the moment of this research, 54 municipalities were using the SINAT system for their everyday work activities, according to information received during the focused group discussion at the MAG.

In summary, we can see that many land administration activities are happening in Ecuador and the government has an ambitious strategy to implement a national cadaster on the full territory in a short time. First registration in urban and rural was conducted differently by different actors, and when public presentation and verification of the land maps and records was finished, they were transferred to municipalities for further maintenance and dissemination. It is unclear what is the quality control process within municipalities and whether the updates have a centralized backup and national database on the internet for data display and dissemination. In the next section, the results from this section are discussed against the fit-for-purpose frameworks.

5. Discussions

In this section, the analyses of the results have been derived based on discussing the results from this research against the basic concept of fit-for-purpose land administration utilizing the three frameworks. The following discussions intend to perform the assessment of land administration in Ecuador based on the fit-for-purpose approach and derive a score table identifying principles that need to be addressed when implementing the fit-for-purpose approach.

Observing the land administration in Ecuador, it is evident that many land-related activities are happening, various legislations are being developed, and organizations for implementation are appointed and created. Looking at the aim to have a national land administration system and full land administration coverage of the territory of Ecuador in a short period of 4–5 years (Section 2.4), it is very positive that there is a commitment from the highest political level, with the idea to perform this as a project. On the other hand, this goal will only be possible with a fit-for-purpose land administration approach [1] because with the current conventional method of land administration it would take more time than planned. Observing the long-term vision used for the SIGTERRAS projects (in Section 4.1), it can be derived that the government is moving the focus from mainly a tax cadaster toward a legal cadaster.

5.1. Discussion of the Results against the Principles of the Fit-for-Purpose Approach

Regarding the institutional framework, one of the principles is "good land governance rather than bureaucratic barriers" [1]. From the expert consultation, as in Section 4.1, it is clear that there is good collaboration between all actors that are involved in land administration in Ecuador; however, the technical standards and protocols for data sharing and exchange still need improvement for the implementation in practice. For example, the three so-called isolated islands data sets in the municipalities appear not to be integrated and still operate separately. Relating this to the next principle "integrated institutional framework rather than sectorial silos", it can be derived that this point needs attention and appropriate integration of the mentioned data sets would improve the efficiency and effectiveness of all stakeholders. Another principle is "flexible ICT approach rather than high-end technology solutions"; based on the expert consultation and focused group discussions, see Section 4.1, the overall information system and the database structure of SINAT satisfy the requirements of the current processes. The last principle in the institutional framework is "transparent land information with easy and affordable access for all" [1]. Linking the last principle to the fact that a big number of attributes are collected in the field during the land survey of fixed boundaries (see Appendix A), this leads us to derive that many elements during collection require more expensive creation and maintenance of the database [12]. All these newly suggested fit-for-purpose opportunities would require adequate capacities and resources development (Chapter 11 of [5]).

Concerning the legal framework, the first principle is "a flexible framework designed along the administrative rather than juridical lines" [1]. This principle should be taken into consideration when planning to speed up the registration and maintenance procedures. It is very important to follow the first principle if/when any changes are planned in the current way of implementing land administration in practice, as those should be backed up with appropriate legal background, bylaws or new regulations. The registration and maintenance procedures in Ecuador are proven as massive and time-consuming, as described in Section 4. The second principle, "a continuum of tenure rather than just individual ownership", can be linked with the law [22], which includes the six legally and socially accepted land tenure types: (1) state property, (2) private property, (3) associative property, (4) cooperative ownership, (5) mixed ownership, and (6) community property. What is missing are some elements of the "continuum of land rights", e.g., certificates of occupancy. Ensuring gender equity for land and property rights (see Section 4.2) is in line with the fourth principle, "ensuring gender equity for land and property rights" [1]. Looking at the third principle "flexible recordation rather than one register", this is already present in Ecuador with land administration in urban (Section 4.2) and in rural (Section 4.2) areas, and maintenance of the three data sets addressed in Section 4.2. Observing the four principles of the legal framework in the context of the Ecuadorian case, we can derive that the first principle needs the most attention.

With regard to the spatial framework, one of the principles is "visible (physical) boundaries rather than fixed boundaries" [1]. This principle is in the opposite situation for the case of Ecuador because according to the current rules and regulations a method of fixed boundaries is mandatory. This is supported with the statement " ... including exhaustive data collection in the field" as expressed during the expert consultation and focus group discussions in the MIDUVI in Section 4.3. The next principle of the fit-for-purpose land administration approach is "aerial/satellite imagery rather than field surveys", again an opposite situation in the case of Ecuador where a field survey is mandatory (as described in both urban and rural areas in Section 4.3). However, it is positive as it is mentioned and demonstrated that orthophotos are used in many ways by multiple stakeholders. The third principle in the spatial framework is "accuracy related to the purpose rather than technical standards". With the regard to the third principle, we can relate to the statement " ... comprehensive level of detail ... "in Section 4.3 and usage of different land survey equipment in Section 4.3, and can conclude that it is not a case in Ecuador; especially for collection of textual data the rural areas as in Appendix A. The fourth principle of the spatial framework is "demand for updating and opportunities for upgrading and ongoing improvement" [1]. From the expert consultation, there is an impression that Ecuador is still in the phase of finalizing the full coverage with land administration, where the demand for updating is in the next steps of implementation. This is partly because this activity is a mandate of the municipal cadaster offices, and they have their own resources and quality control mechanisms. During the expert consultation and the write shop with representatives from the MIDUVI and SOT regarding this fourth principle, there was an opinion that this area needs improvement and should be addressed in the near future.

This research found out that one million parcels have been completed in the rural cadaster in Ecuador in a period of 5 years. This was part of a bigger national mapping project, SIGTER-

RAS, representing 25% of the rural territory (see Section 4.3). These one million parcels in Ecuador are precisely measured with fixed boundaries and linked with around 55–60 text attributes per parcel. It appears that for this project the same rules and regulations as for the urban cadaster were considered. The following question regard these findings: what is the rationale/purpose behind measuring precisely fixed boundaries in rural areas and collecting such a number of attributes? We can derive that a political decision is needed and rethinking of the purpose when the rest of the approximately three million rural parcels are planned to be registered and mapped in a short period of time. Secondly, this could be linked with the multiplier effect: if many attributes are collected, then many attributes have to be stored and maintained [12]; this has a strong implication on the costs and time. Thirdly, based on the SIGTERRAS project description in 4.3, complex logistics, outsourcing, and experts in different areas were required to manage, organize, and administer the work of hundreds of teams; from here we can derive that there is an estimation of a lack of human resources, especially if a national land administration is planned to be implemented by the MIDUVI. Fourthly, as some of the principles are in alignment with the land administration in Ecuador, an adaptation of other principles of fit-for-purpose land administration could be applied. In the results and discussions, the precise measurement of fixed boundaries and a large number of text attributes are considered not in alignment with for-for-purpose principles; it can be derived that interventions could be possible in that regard. An internal MIDUVI assessment on the need and effective use of the precise maps and attributes can determine where these interventions best fit. Our assessment suggests interventions with reducing a number of attributes regarding personal data, legal data, and land use type, and alternative use of imagery.

Additional piloting of the fit-for-purpose approach in several regions in Ecuador to support the suggested interventions was planned under the umbrella cooperation of ESPE University and the School for Land Administration Studies, but because of the COVID-19 outbreak, this was postponed. When presenting the fit-for-purpose land administration concept during the focus group discussions in the MIDUVI (4.1), this concept was discussed further as one of the ways for future implementation of their ambitious aim to have a national cadaster on the full territory in a short time period.

5.2. Score Table for Fit-for-Purpose Land Administration for Ecuador

In continuation, we present a score table for fit-for-purpose land administration for Ecuador based on our assessment and discussions. Because of the qualitative nature of the research, we chose the following scores: H—high level of alignment, M—medium level of alignment, and L—low level of alignment of the results from Ecuador with the principles. The criteria used to apply an appropriate score to the appropriate principles was based on the discussions section where results from the Ecuador case were discussed against the principles of the fit-for-purpose approach. This score table (Table 3) identifies principles that should be addressed if rapid mapping and land registration are desired by the government of Ecuador to be implemented on the whole territory in a short time period. Here follows a fit-for-purpose land administration score table for Ecuador:

Table 3. A score table based on the assessment of the fit-for-purpose land administration for Ecuador.

Key Principles					
Spatial Framework	**Score**	**Legal Framework**	**Score**	**Institutional Framework**	**Score**
Visible (physical) boundaries rather than fixed boundaries	L	A flexible framework designed along administrative rather than juridical lines	M	Good land governance rather than bureaucratic barriers	M
Aerial/satellite imagery rather than field surveys	L	A continuum of tenure rather than just individual ownership	M	Integrated institutional structures rather than sectorial silos	L

Table 3. *Cont.*

Key Principles					
Spatial Framework	**Score**	**Legal Framework**	**Score**	**Institutional Framework**	**Score**
Accuracy related to the purpose rather than technical standards	L	Flexible recordation rather than only one register	M	Flexible ICT approach rather than high-end technological solutions	M
Demands for updating and opportunities for updating and ongoing improvement	M	Ensuring gender equity for land and property rights	H	Transparent land information with easy and affordable access for all	M

From the score table, we can identify which principles are with a low and medium level of alignment with the fit-for-purpose land administration principles as stated in [1]. These could be leading indicators for the Ecuadorian government for areas that need attention when a plan is developed to address rapid mapping and land registration on the whole territory in a short time period. This should be followed with an appropriate strategy for capacity development, available resources, and following the country implementation guide as in [2].

6. Conclusions and Recommendations

The objective of this paper was the assessment of land administration in Ecuador based on the fit-for-purpose approach as an assessment framework. The research was based on a literature review and expert consultation during the data collection activities of primary and secondary data, via interviews and focused group discussions. The analyses of the results and discussions are based on discussing the results against the basic concepts of fit-for-purpose land administration in the structure of the three frameworks and a score table.

We can conclude that Ecuador has established a cadaster in urban areas with approximate 70–75% of municipalities and finished 25% of the area of the rural cadaster. If Ecuador continues at the same pace and speed, it will take many years to finish its full land administration coverage. Land administration in Ecuador is assessed in this paper using the fit-for-purpose land administration approach. This assessment resulted in detailed discussions and a score table. From the score table, we identified principles with a low and medium level of alignment with the land administration in Ecuador. The identified principles are within the spatial and institutional frameworks.

The Ecuadorian government has an ambitious strategy to implement a national cadaster on the full territory in a short period of time and move the focus from mainly a tax cadaster to a legal or fit-for-purpose land administration. Based on our assessment, an adaptation of some of the principles of fit-for-purpose land administration could be utilized. We recommend that identified principles in our score table (with a low and medium alignment) need to be addressed and adopted to the fit-for-purpose approach; specifically with interventions in the precise measurement of fixed boundaries and a large number of text attributes collected in rural areas. For this, an appropriate strategy for implementation should be adopted (as described in [2]) and available resources secured. For the land administration in rural areas, we recommend downsizing the number of attributes collected during the field survey (with reducing of the number of attributes regarding personal data, legal data, and land use type) and alternative use of imagery. Implementing land administration with these interventions would speed up the land administration mapping and land registration in Ecuador.

Author Contributions: Conceptualization, D.T.; Data curation, D.T. and R.S.; Formal analysis, D.T.; Methodology, D.T.; Project administration, D.T. and R.S.; Supervision, D.T.; Writing—original draft, D.T., R.S. and G.J.; Writing—review & editing, D.T., R.S. and G.J. All authors have read and agreed to the published version of the manuscript.

Funding: The funding of the publication costs for this article were kindly provided by the School of Land Administration Studies, Faculty ITC from the University of Twente, in combination with Kadaster International, The Netherlands.

Acknowledgments: The authors would like to express gratitude to the land professionals from Ecuador that provided time and valuable responses during interviews and focus group discussions. Land professionals were from the Ministry for Housing, Ministry of Agriculture, and municipal cadastral office of the capital city of Ecuador, Quito, one municipality which was affected by the 2016 earthquake and University ESPE. The authors would also like to thank anonymous reviewers for their helpful comments, which significantly improved the paper quality.

Conflicts of Interest: The authors declare no conflict of interest.

Appendix A

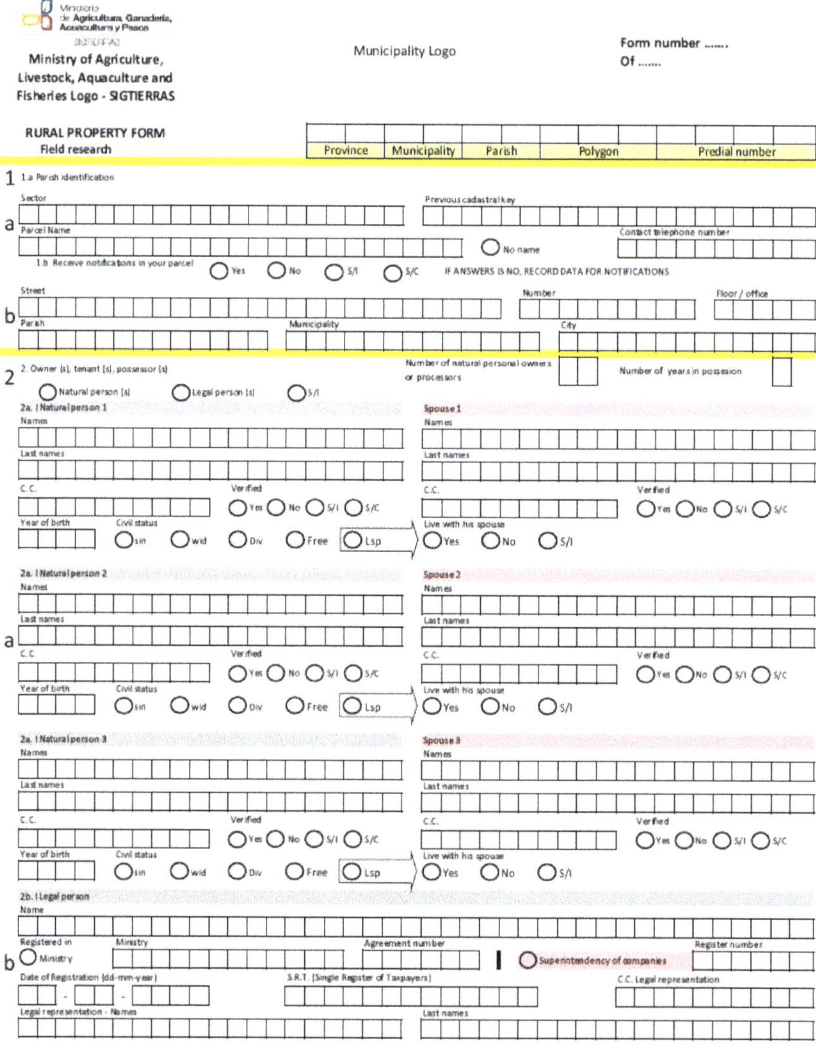

Ministerio de Agricultura, Ganadería, Acuacultura y Pesca

Ministry of Agriculture, Livestock, Aquaculture and Fisheries Logo - SIGTIERRAS

Municipality Logo

Form number ……. Of …….

RURAL PROPERTY FORM
Field research

Province	Municipality	Parish	Polygon	Predial number

1 1.a Parish identification

a Sector — Previous cadastral key

Parcel Name — No name — Contact telephone number

1.b Receive notifications in your parcel: Yes / No / S/I / S/C — IF ANSWERS IS NO, RECORD DATA FOR NOTIFICATIONS

b Street — Number — Floor / office

Parish — Municipality — City

2 2. Owner (s), tenant (s), possessor (s) — Number of natural personal owners or processors — Number of years in possesion

Natural person (s) / Legal person (s) / S/I

a
2a. I Natural person 1 — Spouse 1
Names — Names
Last names — Last names
C.C. — Verified: Yes / No / S/I / S/C — C.C. — Verified: Yes / No / S/I / S/C
Year of birth — Civil status: sin / wid / Div / Free / Lsp — Live with his spouse: Yes / No / S/I

2a. I Natural person 2 — Spouse 2
Names — Names
Last names — Last names
C.C — Verified: Yes / No / S/I / S/C — C.C. — Verified: Yes / No / S/I / S/C
Year of birth — Civil status: sin / wid / Div / Free / Lsp — Live with his spouse: Yes / No / S/I

2a. I Natural person 3 — Spouse 3
Names — Names
Last names — Last names
C.C. — Verified: Yes / No / S/I / S/C — C.C. — Verified: Yes / No / S/I / S/C
Year of birth — Civil status: sin / wid / Div / Free / Lsp — Live with his spouse: Yes / No / S/I

b
2b. I Legal person
Name
Registered in: Ministry — Ministry — Agreement number — Superintendency of companies — Register number
Date of Registration (dd-mm-year) — S.R.T. (Single Register of Taxpayers) — C.C. Legal representation
Legal representation - Names — Last names

Figure A1. *Cont.*

3 3. Legal Information

Parcel with Title | Parcel without Title | W/I

3a. Parcel with Title - Form of acquisition

a Trading | Exchange | Adjudication | Grant | Inheritance | Acquisitive prescription | Site and mountain legal rights and actions | Other

Data according to deeds

Deeds: Yes verified | Not verified | Does not have | W/I | W/C

Requires legal improvement: No | Yes | Not clear

Years without improvement

Certain body: Yes | No

Surface unit according to deed: m² | ha | Block | Tarea | Other

(Record other measure unit)

Value in other measure unit

Trader / Exchanger / Granted / Inheritor - Names

Last names

Through document: Private | Public (deed)

Celebrated before: Public Notary | Judge

Municipality

Date (dd.mm-yyyy)

Registration in the Land Registry: Yes | No | W/I

Municipality

Date (dd.mm-yyyy)

3b. Parcel without Title

b

Land tenure situation	History - Last 10 years
Transfer of possession, crops or improvements	Since (yy) trough (yy) 1 - ; 2 - · Names and last names of previos land possessors (last 10 years) · Years in possession
Succession of possession	Since (yy) trough (yy) · Names and last names of previos land possessors (last 10 years) · Years in possession · Parents · Grandparents · Parents in law · Brothers · Uncles · Other
Individual possession	Desde (aa) · Years of possession
Collective possession	Desde (aa) · Years of possession
Ancestral possession	Desde (aa) - last change · Indigenous village name / ethnicity
Affected by agrarian laws or special decrees	Year · Details
Without identified possession; State	Details
Other	Details

4 4a. Land use

a (REGISTER UP TO 3 MORE IMPORTANT)

Residential (R)	Private residence	Collective residence	Does not have			
Crop	Technified	Traditional	Subsistence	Family garden		
Livestock	Bovine	Goat	Pork	Poultry	Other	Intensive · Extensive · Subsidised
Forestry	Wood	Pulp	Firewood, charcoal			
Aquaculture	Shrimp farm	Other				
Conservation	Natural reserve	Conservation	Other			
NO use	Without use	Not usable				
Other productive	Commerce	Tourism	Industry	Mining	Petroleum	
Social	Education	Health	Worship	Cemetery	Recreational	Public space · Communal
Other productive	Special infrastructure	Other (DESCRIBE IT)				

b 4b. (I) Predominant natural coverage: Arboreal | Shrub | Herbaceous

(II) Relevant ecosystem: Moorland | Wetland | Mangrove | Primary/secondary forest | Does not have

Figure A1. Fieldwork form for rural land survey textual data collection. Source: SIGTERRAS—secondary data collected during fieldwork.

References

1. Enemark, S.; Bell, K.C.; McLaren, R.; Lemmen, C. *Fit-for-Purpose Land Administration; FIG Publication No. 60*. International Federation of Surveyors and the World Bank. 2014. Available online: https://www.fig.net/resources/publications/figpub/pub60/Figpub60.pdf (accessed on 14 January 2021).
2. Enemark, S.; McLaren, R.; Lemmen, C. *Fit-for-Purpose Land Administration: Guiding Principles for Country Implementation*; UN-HABITAT: Nairobi, Kenya, 2016. Available online: https://gltn.net/download/fit-for-purpose-land-administration-guiding-principles-for-country-implementation/ (accessed on 8 December 2020).
3. *Land Administration Guidelines*; UNECE: New York, NY, USA; Geneva, Switzerland, 1996. Available online: http://www.unece.org/fileadmin/DAM/hlm/documents/Publications/land.administration.guidelines.e.pdf (accessed on 7 December 2020).
4. Dale, P.; McLaughlin, J. *Land Administration*; Oxford University Press: Oxford, UK, 1999.
5. Williamson, I.; Enemark, S.; Wallace, J.; Rajabifard, A. *Land Administration for Sustainable Development*; ESRI Press: Redlands, CA, USA, 2010.
6. www.unhabitat.org. *Framework for Evaluating Continuum of Land Rights Scenarios; Securing Land and Property Rights for All*; UNHABITAT: Nairobi, Kenya, 2016. Available online: https://unhabitat.org/sites/default/files/download-manager-files/Framework%20for%20Evaluating%20Continuum%20of%20Land%20Rights%20Scenarios_English_2016.pdf (accessed on 19 May 2021).
7. Zevenbergen, J.; Augustinus, C.; Antonio, D. *Designing a Land Records System for the Poor*; UNHABITAT: Nairobi, Kenya, 2012. Available online: https://unhabitat.org/designing-a-land-records-system-for-the-poor-secure-land-and-property-rights-for-all (accessed on 7 December 2020).
8. *Voluntary Guidelines on the Responsible Governance of Tenure of Land, Fisheries and Forests in the Context of National Food Security*; FAO: Rome, Italy, 2012. Available online: http://www.fao.org/3/a-i2801e.pdf (accessed on 7 December 2020).
9. *Land Governance Assessment Framework*; World Bank: Washington, DC, USA, 2015. Available online: https://www.worldbank.org/en/programs/land-governance-assessment-framework (accessed on 7 December 2020).
10. UN-GGIM. Framework for Effective Land Administration (p. 30). p. 30. Available online: https://ggim.un.org/meetings/GGIM-committee/9th-Session/documents/E_C.20_2020_10_Add_1_LAM_background.pdf (accessed on 23 April 2021).
11. Zevenbergen, J.; de Vries, W.; Bennett, R. (Eds.) *Advances in Responsible Land Administration*; CRC Press (Tylor and Francis Group): Boca Raton, FL, USA, 2016.
12. Lemmen, C. A Domain Model for Land Administration. Ph.D. Thesis, Technical University Delft, Delft, The Netherlands, 5 July 2012.
13. Enemark, S.; McLaren, R. Making FFP Land Administration Compelling and Work in Practice. In Proceedings of the FIG Commission 7 International Commission 7 2018 International Seminar, Bergen, Norway, 24–28 September 2018.
14. Ho, S.; Choudhury, P.R.; Haran, N.; Leshinsky, R. Decentralization as a Strategy to Scale Fit-for-Purpose Land Administration: An Indian Perspective on Institutional Challenges. *Land* **2021**, *10*, 199. [CrossRef]
15. Molendijk, M.; Morales, J.; Lemmen, C. Light Mobile Collection Tools for Land Administration; Proof of Concept for Colombia. GIM International, Online Magazine. 2015. Available online: https://www.gim-international.com/content/article/light-mobile-collection-tools-for-land-administration (accessed on 4 February 2021).
16. Todorovski, D.; Salazar, R.; Jacome, G.; Bermeo, A.; Orellana, E.; Zambrano, F.; Teran, A.; Mejia, R. Land Administration in Ecuador; Current Situation and Opportunities with Adoption of Fit-for-Purpose Land Administration Approach. In Proceedings of the World Bank Conference on Land and Poverty 2018: Land Governance in an Interconnected World, Washington, DC, USA, 19–23 March 2018.
17. Thomas, R.M. *Blending Qualitative and Quantitative Research Methods in Theses and Dissertations*; Corwin Press: Thousand Oaks, CA, USA, 2003.
18. Yin, R.K. *Case-Study Research–Design and Methods (Vol. 5)*; Sage Publications: Thousand Oaks, CA, USA, 2003.
19. Ministerial Agreement No. 029-16, 2016. *National Technical Standards for the Cadaster of Urban-Rural Real Estate and Property Appraisals*; MIDUVI: Quito, Ecuador, 2016. Available online: https://www.habitatyvivienda.gob.ec/wp-content/uploads/downloads/2016/08/Acuerdo-Ministerial-No-0029-16-Normas-Tecnicas-Nacionales-para-el-Catastro-de-Bienes-Inmuebles-Urbanos-Rurales-y-Avaluos-de-Bienes-Operacion-y-Calculo-de-Tarifas-de-la-Dinac.pdf (accessed on 16 August 2021).
20. Constitution of the Republic of Ecuador. 2008. Available online: http://pdba.georgetown.edu/Constitutions/Ecuador/english08.html (accessed on 11 December 2020).
21. Presidential Decree, No. 410 for the creation of the National Committee of Cadastre, Quito-Ecuador, Presidency of Ecuador. Available online: https://www.registroficial.gob.ec/index.php/registro-oficial-web/publicaciones/registro-oficial/item/3390-registro-oficial-no-410.html (accessed on 16 August 2021).
22. Law 1196. *Law for Territorial Ordering and Land use Management*. National Parliament of the Republic of Ecuador. 2016. Available online: http://www.habitatyvivienda.gob.ec/wp-content/uploads/downloads/2016/08/Ley-Organica-de-Ordenamiento-Territorial-Uso-y-Gestion-de-Suelo1.pdf (accessed on 11 December 2020).
23. Ministerial Agreement 160. *Establishment of MAGAP Execution Unit for Implementation National Rural Land Management and Information Program-SIGTIERRAS*; MAG: Quito, Ecuador. Available online: http://servicios.agricultura.gob.ec/mag01/pdfs/aministerial/2008/2008_160.pdf (accessed on 16 August 2021).

24. Manual Urban, Guidelines for urban cadastral projects Quito-Ecuador, Banco de Desarrollo del Ecuador. 2015. Available online: http://bde.fin.ec/asistencia-tecnica/ (accessed on 15 February 2018).
25. Manual for Data Collection for Rural Parcels. Available online: http://www.sigtierras.gob.ec/descargas/ (accessed on 8 December 2020).